HAGFISH

B I O L O G Y

CRC
MARINE BIOLOGY
SERIES

The late Peter L. Lutz, Founding Editor
David H. Evans and Stephen Bortone, Series Editors

PUBLISHED TITLES

Biology of Marine Birds
  E.A. Schreiber and Joanna Burger

Biology of the Spotted Seatrout
  Stephen A. Bortone

Early Stages of Atlantic Fishes: An Identification Guide for the Western Central North Atlantic
  William J. Richards

Biology of the Southern Ocean, Second Edition
  George A. Knox

Biology of the Three-Spined Stickleback
  Sara Östlund-Nilsson, Ian Mayer, and Felicity Anne Huntingford

Biology and Management of the World Tarpon and Bonefish Fisheries
  Jerald S. Ault

Methods in Reproductive Aquaculture: Marine and Freshwater Species
  Elsa Cabrita, Vanesa Robles, and Paz Herráez

Sharks and Their Relatives II: Biodiversity, Adaptive Physiology, and Conservation
  Jeffrey C. Carrier, John A. Musick, and Michael R. Heithaus

Artificial Reefs in Fisheries Management
  Stephen A. Bortone, Frederico Pereira Brandini, Gianna Fabi, and Shinya Otake

Biology of Sharks and Their Relatives, Second Edition
  Jeffrey C. Carrier, John A. Musick, and Michael R. Heithaus

The Biology of Sea Turtles, Volume III
  Jeanette Wyneken, Kenneth J. Lohmann, and John A. Musick

The Physiology of Fishes, Fourth Edition
  David H. Evans, James B. Claiborne, and Suzanne Currie

HAGFISH

BIOLOGY

Edited by

Susan L. Edwards

Appalachian State University, Boone, North Carolina, USA

Gregory G. Goss

University of Alberta, Edmonton, Alberta, Canada

CRC Press
Taylor & Francis Group
Boca Raton London New York

CRC Press is an imprint of the
Taylor & Francis Group, an **informa** business

CRC Press
Taylor & Francis Group
6000 Broken Sound Parkway NW, Suite 300
Boca Raton, FL 33487-2742

First issued in paperback 2020

ISBN 13: 978-0-367-57551-9 (pbk)
ISBN 13: 978-1-4822-3345-2 (hbk)

This book contains information obtained from authentic and highly regarded sources. Reasonable efforts have been made to publish reliable data and information, but the author and publisher cannot assume responsibility for the validity of all materials or the consequences of their use. The authors and publishers have attempted to trace the copyright holders of all material reproduced in this publication and apologize to copyright holders if permission to publish in this form has not been obtained. If any copyright material has not been acknowledged please write and let us know so we may rectify in any future reprint.

Library of Congress Cataloging-in-Publication Data

Hagfish biology / editors, Susan L. Edwards and Gregory G. Goss.
    pages cm. -- (CRC marine biology series ; 18)
    Includes bibliographical references and index.
    ISBN 978-1-4822-3345-2 (alk. paper)
    1. Hagfishes. I. Edwards, Susan L. II. Goss, Gregory G.

QL638.14.H34 2016
597'.2--dc23
                                          2015010465

Visit the Taylor & Francis Web site at
http://www.taylorandfrancis.com

and the CRC Press Web site at
http://www.crcpress.com

Contents

Preface

The hagfishes are a cosmopolitan group of craniate chordates, inhabiting almost all the oceans of the world, other than Antarctica. Hagfishes as craniates do not have vertebrae but do possess cranial bones and are considered to be the most ancient of the jawless fishes, having diverged from the main vertebrate lineage more than 500 million years ago. That hagfishes are physiologically unique in that they are the only living craniates to maintain their plasma NaCl concentration almost the same as that of seawater is a feature that has been known for almost 100 years. However, new high-throughput sequencing technologies allowing for improved genetic information, advanced microscopy techniques, the recent description of hagfish embryology, and the development of techniques to understand ancient evolutionary relationships have together led to a resurgence in interest in the hagfish as a key species to understand the evolution of the vertebrates.

These advances have resulted in new perspectives in hagfish research and this has led to the development of this new compendium of information on the hagfishes. An important part of putting this publication together was the gathering of hagfish biologists from all over the world in a symposium at the International Congress on the Biology of Fishes held in Edinburgh, Scotland, August 2014. This book aims to build on our previous knowledge and further expand the scientific interest for this fascinating yet understudied key evolutionary species. We hope that this book leads to a larger conversation and provides the impetus for new research avenues to develop.

Editors

Susan L. Edwards, PhD, is a professor of biology and chairperson at Appalachian State University in Boone, NC. She received her BSc in biology from Deakin University in 1994, her MSc in neuroscience from the University of Melbourne in 1997, and earned a PhD in comparative physiology from Deakin University in 2000. Her postdoctoral studies were conducted with Dr. James B. Claiborne at Georgia Southern University. In 2002, she took her first academic position in Queensland, Australia at James Cook University. However, in 2006, after many years of commuting from Australia to the United States for research, she decided to relocate to the United States to be closer to her model organisms.

Dr. Edwards' research program is focused on the identification and localization of ion transport mechanisms associated with osmotic balance, acid/base homeostasis, and more recently nitrogenous waste excretion in fishes. She is an active member of the American Fisheries Society and is the president-elect of the physiology section. She is also a member of the Canadian Society of Zoology, Society for Experimental Biology, Australian and New Zealand Society of Comparative Physiology and Biochemistry, and is a life member of the Mount Desert Island Biological Laboratory. She serves on the editorial board of *Comparative Biochemistry and Physiology* and has served on a number of panels for the National Science Foundation.

Gregory G. Goss, PhD, is a professor in the Department of Biological Sciences at the University of Alberta and is cross-appointed to the School of Public Health and a Fellow of the National Institute of Nanotechnology. Dr. Goss was appointed assistant professor in July 1997 and held ranks of associate and full professor in July of 2002 and 2005, respectively. Prior to his faculty position, Dr. Goss held two postdoctoral appointments at Hospital for Sick Children, Toronto, Ontario, and Beth Israel Hospital, Harvard Medical School, Boston, MA. He completed his BSc (1986) and MSc (1988) at McMaster University with Professor Chris M. Wood and obtained a PhD (1993) at the University of Ottawa under the tutelage of Professor Steve Perry. He is the past winner of the Petro-Canada Young

Innovator Award, the Canadian Society of Zoologists Early Investigator Award, the American Physiological Society Young Investigator Award, the McCalla award for teaching and research, and was awarded a Killam Annual Professorship in 2009–2010.

Dr. Goss' research is focused on the twin areas of toxicology and comparative physiology in a variety of fish species. In addition to his research at the University of Alberta, he regularly conducts his research program at Bamfield Marine Sciences Centre (BMSC), where he leads a productive program aimed at understanding the physiology of ion, acid–base, and solute transport regulation in trout, dogfish, and hagfish. It is at BMSC where Dr. Goss became acutely interested in hagfish physiology and thus formed the impetus for this book. Dr. Goss has served as president of the Canadian Society of Zoologists and serves on the council for numerous national and international societies. He is an associate editor of the *Canadian Journal of Zoology* and on the editorial boards for *Nanotoxicology and Environmental Science: Nano*.

Contributors

William C. Aird
Department of Medicine
Beth Israel Deaconess Medical
    Center
Harvard Medical School
Boston, Massachusetts

Carol Bucking
Department of Biology
York University
Toronto, Ontario, Canada

Nic R. Bury
Diabetes and Nutritional Sciences
    Division
King's College London
London, United Kingdom

Alexander M. Clifford
Department of Biological
    Sciences
University of Alberta
Edmonton, Alberta, Canada

Michael I. Coates
Department of Organismal
    Biology and Anatomy
University of Chicago
Chicago, Illinois

Shaun P. Collin
School of Animal Biology and the
    Oceans Institute
The University of Western Australia
Crawley, Western Australia,
    Australia

Max D. Cooper
Department of Pathology and
    Laboratory Medicine
Emory University School of
    Medicine
Atlanta, Georgia

Sabyasachi Das
Department of Pathology and
    Laboratory Medicine
Emory University School of
    Medicine
Atlanta, Georgia

William Davison
School of Biological Sciences
University of Canterbury
Christchurch, New Zealand

Ann M. Dvorak
Department of Pathology
Beth Israel Deaconess Medical
    Center
Harvard Medical School
Boston, Massachusetts

Susan L. Edwards
Department of Biology
Appalachian State University
Boone, North Carolina

Douglas S. Fudge
Department of Integrative Biology
University of Guelph
Guelph, Ontario, Canada

Chris N. Glover
School of Biological Sciences
University of Canterbury
Christchurch, New Zealand

Gregory G. Goss
Department of Biological Sciences
University of Alberta
Edmonton, Alberta, Canada

Scott M. Grant
Centre for Sustainable Aquatic
   Resources
Marine Institute of Memorial
   University of Newfoundland
St. John's, Newfoundland and
   Labrador, Canada

Julia E. Herr
Department of Integrative Biology
University of Guelph
Guelph, Ontario, Canada

Brantley R. Herrin
Department of Pathology and
   Laboratory Medicine
Emory University School of
   Medicine
Atlanta, Georgia

Masayuki Hirano
Department of Pathology and
   Laboratory Medicine
Emory University School of
   Medicine
Atlanta, Georgia

Philippe Janvier
Muséum National d'Histoire
   Naturelle
UMR7207 (CNRS, Sorbonne
   Universités, UPMC, MNHN)
Paris, France

Trevor D. Lamb
Department of Neuroscience and
   ARC Centre of Excellence in
   Vision Science
Australian National
   University
Canberra, Australia

Jianxu Li
Department of Pathology and
   Laboratory Medicine
Emory University School of
   Medicine
Atlanta, Georgia

Tetsuto Miyashita
Department of Biological
   Sciences
University of Alberta
Edmonton, Alberta, Canada

Masumi Nozaki
Sado Marine Biological Station
Niigata University
Niigata, Japan

Jinae N. Roa
Marine Biology Research Division
University of California San Diego
La Jolla, California

Robert S. Sansom
Faculty of Life Sciences
University of Manchester
Manchester, United Kingdom

Stacia A. Sower
Department of Molecular, Cellular
    and Biomedical Sciences
University of New Hampshire
Durham, New Hampshire

Martin Tresguerres
Marine Biology Research Division
University of California
    San Diego
La Jolla, California

Alyssa M. Weinrauch
Department of Biological Sciences
University of Alberta
Edmonton, Alberta, Canada

Timothy M. Winegard
Department of Integrative Biology
University of Guelph
Guelph, Ontario, Canada

chapter one

Anatomy of the Pacific hagfish (*Eptatretus stoutii*)

Alyssa M. Weinrauch, Susan L. Edwards,
and Gregory G. Goss

Contents

Introduction

Collectively, hagfishes are among the earliest extant members of the vertebrate lineage. Although there have been multiple reports detailing the histology and gross morphology of various tissues, a comprehensive atlas

1

of the hagfish that includes gross morphology, histology, and electron microscopy has never been created before. This chapter will provide a source for understanding the fundamental histology of cells and tissues in the Pacific hagfish (*Eptatretus stoutii*). By combining physiological studies with the described anatomy, we will be able to better elucidate the evolution of physiological and morphological traits.

Tissue histology

Integument

Hagfish lack scales and an outer *stratum corneum* (Figure 1.1a). Instead, the outer layer is homologous to the *stratum germinitavium* found in vertebrates. Resting atop a layer of skeletal muscle are the epidermis (E), dermis (D), and hypodermis (Andrew and Hickman, 1974). The epidermal layer is 75–200 μm thick, with variances depending on species, nutritional state, and location along the body (Andrew and Hickman, 1974). It is comprised of four cell types, held together by desmosomes (Spitzer and Koch, 1998): undifferentiated cells (UC), small mucous cells (SMC), large mucous cells (LMC), and epidermal thread cells (ETC) (Figure 1.1b) (Blackstad, 1963; Andres and von Düring, 1993; Spitzer and Koch, 1998). Mucous cells are thought to be homologous to goblet cells and are constitutively expressed in the epidermis, forming near the basement membrane (BM) and migrating to the epidermis to release their contents as they mature (Andrew and Hickman, 1974). Small mucous cells are regular in shape (diameter averaging ~12.4 μm), whereas large mucous cells are ellipsoidal, becoming more elongated as they reach the epidermal surface (diameter averaging ~31.2 μm). The nucleus (N) is large, with cytoplasmic filaments compacted to the basal and lateral sides of differentiated cells and prominent nucleoli (Ni) observed throughout all cell types (Figure 1.1c). The large mucous cells nuclei stain darkly but the diffuse cytoplasmic components stain weakly highlighting contrast, whereas the scanning electron microscopy (SEM) enhances the fusion and release of mucous contents from the clustering mucous cells, as well as the ~0.25–0.55 μm long projections on the surface (microplicae (MP)) (Figures 1.1c–e). The functions of the epidermal thread cells (diameter averaging ~19.8 μm) are undefined to date, however some speculate that the threads and mucous cells act like the slime glands to yield additional slime (Spitzer and Koch, 1998). Staining techniques employed in our study of *E. stoutii* have appeared to unravel the threads, leaving an acidic granular product within the epidermal thread cells (Figure 1.1b1). When different staining protocols are used, the thread-like nature of *Myxine glutinosa* epidermal thread cells can be observed (Figure 1.1b2; kindly provided by Sarah Schorno, Tim Winegard, and Dr. Douglas Fudge).

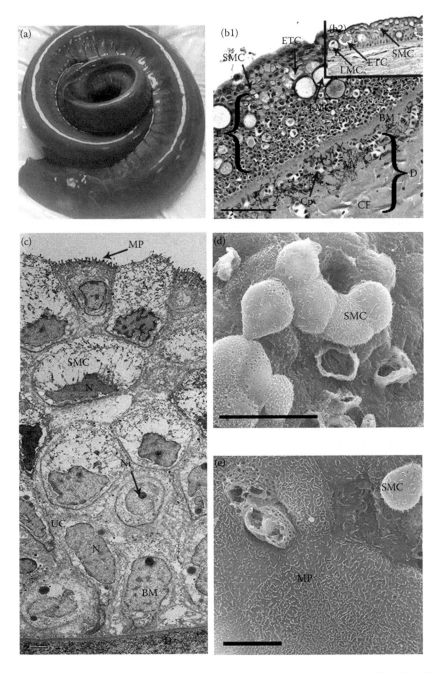

Figure 1.1 (**See color insert.**)   (*Continued*)

Specialized cell types include the flask-shaped sensory Schreiner organs and ionocytes, which, although not shown here, are infrequently dispersed throughout the skin (Braun, 1998; Spitzer and Koch, 1998). The precise sensory function of the Schreiner organ is yet to be determined. A unique basement membrane with a basal lamina on both sides separates the dermis from the epidermis, and housed within this layer are hormonally controlled melanocytes (Me) containing brown/black melanin granules (Coonfield, 1940; Andrew and Hickman, 1974). Both the dermo-epidermal and the dermo-hypodermal junctions contain elastic-like fibers that are difficult to stain. The dermis is comprised of 5–10 μm thick collagen fibers (CF), providing the elasticity necessary for undulation during swimming (Andrew and Hickman, 1974; Welsch et al., 1998). Numerous capillaries (Cp) filled with erythrocytes (RBC) occur throughout the dermis, permitting gas exchange and possibly supplying mucous cells with mucous composites (Figure 1.1b) (Potter et al., 1995). Reduced vascularization is observed in *Myxine* sp. when compared with *Eptatrids* sp., likely because of their burrowing behavior. The hypodermis is comprised of loose connective tissues (CT), blood vessels, and glycogen-rich adipose tissue (Blackstad, 1963; Andrew and Hickman, 1974).

Sensory barbels

Hagfish have limited visual capacity, having only eye spots rather than compound eyes (see Chapter 6). Thus, they rely heavily on mechanosensation using their barbels to perceive their surroundings (Greene, 1925),

---

Figure 1.1 (Continued) Integument. (a) Gross morphology of *E. stoutii* demonstrating the scale-less nature of hagfish skin. (b) Light micrograph demonstrating the cellular morphology of the dermis and epidermis of hagfish integument. (b1) A 75–200 μm thick epithelium (E) is comprised of four distinct cell types: undifferentiated cells (UC), small mucous cells (SMC), large mucous cells (LMC), and epidermal thread cells (ETC). These rest upon a basement membrane (BM), which sits atop the dermis (D) comprised of collagen fibers (CF) and housing the pigment-containing melanocytes (Me) and numerous capillary (Cp) networks. H&E stain. Scale bar = 100 μm. (b2) Light micrograph of *M. glutinosa* epidermis highlighting the coiled, thread-like nature of the ETCs that were not found to be as distinct in *E. stoutii*. Scale bar = 25 μm. Photograph kindly provided by Sarah Schorno, Tim Winegard, and Dr. Douglas Fudge. (c) Transmission electron micrograph of *E. stoutii* skin highlighting the microplicae (MP) on the epithelial surface along with the prominent nucleoli (Ni) within the basolaterally located nuclei (N) and diffuse surrounding cytoplasm. UCs are strikingly different with a central nucleus and more compact cytoplasm. Scale bar = 2 μm. (d) Scanning electron micrograph of *E. stoutii* epidermis. The clustering of SMCs is apparent alongside cells that have burst to release their contents. Scale bar = 20 μm. (e) Scanning electron micrograph of *E. stoutii* integument demonstrating the ubiquitous distribution of MP around and on the SMCs. Scale bar = 10 μm.

although a role for chemosensation cannot be ruled out. Notably, when the hagfish coils, the head is consistently on the outside of the coil exposing the sensory barbels (SB), of which there are three pairs in *Eptatretus* sp. that surround the nasohypophysial opening (NO) (Figure 1.2a) (Greene, 1925). These barbels oscillate both while swimming and at rest, suggesting a mechanosensory function that is sensitive throughout the barbel (Worthington, 1905; Greene, 1925; Blackstad, 1963; Andrew and Hickman, 1974). Hagfish barbels are similar to catfish tentacles, which contain numerous goblet cells and have been characterized using various stains (Yan, 2009). Catfish tentacles are present only in males as a result of sexual dimorphism, with the mucous cell components thought to provide nourishment for the larvae as the males alone care for the young (Yan, 2009), although no apparent sexual dimorphisms exist in hagfishes. Morphologically, the sensory barbel contains an outer epithelium (E) with distinct basolateral congregation of nuclei within each cell type: as in the skin, the epidermis is comprised of small mucous cells and undifferentiated cells resting on a basement membrane, whereas large mucous cells and epidermal thread cells are not observed. Also consistent with other integument is the presence of microplicae on the outer surface. The underlying dermis (D) is consistent with the aforementioned description, complete with abundant capillary (Cp) networks and melanocytes, however the underlying layer contains muscular tissue (Ms) and connective tissue, along with a small pocket of cartilage (Crt), permitting rigidity, control, and oscillation of the tentacles (Figures 1.2b–e). This could allow the tentacles to remain sentient when facing a strong force, for example, during active predation (Zintzen et al., 2011). Sensory Schreiner organs are evident and abundant, as long, conical cells located in sensory buds (SBd) deep in the epidermis and extending to the surface (Blackstad, 1963; Braun, 1998), with flagellated structures observable in the SEM images (Figures 1.2f1 and f2) that seem to be lost during other fixation procedures (e.g., TEM, light microscopy).

Dental plate

Hagfish are demersal opportunistic scavengers known to feed on a wide variety of decaying animals ranging from polychetes and mollusks to mammals (Martini, 1998; Clark and Summers, 2007, 2012; Glover et al., 2011). Recent video evidence from Zintzen et al. (2011) also demonstrates a predatory behavior in some species (*Neomyxine*), where possible suffocation by slime exudate occurs within burrows. Although this behavior could be universal across the *Myxine*, it is currently unknown how many hagfish species utilize active predation and if used, how often in comparison to decaying ocean fall (Zintzen et al., 2011). Morphological similarities between juvenile and adult hagfish

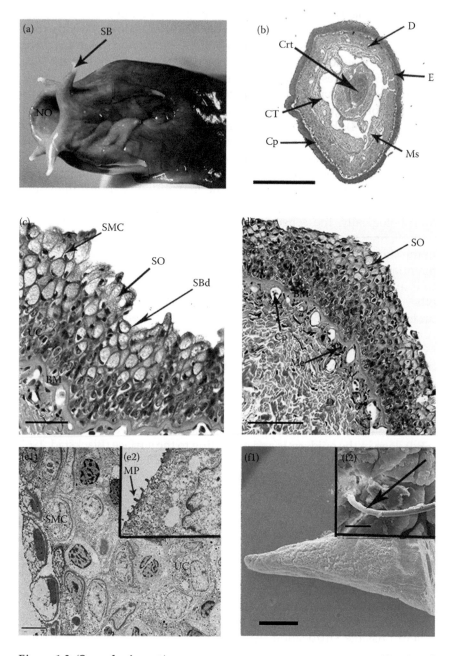

Figure 1.2 (**See color insert.**)

(*Continued*)

teeth are suggestive of similar diets throughout the hagfish life history, as well as between species (*Myxine* and *Eptatrids*) (Clark and Summers, 2012). Detection of food appears to occur via smell and touch with the sensory barbels (Clark and Summers, 2007).

Continual protraction and retraction of a heart-shaped dental plate, housing two rows of conical teeth (T), tear tissues from carrion and permits entry into the body cavity (Figure 1.3a) (Dawson, 1963; Krejsa et al., 1990; Clark and Summers, 2007, 2012; Chiu et al., 2011). Recent studies determined that hagfish acquire nutrients/ions via extraintestinal means such as the skin (Glover et al., 2011; Schultz et al., 2014), and perhaps entry into carrion enhances uptake.

Plate movement occurs as a result of protractor and retractor muscles (see pharyngeal muscle) with produced force being equivalent to that of some gnathostomes despite the lack of bone used for muscle attachment (Dawson, 1963; Krejsa et al., 1990; Clark and Summers, 2007, 2012; Chiu et al., 2011). The force of penetration, as well as tooth stress, alters the tearing ability, whereas the ability to grasp prey is proportional to tooth number and surface area (Clark and Summers, 2012).

The dental plate houses a single palatine tooth responsible for continued hold on prey, as well as two rows of bilaterally symmetrical lingual teeth (7–9 per row) for rasping (Dawson, 1963; Slavkin et al., 1983; Rice et al., 1994; Clark and Summers, 2007, 2012). Hagfish ranging 98–202 mm in length have palatal teeth 0.5–1.5 mm in length, whereas the outer and inner rows of lingual teeth are 1.6–3.8 and 1.9–4.5 mm in length, respectively (Krejsa et al., 1990). The lingual teeth change orientation and fold into one another during plate retraction, rasping at decaying flesh (Clark and Summers, 2012).

Figure 1.2 (Continued) Sensory barbels (SB). (a) A gross morphological representation of the three pairs of SBs surrounding the nasohypophysial opening (NO) in Pacific hagfish. (b) Transverse section of SB demonstrating the layers within. The outer epithelial layer (E) rests atop a dermis (D) housing the capillaries (Cp). Beneath this layer are layers of connective tissue (CT) and muscle (Ms), surrounding a small central pocket of cartilage (Crt). H&E Stain. Scale bar = 1 mm. (c) Sagittal section of hagfish SBs. Resting atop a basement membrane (BM) are undifferentiated cells (UC), which migrate to the surface and differentiate into small mucous cells (SMC). Also abundant are sensory Schreiner organs (SO), which form sensory buds (SBd) thought to hold flagella. H&E stain. Scale bar = 50 µm. (d) Sagittal section of hagfish barbel demonstrating the cellular composition of the epidermis (e) containing UCs, SMCs, and SOs. Beneath the BM is the dermis (d) housing the capillaries (Cp) and melanocytes (Me). H&E stain. Scale bar = 100 µm. (e) Transmission electron microscopy detailing the differentiated SMCs atop the UC (e1) and microplicae (MP) (e2). Scale bar = 5 µm. (f) Scanning electron micrograph of the SB highlighting the rough surface (f1). Scale bar = 500 µm. Long flagellated structures (arrow) were observed using SEM and are possible components of the SOs (f2). Scale bar = 50 µm.

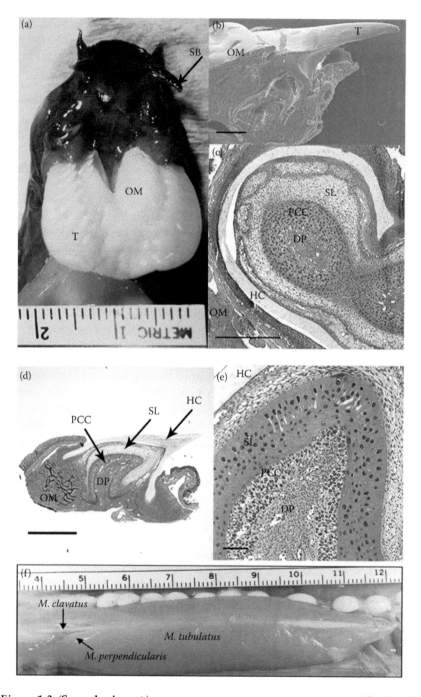

Figure 1.3 (See color insert.)   (*Continued*)

The tooth is comprised of multiple "cone" layers, making it conducive to postprandial tooth replacement (Dawson, 1963; Krejsa et al., 1990; Clark and Summers, 2007, 2012). The tooth replacement process is similar to squamate integument shedding and also follows typical vertebrate tooth formation until crown formation where the dentin–protein complex forms. This is because cyclostomes lack calcified structures and thus enamel and dentin are not present; however, an enamel-like protein of unknown function has been discovered and is suggestive of the conservation of enamel protein structure throughout vertebrate evolution (Slavkin et al., 1983; Krejsa et al., 1990; Clark and Summers, 2012). The layers composing the tooth include the central dental papilla (DP), pokal cell cone (PCC), stellate layer (SL), and outer horn cap (HC), all of which rest in the oral mucosa (OM) (Figure 1.3a–e) (Dawson, 1963; Krejsa et al., 1990; Clark and Summers, 2007, 2012).

The dental papilla is the site of new tooth formation (Krejsa et al., 1990), which eventually gives rise to the pokal cell cone, or functional tooth. This area is rich in lipids and phospholipids, providing the structural support for the tooth. Moving outward, the stellate layer is incredibly vacuolated and these cells produce the keratinous outer horn cap and separate the outer shedding tooth from the underlying replacement (Dawson, 1963). These cells are also found in lamprey and are thought to be synonymous to the stellate reticulum of gnathostomes (Krejsa et al., 1990). Some adults have multiple stellate layers and pokal cell cones to increase the rapidity of tooth replacement (Krejsa et al., 1990). The keratinous horn cap provides the strength and shape of the tooth, with elongated cells at the apex representing pokal cell remnants (Appy and Anderson, 1981; Krejsa et al., 1990; Clark and Summers, 2012).

Figure 1.3 (Continued) Dental plate. (a) Gross morphology of the hagfish dental plate. The sensory barbels (SB) surrounding the nasopharyngeal opening and mouth presumably detect food sources. The hagfish then repeatedly everts and retracts the dental plate where two rows of teeth (T) sit in an oral mucosa (OM). (b) Scanning electron micrograph of a hagfish tooth (T) still intact with the OM. Scale bar = 500 μm. (c) Transverse section of a hagfish tooth demonstrating the "cone within a cone" makeup of the tooth within the OM. The central dental papilla (DP) initiates tooth growth into the pokal cell cone (PCC) otherwise known as the functional tooth. Residing atop this layer is the stellate layer (SL), which produces the outer keratinous horn cap (HC). H&E stain. Scale bar = 500 μm. (d) Sagittal representation of the hagfish tooth resting in OM. The central DP forms the functional tooth known as the PCC. Atop this rests the SL, which produces the outer keratinous horn cap (HC). H&E stain. Scale bar = 1 mm. (e) Higher magnification of the pokal cell layers, as above. H&E stain. Scale bar = 100 μm. (f) Gross anatomical portrait of the pharyngeal muscles. It is comprised of two muscles in concentric rings: the outer *M. tubulatus* and the inner *M. clavatus*. Also evident is the *M. perpendicularis* running between the paired *M. clavatus* muscle.

Dental muscle (pharyngeal muscle)

The dental plate is maneuvered using a set of protractor and retractor muscles. Specifically, the retractor muscles include the M. tubulatus, M. elevatus, and M. perpendicularis, all enclosed within a sheath of connective tissue (Figure 1.3f). M. tubulatus is a paired muscle that forms an outer ring of muscle surrounding the inner M. clavatus, which attaches to the posterior edge of the dental plate (Figure 1.4a) (Dawson, 1963; Baldwin et al., 1991). The muscle attaches to the cartilaginous cranium (Clark and Summers, 2012), and transference of force to the dental plate occurs via the clavatus tendon, which has properties similar to those of some gnathostomes (Clark and Summers, 2007). It is comprised of thin white muscle only, as opposed to the white, red, and intermediate fibers that make up the somatic muscle (Clark and Summers, 2007; Chiu et al., 2011). The white muscle can exert rapid and powerful force in anaerobic conditions, whereas the red muscle contracts for extended periods of time with little force (Dawson, 1963; Baldwin et al., 1991). Furthermore, the somatic and pharyngeal muscles differ in enzymatic activity, with the dental muscle having a higher overall activity, apart from those enzymes involved in aerobic catabolism of carbohydrates and lipids (Chiu et al., 2011). This striated muscle (StM) contains few nuclei, however glycogen granules are abundant (Figure 1.4b).

Somatic muscle

Muscles are necessary for movement and therefore feeding and ventilation, however they also act in storage and as energy reserves. The muscle tissue of hagfish is iso-osmotic to seawater, holding concentrations of inorganic ions below that of the plasma (Currie and Edwards, 2010). Hagfish have three distinct muscular groups: M. parietalis, M. obliquus, and M. rectus, with M. parietalis constituting the majority of the body, running from nasal to caudal end and extending laterally from the midline to the slime glands (Jansen and Andersen, 1963; Korneliussen and Nicolaysen, 1973). It is comprised of W-shaped metameric segments, as opposed to the common V-shape of other teleosts. Less-developed metameric segments account for the reduced swimming speeds demonstrated by the hagfish (Jansen and Andersen, 1963; Andrew and Hickman, 1974).

Striated muscle fibers (StM), themselves made up of myofibrils, contain numerous nuclei (Figure 1.4e) (Jansen and Andersen, 1963; Hardisty, 1979) and are found in the ~3 mm long myotomes. The myotomes (Mto) are segmented by myosepta (Msp) (tissues containing vascular and nervous elements) (Figures 1.4c and g) (Jansen and Andersen, 1963; Mellgren and Mathisen, 1966) into compartments, with lateral and ventral linings of red fibers and white fibers in the remaining space (Mellgren and Mathisen, 1966). The makeup of the muscle is consistent with that

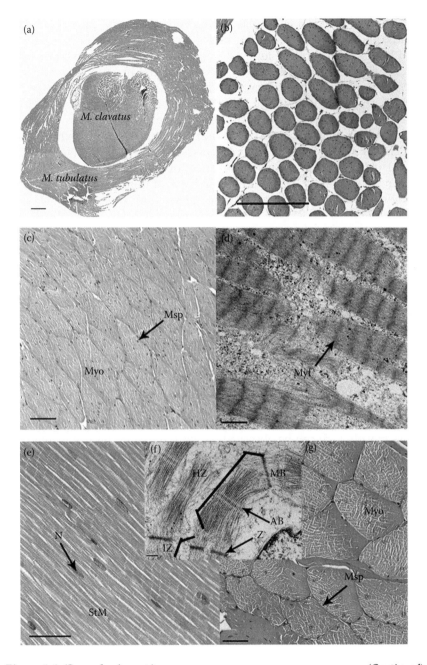

Figure 1.4 (**See color insert.**)   (*Continued*)

typically observed in vertebrates being comprised of Z lines, I bands, A bands, and M and H zones (Figures 1.4d and f). These myotomes contain both small, aerobic slow red fibers and large, anaerobic fast white fibers, with 14% of the red fiber surface covered in capillaries and only 0.4% of the white fibers (Jansen and Andersen, 1963; Korneliussen and Nicolaysen, 1973; Baldwin et al., 1991). Within the red fibers are extensive lipid and glycogen deposits (Baldwin et al., 1991), demonstrating the energy reserve capacity of the tissue as opposed to the white tissue, which lack these structures (Mellgren and Mathisen, 1966). Not pictured are *M. obliquus* which forms a thin sheet over *M. parietalis* in the cloacal region, as well as *M. rectus*; a narrow band of muscle running between the slime glands and ventral midline (Jansen and Andersen, 1963).

Intestine

The cyclostome intestine is thought to have derived from the microphagous feeding of ancestral craniates (Appy and Anderson, 1981). As in some fish, hagfish do not have a defined stomach or a differentiated small and large intestine (Andrew and Hickman, 1974; Menke et al., 2011), thus digestion occurs directly in the straight intestinal tube via lipases (Adam, 1963). The intestine is anchored to the body wall, with mesentery containing nerves, blood, and lymph vessels, which is removed in the photograph (Figure 1.5a) (Adam, 1963). Hemocytopoietic tissue and adipose submucosa constitute further portions of the alimentary canal, along with other derivatives such as the liver, biliary system, and pancreas (Jordan and Speidel, 1930; Papermaster et al., 1962; Clark and Summers, 2007). Differences in external gross morphology can be observed between the pharyngocutaneous duct (foregut) and the hindgut, which are separated in the region of the portal

---

Figure 1.4 (Continued) Musculature. (a) Light micrograph showing a sagittal section of the pharyngeal muscle that controls the dental plate. The muscles are paired and align in concentric circles with the *M. tubulatus* surrounding the *M. clavatus*. H&E stain. Scale bar = 1 mm. (b) Transverse section of the pharyngeal muscle demonstrating the longitudinal band arrangement. H&E stain. Scale bar = 500 μm. (c) Transverse section of the parietal muscle. Individual multinucleated myomeres (Myo) are tightly joined by myosepta (Msp). H&E stain. Scale bar = 100 μm. (d) Transmission electron micrograph of myofibrils (Myf) of the pharyngeal muscle demonstrating the striated nature and collagenous matrix between muscle fibers. Scale bar = 1 μm. (e) The striated muscle (StM) contains multiple nuclei (N) as in typical vertebrate StM. H&E stain. Scale bar = 50 μm. (f) Transmission electron micrograph of StM fibers and nucleus (N). The striations are clear and demonstrate the typical banding pattern observed in vertebrates including the Z line (Z), M band (MB), A band (AB), I zone (IZ), and H zone (HZ). Scale bar = 200 nm. (g) Cross-sectional light micrograph of StM fibers illustrating the myomeres (Myo) separated by myosepta (Msp). H&E stain. Scale bar = 100 μm.

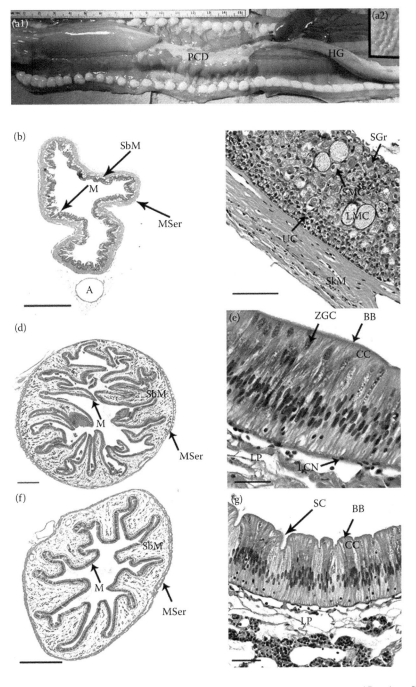

Figure 1.5 (**See color insert.**)   (*Continued*)

heart where cross-striated muscle fibers cross the intestine (Adam, 1963; Andrew and Hickman, 1974). However, both foregut and hindgut are comprised of three layers: mucosa (M), submucosa (SbM), and *Muscularis serosa* (Figure 1.5b,d,f) (Adam, 1963; Andrew and Hickman, 1974). Although not previously acknowledged, there are indeed differences along the length of the hindgut, which will be discussed herein.

Foregut

The foregut contains a distinct cellular makeup, vastly different from the hindgut, but akin to the skin. Resting upon the basement membrane, are undifferentiated cells, which migrate upward to become either mature epithelial cells, small mucous cells, or large mucous cells (Figures 1.5c and 1.6a, b1, and b2). Uniquely, hagfish restrict all mucous cells to the foregut (Appy and Anderson, 1981; Menke et al., 2011). There are also secretory granulocytes (SGr) containing acidophilic granules, which are likely peptidases used to further enhance digestion (Figure 1.5c). The surface of this epithelium is covered with microridges (MRs) with no apparent pattern and as in the skin, function to reduce friction and aid in the progression of mucous and food down the tract (Figures 1.6a, b1, and b2). Irregularly shaped epithelial cells have clear borders

---

Figure 1.5 (Continued) Alimentary canal. (a1) Gross representation of the straight intestinal tube of the Pacific hagfish. Lacking a stomach, the hagfish intestine begins with the pharyngocutaneous duct (PCD), which runs alongside the gills. Posterior to the cardiac striations near the liver, the hindgut (HG) appears with distinct morphology. (a2) Mucosal surface of the hindgut. Large, permanent zigzag folds are used to enhance the surface area and may change color with nutritional status. (b) Cross-section of the PCD and corresponding intestinal artery (a). The intestine is composed of three layers: the outer *M. serosa* (MSer), a submucosa (SbM), and the mucosal layer (M). H&E stain. Scale bar = 1 mm. (c) Sagittal section of the PCD, displaying the SbM and the cellular components of the mucosa, which like the skin include, undifferentiated cells (UC), small mucous cells (SMC), large mucous cells (LMC), and secretory granules (SGr). H&E stain. Scale bar = 100 μm. (d) Cross-section of the anterior hindgut displaying the three layers: mucosa (M), submucosa (SbM), and *M. serosa* (MSer). H&E stain. Scale bar = 1 mm. (e) Sagittal plane of the anterior hindgut with a distinct morphology in comparison to the PCD. Atop a lamina propria (LP) rest the columnar cells (CC) with a brush border (BB). Interspersed among these cells are zymogen granule cells (ZGC) with acidic digestive enzymes. Also present are lymphatic cell nuclei (LCN) for immunity. H&E stain. Scale bar = 100 μm. (f) Transverse section of the posterior hindgut. Again the three layers are present [*M. serosa* (MSer), submucosa (SbM), and mucosa (M)], but with reduced infolding in comparison to the anterior hindgut. H&E stain. Scale bar = 1 mm. (g) Transverse section of the posterior hindgut. Resting upon an LP are the CCs with a BB, however also present are secretory cells (SC) of unknown function. H&E stain. Scale bar = 100 μm.

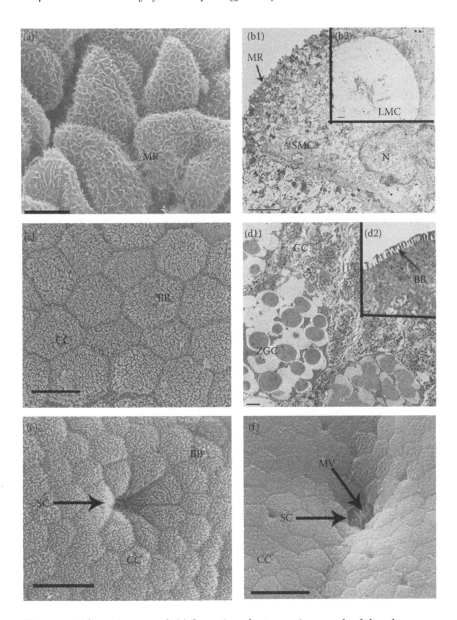

Figure 1.6 Alimentary canal. (a) Scanning electron micrograph of the pharyngo-cutaneous duct demonstrating the abundance of microridges (MR) on irregularly shaped epithelial cells. Scale bar = 5 μm. (b1) Transmission electron micrograph of the MRs atop a small mucous cell (SMC) with the typical basolaterally located nucleus (N). Scale bar = 2 μm. (b2) The disperse nature of the cytoplasm within the large mucous cells (LMC) of the PCD. Scale bar = 2 μm.   (*Continued*)

with distinct cell–cell junctions and deeper furrows occur between cells, unlike the typical squamous cell covering of other vertebrate foreguts (Figure 1.6b1).

Beneath the basement membrane (BM) is a *lamina propria* (LP), an area rich in vasculature and adipose tissue. The submucosa (SbM) contains connective tissue (CT), lymph, and blood vessels, as well as varying amounts of adipose depending on the nutritional state (Adam, 1963; Andrew and Hickman, 1974) and dispersed lymph vessels that have sometimes been referred to as a diffuse spleen (Andrew and Hickman, 1974). The submucosa functions to provide turgidity to the intestinal walls, whereas the *M. serosa* aids in transport of nutrients by applying peristaltic pressures, with assistance required from body undulation (Andrew and Hickman, 1974). Not pictured here is the transition from foregut to hindgut, which lacks all specialized cells and contains only transitional epithelium.

Hindgut

The diameter of the hindgut measures 1–2 cm, becoming thinner toward the posterior end and changing color with nutritional status (Adam, 1963). It contains approximately 10 large, permanent zigzag folds which become somewhat diffuse in the preanal or anal regions (Figure 1.5a2). This is most likely due to the fact that absorption is reduced in these areas, as most of the nutrients have been absorbed and waste remains. The cells of the mucosal region are simple, columnar cells (CC) up to 250 μm in length (Figures 1.5e and g). They exhibit palpable polarization and a brush border (BB) to enhance surface area and aid in passage of food (Figures 1.6c, d1, d2, e, and f). Constituents that form peritrophic sacs are also secreted in this segment, which prevent bacterial penetration while allowing enzymatic penetration. These sacs enshroud the waste to be excreted (Adam, 1963; Hardisty, 1979). Interspersed among the columnar cells are zymogen granule cells (ZGCs) containing large, acidophilic secretory granules which have been likened to pancreatic enzymes and therefore, all digestion is thought to occur in the hindgut (Adam, 1963; Andrew and Hickman, 1974). They are easily identifiable

---

Figure 1.6 (Continued) (c) Scanning electron micrograph of the anterior hindgut displaying the extensive brush border (BB) atop the columnar cells (CC). Scale bar = 5 μm. (d1) Transmission electron microscopy of the zymogen granule cells (ZGC) among the CCs with a BB. Scale bar = 2 μm. (d2) Transmission electron microscopy detailing the fine BB lining the cells of the hindgut. Scale bar = 0.5 μm. (e) Scanning electron micrograph of a secretory cell (SC) surrounded by CCs covered with a BB. Scale bar = 10 μm. (f) Scanning electron micrograph of the various-sized pores within the CCs of the posterior hindgut. The central secretory cell (SC) appears to house microvilli (MV) of unknown function. The smaller pores are predicted to play a role in water regulation. Scale bar = 10 μm.

due to a single nucleolus in the nucleus along with a tapered apex, all of which is contained within a pit reminiscent of the mammalian pancreatic acinar cell (Adam, 1963; Appy and Anderson, 1981). Two distinct types of granules appear to reside within zymogen granule cell, a lightly stained, less-dense granule that is seemingly more basophilic than the alternative more-dense acidophilic granules (Figure 1.5e). These specialized epithelia rest upon a *lamina propria* (LP) containing dense connective tissue (CT) and numerous capillaries. At the interface between the basement membrane (BM) and lamina propria (LP) are distinct nuclei, which are thought to represent lymphatic cell nuclei (LCN), a probable necessity at an internal/external junction, given the feeding environment of the hagfish (Figure 1.5e). Similar to lamprey, the mucosal region of hagfish contains more absorptive cells than secretory cells (Appy and Anderson, 1981). However, the number of secretory cells (SC) appears to increase as you move down the intestine and become more diffuse toward the cloacal region (Figure 1.5g). These cells appear to group and form a secretory cleft or pit, distinct from the surrounding columnar epithelium (more lightly stained and a reduced BB), appearing pore-like in the SEM images (Figures 1.6e and f). The mucosal mucous lubricates the food and slows down diffusion of nutrients while preventing foreign invasion (Rogers, 1983; Clark and Summers, 2012). The submucosa and *M. serosa* function as mentioned above.

Liver

Located caudal to the portal heart, the liver (L) is contiguous with the gall bladder (GB) via a bile duct (BD) off the smaller, rostral lobe (Figure 1.7a) (Adam, 1963). The hagfish liver is a large, two-lobed tubular organ that comprises ~3.5% of the total body weight and is surrounded by a thin layer of squamous epithelia (Anderson and Haslewood, 1967; Mugnaini and Harboe, 1967; Yousen, 1979; Umezu et al., 2012). It is structurally similar to a mammalian liver, suggesting that it was acquired early on in vertebrate phylogeny (Umezu et al., 2012). Hagfish hepatocytes (HE) have polarity, which is obvious due to apical microvilli (MV) (Mugnaini and Harboe, 1967). Also notable are the lipid deposits (LI) in the liver, giving hepatocytes a vacuolated appearance (Figures 1.7b and c). This fatty status corresponds well with the elasmobranch liver, which is utilized in flotation and intermediary metabolism (Inui and Gorbman, 1978).

The hepatocytes are exocrine, producing bile and passing it through multiple canaliculi (Ci), which merge to form the bile duct (BD), leading to the gall bladder (GB) for storage (Figure 1.7c). A close-up of the caniliculus (Figure 1.7d) shows a highly interdigitated surface epithelium with numerous microvilli (MV), with secretory products (SPs) in the lumen (L). A separate bile duct lined with a columnar epithelium (CE) and surrounded by

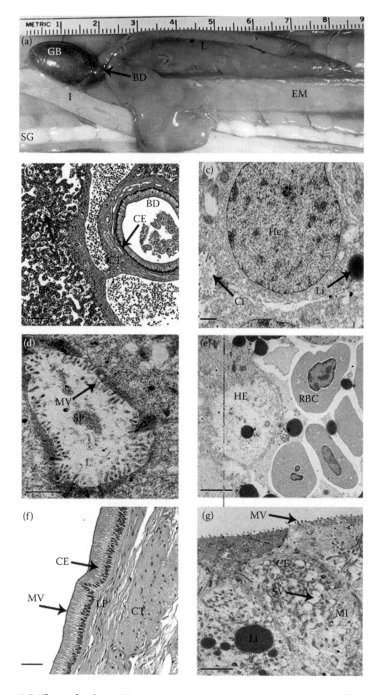

Figure 1.7 **(See color insert.)**   (*Continued*)

connective tissue permits passage into the intestine (Figure 1.7b) (Adam, 1963; Umezu et al., 2012). Whether any connection exists between bile ducts or whether the portal triad exists is unknown to date (Umezu et al., 2012). The bile itself contains a bile alcohol, myxinol disulfate, which is less effective than vertebrate bile alcohols (Anderson and Haslewood, 1967).

Hepatocytes are also responsible for the regulation of blood glucose and proteins, with insulin being released from the liver. However, hepatectomized hagfish live 5–30 days, with no lowering of blood glucose, suggesting additional regulation and quite possibly intestinal supplementation (Fänge et al., 1963; Inui and Gorbman, 1978; Malte and Lomholt, 1998). Only one enzyme of the ornithine-urea cycle (arginase) has been detected in hagfish liver tissue (Read, 1975). While urea is seemingly unable to be produced, the liver is speculated to be excretory due to transport of organic acids from the blood into the bile, a task that the primitive archinephric duct is incapable of. Blood stems from the *vena supra-intestinalis* and right jugular vein into three types of vessels: bile-ductule-associated, nonassociated (such as portal veins lined with epithelium), and sinusoids (Si) draining to the central vein (Figure 1.7b) (Fänge et al., 1963). Red blood cells (RBC) are notable within these vessels where exchange of material occurs over a thin endothelium and peritubular connective tissue with foot processes for an increased surface area (Figure 1.7e) (Malte and Lomholt, 1998).

Gall bladder

The gall bladder is responsible for the storage of bile received from the liver via hepatic ducts, until release is prompted by cholinergic signals

---

*Figure 1.7 (**Continued**)* Liver and gall bladder. (a) Gross morphological details of the bi-lobed liver (L) and gall bladder (GB), attached via a bile duct (BD). Ducts also lead to the intestine (I) for bile distribution. EM, epaxial muscle; SG, slime gland. (b) Light micrograph of the hepatocytes (He) and sinusoids (Si) within, alongside a BD lined with columnar epithelium (CE) and supported by connective tissue (CT). H&E stain. Scale bar = 100 μm. (c) Transmission electron micrograph of an HE. Spherically shaped they are surrounded by lipid deposits (LI) and canaliculi (Ci) for transport. Scale bar = 1 μm. (d) A single, microvilliated (MV) canaliculus within a hagfish liver. The lumen (L) contains multiple secretory products (SP). Scale bar = 1 μm. (e) Transmission electron micrograph displaying the exchange capacity within the liver as the hepatocytes (HE) are surrounded by capillaries filled with erythrocytes (RBC). Scale bar = 5 μm. (f) GB wall. A simple CE topped with MV resting on an lamina propria (LP) and CT layer. Beneath this is the muscular layer necessary to contract the GB and expel bile. H&E stain. Scale bar = 100 μm. (g) Epithelial surface of the GB. The CE houses MV. Within the cells themselves, organelles are not present near the surface but include mitochondria (MI), lipid droplets (LI), and numerous vacuoles and vesicles (SV). Scale bar = 2 μm.

(Fänge et al., 1963; Anderson and Haslewood, 1967; Malte and Lomholt, 1998). Located caudal to the portal heart, it is obvious due to its bluish-green coloration (Figure 1.7a). The inside wall is lined with the tall, simple columnar epithelium resting on a basement membrane, which in combination with the underlying lamina propria composes the mucosal layer. Beneath the mucosa is the submucosa consisting of connective tissue and blood vessels, all of which are surrounded by a well-developed smooth muscle used for the propulsion of bile (Inui and Gorbman, 1978; Morrison et al., 2006). The perimuscular layer is again connective tissue and the surrounding serosal layer is continuous with the peritoneum and physically attaches the gall bladder to the liver and epaxial muscle.

Ultrastructurally, the apical surface of the columnar cells contains a brush border of ~0.175 µm long microvilli appearing consistently (Figures 1.7f and g). Apart from obviously increasing the surface area, the microvilli do not have a defined functional significance. The function of the gall bladder is primarily a storage of bile salts for fat emulsification and also enzyme maturation. The functional significance of hagfish gall bladder is as yet undefined, however with lipid granules (LI) and numerous mitochondria (MI) present, it suggests that the highly microvilliated surface has an absorptive or secretory capacity. The content of these microvilli does not appear to differ from that of the cytoplasm in the rest of the cell. Typically, the cytoplasm is of consistent electron density and the first ~2.5–3.0 µm of the apical region of the cell is devoid of organelles. Below this area is filled with multiple vacuoles and vesicles (SV). Studies conducted in some vertebrates suggest that some vesicles form by pinocytosis for the consumption of water and lipids (LI), whereas other vesicles are contingents of the endoplasmic reticulum used for trafficking to the Golgi apparatus in the supranuclear zone (Johnson et al., 1962; Inui and Gorbman, 1978).

Numerous elongated mitochondria (MI) observed with higher concentrations typically in the infranuclear region, associating with the endoplasmic reticulum. Various-sized lipid deposits (LI) were observed, although not membrane-bound (Figure 1.7g).

Renal system

Hagfish have two retroperitoneal archinephric ducts (AD) placed laterally on either side of the dorsal aorta (DA) and postcardinal vein (Figure 1.8a). Perfusion with Coomassie Brilliant Blue helped highlight distinct very large, independent glomeruli (G) joined by ductile tubule location as distinct circular structures along the archinephric duct (Figure 1.8b). Arising blindly just anteriorly to the liver, these ducts store urine before output into the cloaca via posterior ureters. The urine output is low at 4–10 mL/kg/day and is believed to be related to glomerular filtration rate (Andrew and Hickman, 1974; Hardisty, 1979; Fels et al., 1998; Riegel and Bardack, 1998). The 30–40

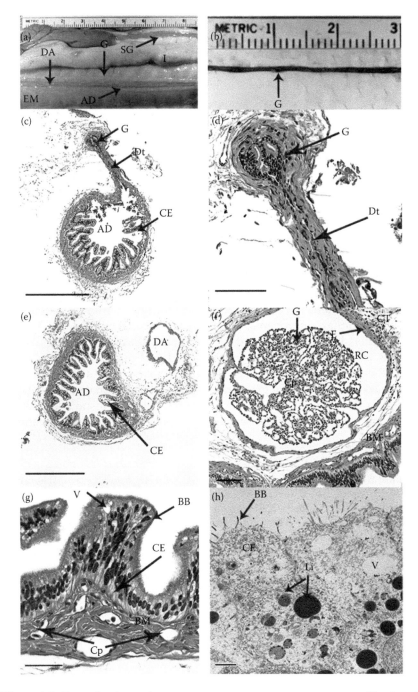

Figure 1.8 (See color insert.) (Continued)

ovoid renal corpuscles (RC) are each individually connected to the archinephric duct via ductile tubules (Dt). The renal corpuscles house the glomeruli (G) and dissipate toward the caudal end (Figures 1.8b, c, d, and f).

The archinephric duct is lined with simple columnar or cuboidal epithelia (CE), resting on a basement membrane (Figure 1.8e) (Yousen, 1979; Fels et al., 1998). There is an obvious brush border of long, straight microvilli (MV) and along with vacuoles (V) aids in the absorption of materials during the extended time when urine is stored (Andrew and Hickman, 1974; Yousen, 1979; Fels et al., 1998). Besides large vacuoles, vesicles, inclusion bodies, and apical pits are observed among a tubular network on the apical surface. Surrounding the duct is a connective tissue network.

The blood supply is entirely arterial, with arteries stemming from the dorsal aorta and thinning into an extensive capillary network (CP) within the glomerulus (Figures 1.8f and g). There are generally three efferent arterioles through which blood can exit the system into the postcardinal vein. The corpuscle itself is lined with mesangial cells, apart from the thinnest capillary, which contains endothelial cells (E) alone (Figure 1.8f). Within the columnar cells are multiple lipid droplets (LI) of varying size and staining intensity (Figure 1.8h). It is possible that these represent different functional components, but this is yet to be elucidated.

Glomeruli function to produce peritoneal fluid using a central mass of cells as a phagocytic filter for the blood and peritoneal fluid (Fänge, 1963; Riegel and Bardack, 1998). Although potassium, magnesium, and sulfate are excreted in the glomerular filtrate, it seems that sodium is able to be resorbed (McInerney, 1974). The renal system is responsible for the maintenance of the ionic and osmotic composition through the secretion

---

Figure 1.8 (Continued) Archinephric duct (AD). (a) Hagfish have two ADs; one on either side of the dorsal aorta (DA). Large, visible glomeruli (G) are seen along the length of the duct. I, intestine; EM, epaxial muscle; SG, slime gland. (b) Excised AD stained with Coomassie Blue to increase glomeruli (G) visibility. (c) Cross-section of the AD and associated glomerulus (G). The duct is lined with a columnar epithelium (CE) and connects to the glomerulus via a small ductule (Dt). H&E stain. Scale bar = 500 μm. (d) Detailed images of the glomerulus (G), connected to the Dt. H&E stain. Scale bar = 100 μm. (e) Cross-section of the CE lined AD and associated DA. H&E stain. Scale bar = 500 μm. (f) Sagittal section of a glomerulus (G) inside the surrounding connective tissue (CT). The glomerulus is housed within a renal corpuscle (RC) lined with an endothelial layer (E) and filled with numerous capillaries (Cp). BM, basement membrane. H&E stain. Scale bar = 100 μm. (g) Sagittal section of the AD highlighting the numerous vacuoles (V) within the CE that are topped with a brush border (BB). The transport capacity is highlighted beneath the BM where multiple capillaries (Cp) are observed. H&E stain. Scale bar = 50 μm. (h) Transmission electron micrograph of the long, thin BB atop cells filled with multiple vacuoles (V) and lipid droplets (LI). Scale bar = 5 μm.

and resorption of materials (Fels et al., 1998). Hagfish are osmoconformers and it is unknown whether the primitive hagfish kidney regulates water content. Resorption and organic acid secretion do not occur in the archinephric duct as they would in the proximal tubules of other vertebrates (Rall and Burger, 1967; Fels et al., 1998; Riegel and Bardack, 1998) and the function of the distal tubule has not yet been investigated in hagfishes.

Gill

E. stoutii have 10–13 mediolaterally flattened gill pouches (GP) encapsulated in individual pouches and individually adjoined to the pharyngocutaneous duct (PCD) via a sphincter-controlled afferent water duct (AWD) (Mallatt and Paulsen, 1986). The convex lens-shaped gill pouches (GPs) demonstrate no obvious differences from one another apart from the disk-shaped first pouch and are each connected to the external environment via the efferent water duct (EWD) that exits through an unpigmented gill slit on the ventrolateral surface (Figures 1.9a and b). The left posterior-most gill slit is enlarged and used for water overflow. The water duct lies on the central axis of each gill and radiating outward from this point are the 40–60 highly branched lamellae (L) of varying lengths (Figures 1.9c and d).

The gills are well vascularized with branchial arteries (BA) encapsulating the pouch and visibly radiating outward from the central axis under the muscular layer (Choe et al., 1999; Evans et al., 2005). Each primary lamella is supplied with an artery that anastomoses first into pillar capillaries and then into the marginal capillaries (MC) in the secondary and tertiary lamellae, respectively (Adam, 1963; Andrew and Hickman, 1974; Mallatt and Paulsen, 1986). Pillar capillaries are generally 8 μm in diameter and are lined by flange-coated pillar cells (P), whereas MCs are 11 μm in diameter and are lined by smooth endothelial cells (Figure 1.9f) (Mallatt and Paulsen, 1986).

The blood–water interfacial distance allows for an efficient counter-current transfer across a thin respiratory surface (~2 cell-layers thick) and endothelium-lined capillaries (Mallatt and Paulsen, 1986; Evans et al., 2005). Water flows from the afferent zone (AZ), over the respiratory zone (RZ) and finally exits over the efferent zone (EZ) (Figure 1.9g). This efficient high-surface-area tissue, as well as a sedentary lifestyle, permits hagfish to have the lowest ventilation rate of any known vertebrate (Munz and Morris, 1965; Forster, 1990). This tissue is comprised of four cell types: basal (B), intermediate (I), pavement (PA), and ionocytes (IO) (Figure 1.9e) (Mallatt and Paulsen, 1986). The basal cells lie along the basal lamina (BL) and differentiate into intermediate cells, which are not present at the respiratory epithelium. The intermediate cells contain intracellular structures of either pavement cells or basal cells, for instance, small versus large Golgi

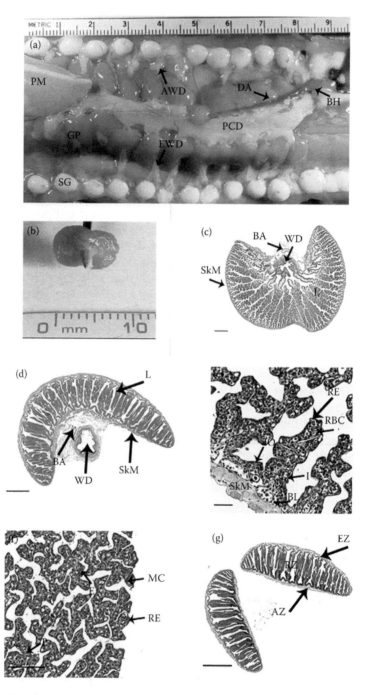

Figure 1.9 (**See color insert.**)   (*Continued*)

complexes (Mallatt and Paulsen, 1986). Pavement cells contain distinctive ellipsoidal secretory vesicles (SG) clustered at the apical surface, staining positive with the periodic acid–Schiff (PAS) stain indicating the presence of acidic mucous. The PAS stain is localized to the afferent zone, in the region where pavement cells dominate (Figure 1.9g). Mucous-containing goblet cells are not present in hagfish gills and it is thought that these vesicles fulfill that role (Mallatt and Paulsen, 1986). The Golgi complex is enlarged, whereas both the smooth endoplasmic reticulum and rough endoplasmic reticulum are regressed (Mallatt and Paulsen, 1986). Apart from respiration, the gill is also capable of acid/base and ion regulation owing to specialized ionocytes within the gill epithelium (see Chapter 11). These ionocytes (or mitochondrion-rich cells) are columnar in nature, containing numerous mitochondria (MI) and are interspersed among the pavement cells but in close proximity or on the filamental epithelium (Figure 1.10a). They have been demonstrated to express a suite of ion transporters including Na^+/H^+ exchanger isoforms 2 and 3 (NHE2, NHE3), Na^+/K^+ ATPase (NKA), V-type H^+-ATPase and Cl^-/HCO_3^- exchangers (Edwards et al., 2001; Choe et al., 2002; Tresguerres et al., 2006; Parks et al., 2008; Braun and Perry, 2010; Currie and Edwards, 2010). Both ionocytes and pavement cells have microplicae (averaging ~0.165 μm in length) on the apical (water facing) surface that can be seen as small projections of the apical surface. It is thought that these microplicae act to reduce friction and drag (Sperry and Wassersug, 1976). Lipid (LI) droplets were observed in all gill cell types, but not in the ionocytes of *Myxine* (Mallatt and Paulsen, 1986) (see Figure 1.10).

Figure 1.9 (Continued) Gill pouches. (a) *In situ* representation of the 10–13 gill pouch pairs (GP) lining the pharyngocutaneous duct (PCD). Adjoined to this duct via afferent water ducts (AWD), the gill exit port is through the efferent water duct (EWD). PM, pharyngeal muscle; SG, slime gland; DA, dorsal aorta; BH, branchial heart. (b) A singular GP. The outer skeletal muscle layer is noticeably encapsulated in a capillary network. (c) Sagittal representation of a GP. Centrally located is the water duct (WD) with lamellae (L) radiating outward from this point. The entire gill is encapsulated in a skeletal muscle (SkM) with branchial arteries (BA). H&E stain. Scale bar = 1 mm. (d) Transverse section of a GP highlighting the aforementioned components. WD, water duct; L, lamellae; SkM, skeletal muscle; BA, branchial artery. H&E stain. Scale bar = 1 mm. (e) Sagittal section of the lamellar base. Radiating upward from the skeletal muscle (SkM) layer are the lamellae stalks. Starting with a basal cell layer (BL), the cells differentiate into ionocytes (IO) and intermediate cells (I). At the respiratory epithelium (RE), fine capillaries contain many erythrocytes (RBC) for exchange. H&E stain. Scale bar = 100 μm. (f) Sagittal section of the respiratory epithelium (RE) at the end of a lamellar process. The pillar capillaries are lined with pillar cells (P) and become marginal capillaries (MC) at the lamellar end. E, endothelial cell. H&E stain. Scale bar = 100 μm. (g) Transverse plane demonstrating the afferent zone (AZ), efferent zone (EZ), and respiratory zone (RZ) of the gill. PAS stain. Scale bar = 1 mm.

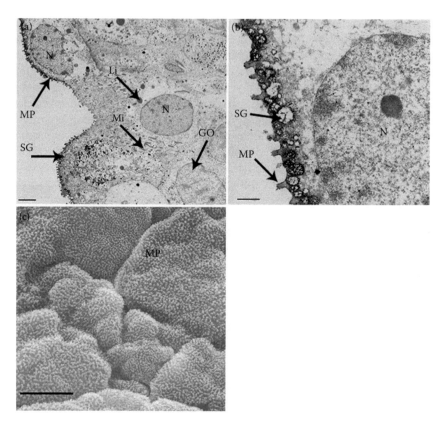

Figure 1.10 Gill pouches (GP). (a) Transmission electron microscopy detailing the ultrastructural components of the GP. The pavement cells are lined with microplicae (MP) while also housing secretory granules (SGs). The central cell in this image is an ionocyte, and numerous mitochondria (MI) are seen. Other apparent organelles include the Golgi complex (GO) and lipid deposits (LI). N, nucleus Scale bar = 2 μm. (b) MP, microplicae; SG, secretory granule; N, nucleus. Scale bar = 0.5 μm. (c) Scanning electron micrograph of the respiratory surface demonstrating the coverage of MP. Scale bar = 5 μm.

An extensive ultrastructure review of the gill of E. *stoutii* was conducted by Mallat and Paulsen (1986), to which the reader is referred for more details.

Cardiac tissue and associated structures

Hagfish have multiple hearts including branchial, portal, cardinal, and caudal. The branchial heart is the principal pump, lying posterior to the last gill pouch with the ventricle (Vt) slightly left of the atrium (At) and sinus venosus (Figure 1.11a) (Johansen, 1963; Jensen, 1966). The heart is

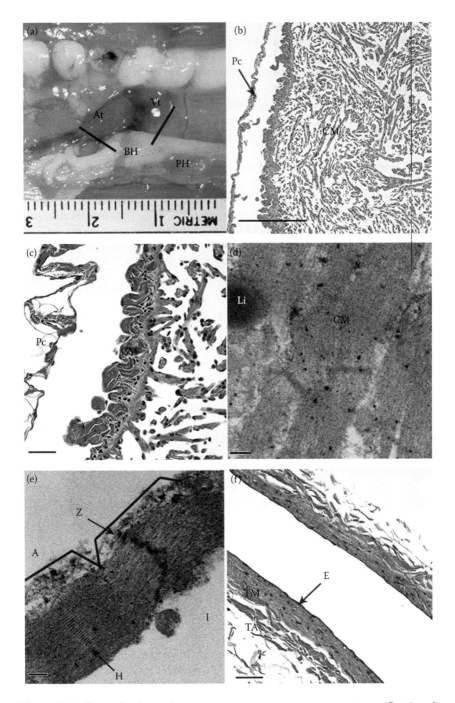

Figure 1.11 (See color insert.)   (*Continued*)

enveloped in a pericardium (Pc) and is incredibly basic and capable of anaerobic metabolism, a feat necessary for the hypoxic or anoxic conditions the hagfish regularly face (Johansen, 1963; Hardisty, 1979). A loose, striated cardiac muscle (CM) comprises both the ventricle and atrium of the branchial heart and contains lipid deposits (LI) as well as high levels of epinephrine and norepinephrine (Figure 1.11b–d) (Forster, 1998). As in expaxial muscles, the hagfish cardiac myofibrils contain typical vertebrate musculature including Z lines (Z) in the I band (I) and H zones (H) within the A band (A) (Figure 1.11e). The branchial and portal hearts lack autonomous regulation, however nerves have been found adjacent to and inside the epicardium of *E. stoutii* (Greene, 1925; Chapman et al., 1963; Hirsch et al., 1964; Farrell, 2007; Cox et al., 2011). The remaining three hearts are accessory in nature, residing within the sinuses that exist instead of capillaries and thus are necessary to force pooling blood back toward the branchial heart (Jensen, 1966). Further assistance can be provided by muscles, for instance, the gills which are known to pulse with ventilation (Jensen, 1966). The portal heart receives blood from the head as well as the gut and gonad via two openings, pushing it into the liver. The cardinal heart lies within the head and pulses due to skeletal muscle surrounding the pair of sacs (Jensen, 1966). The caudal heart is a pair of flattened sacs that contract due to external skeletal muscles that interact with the heart (Chapman et al., 1963) and therefore is not innervated itself. A cartilaginous shaft lies between the sacs and as the animal swims, this shaft compresses one sac to force blood out, whereas the other pools with blood (Jensen, 1966). Hagfish have the lowest cardiac output of all fish, similar to that of sedentary teleosts and elasmobranchs (Forster, 1998). This low output may hinder kidney function, but has a high glycolytic potential and a means by which to deal with anaerobic wastes (Forster, 1998; Cox et al., 2010).

The dorsal aorta leads oxygenated blood from the heart to the other tissues. Arteries are comprised of three layers: *tunica intima, tunica media,*

Figure 1.11 (Continued) Cardiac tissues. (a) Gross morphology of the true hearts: the branchial heart (BH) comprised of an atrium (At) and ventricle (Vt), and the portal heart (PH) leading to the liver. (b) Sagittal section of the cardiac muscle (CM) and surrounding pericardium (Pc) of the branchial heart. H&E stain. Scale bar = 500 μm. (c) Pc, pericardium; CM, cardiac muscle. H&E stain. Scale bar = 100 μm. (d) Transmission electron micrograph of the branchial heart cardiac muscle (CM). Lipid deposits (LI) were found within this tissue. Scale bar = 200 nm. (e) Detailed cardiac myofibril structure highlighting the dark Z line (Z) in the I band (I) and the H zone (H) in the A band (A). Scale bar = 200 nm. (f) Longitudinal section of the dorsal aorta. The inner lining is comprised of endothelial cells (E). Beneath this layer is the *Tunica media* (TM) and the *Tunica adventitia* (TA). H&E stain. Scale bar = 100 μm.

and *tunica adventitia*. The *tunica intima* contains an endothelium (E) of varying thickness as well as connective tissue (Worthington, 1905). The *tunica media* (TM) contains smooth muscle, unique elastic microfibrils (Forster, 1998), and is lined with endothelium (E), while the *tunica adventitia* (TA) is comprised of connective tissue (Figure 1.11f). The veins contain an endothelial layer atop connective tissue. Monomeric hemoglobins lacking immunoglobins comprise the blood, as the immune system instead produces a complement-like factor when faced with antigens, and injection of vertebrate antibodies yields no response (Forster, 1998). A 95% saturation is reached at oxygen levels lower than 40 mmHg, yet oxygenated blood is only about 50% saturated which suggests hagfish can live in 10–20 mmHg oxygen environments (Chapman et al., 1963; Johansen, 1963). This low oxygen-carrying capacity indicates a great extraction capacity (Forster, 1998). Red blood cells are reproduced in the bloodstream from large lymphocytes like the hemocytoblast of higher vertebrates, however, variable hematocrit levels make it difficult to calculate total blood volume (Andrew and Hickman, 1974; Forster, 1998).

Slime glands

Hagfish are notorious for the secretion of copious amounts of colorless slime when agitated (Leppi, 1968). Aiding in predator evasion via a gill clogging mechanism, slime can comprise 3–4% of total body weight (Blackstad, 1963; Downing et al., 1981; Fudge et al., 2005; Lim et al., 2006; Winegard and Fudge, 2010). It is produced by approximately 150 epidermally derived slime glands (SG) (Figure 1.12a) that line the entire length of the body in two rows at the border of the myomere and are ~0.5 cm long (Blackstad, 1963; Leppi, 1968; Lim et al., 2006). These glands are covered in connective tissue and a striated muscle (StM) layer (Figures 1.12b and c), which can produce localized contraction and expulsion of the gland contents through a 0.5 mm slit (Blackstad, 1963; Downing et al., 1981, 1984). Residing within the gland are two distinct cell types: thread cells (TCs) and mucous cells (MC) (Leppi, 1968; Terakado et al., 1975). Long, coiled, proteinaceous threads (Th) (1–60 cm in length) are within the 180 × 80 µm ellipsoidal thread cells (Figures 1.12c–d). These threads stain acidophilic and contain a compacted, spindle-like nucleus (N) shifted to one pole (Blackstad, 1963; Terakado et al., 1975; Downing et al., 1981, 1984; Fudge et al., 2005; Winegard et al., 2014). The mucous cells contain mucins, which upon ejection are believed to adhere to the thread cells and cause uncoiling. Agitation of the surrounding water is required to aid in the uncoiling as well as to cause swelling of the mucins, which become entangled with one another, producing slime that is 99% water (Leppi, 1968; Lim et al., 2006; Winegard and Fudge, 2010). See Chapter 15 for an extensive look at the slime glands.

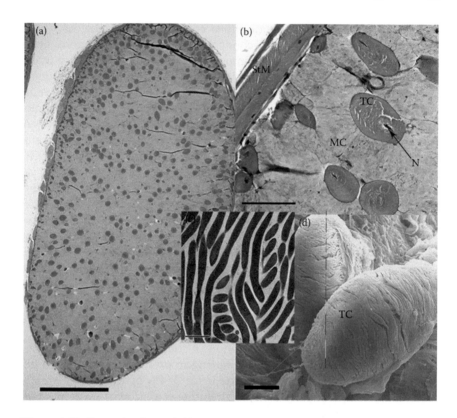

Figure 1.12 (**See color insert.**) Slime glands (SG). (a) Sagittal section of a whole SG. H&E stain. Scale bar = 1 mm. (b) The components of the SG. An outer skeletal muscle (SkM) contracts to expel the inner mucin vesicles (MC) and nucleated (N) thread cells (TC). H&E stain. Scale bar = 100 μm. (c) Transmission electron micrograph detailing the coiling nature of the TC within the SG. Scale bar = 2 μm. (d) Scanning electron micrograph of the tightly coiled TCs. Scale bar = 20 μm.

Reproductive organs

Little is known on the subject of hagfish reproduction and embryology, as they do not typically reproduce in captivity, and fertilized eggs are difficult to come across, with the last *E. stoutii* collection being in 1905 (Tsuneki and Gorbman, 1977). An exception is the unique *Eptatretus burgeri* that have seasonal migratory reproductive behavior (Fernholm, 1974) and have recently been bred in captivity, thereby opening a new topic of hagfish study. The gonads (GON) are located in the peritoneal cavity, attached via mesentery to the intestine (Figure 1.13a) (Koch et al., 1993). There are contradictory reports on the fertilization process, however, no copulatory organs have been found and thus, it is predicted that fertilization is an external process (Gorbman, 1990; Koch et al., 1993; Morisawa, 1999). Hagfish are said to be dioecious, with the posterior portion being male and the anterior female (Andrew and Hickman, 1974; Hardisty, 1979). The testes can develop while the ovaries regress to produce a male, or conversely, the testes regress and the ovaries develop to produce a female. Finally, the testes can develop along with the ovary to produce a hermaphrodite (Hardisty, 1979), although the proportion of functional hermaphrodites is unknown (Powell et al., 2005). The testes contain lobules (LB) up to 0.5 mm in diameter, which are surrounded by a single layer of epithelia (Figure 1.13b,c) (Andrew and Hickman, 1974; Jespersen, 1975). The outer edge of each lobule contains vegetative Sertoli cells and germinating spermatozoan (Jespersen, 1975; Gorbman, 1990). Early spermatids are spherical, averaging 14 μm in diameter, whereas mature sperm are about 40–60 μm long (Jespersen, 1975).

Hagfish ovaries are adjoined to the intestine, however, due to their thin membranous structure, they are not easily distinguished from the mesentery (Tsuneki and Gorbman, 1977). The hagfish produces 12–40 round eggs of 8 × 28 mm in size via spontaneous ovulation (Fernholm, 1975; Koch et al., 1993). During development, these eggs become ellipsoidal as they fill with yolk platelets (YPs) and house anchoring filaments (micropyles; Mpy) at each end to string the eggs together (Figure 1.13d,e) (Morisawa, 1999; Nozaki et al., 2000; Powell et al., 2005). The flattened egg surface is comprised of theca cells (TCs) embedded in connective tissue of collagen fibrils, which overlay the columnar granulosa layer (Andrew and Hickman, 1974; Tsuneki and Gorbman, 1977). A jelly coat containing acidic and neutral mucopolysaccharides and proteins also lines the egg (Koch et al., 1993). Brown bodies akin to a *corpus luteum* have been found and are indicative of a seasonal reproductive cycle in *Myxine* (Powell et al., 2005). The low number of eggs in a gravid female, coupled with a low number of sexually mature hagfish, may indicate a limited reproduction potential (Powell et al., 2004).

Recently, hormones involved in reproduction have been the subjects of study. Females demonstrate increased estradiol concentrations with the development of the ovaries, whereas hagfish without ovarian development have higher levels of testosterone and progesterone (Nozaki, 2013). Furthermore, the production of the egg protein vitellogenin can be prompted with estrogenic signals (Yu et al., 1981). Testicular development and hormones, however, are not so clear-cut. There is no observable correlation between development and plasma estradiol or testosterone, however progesterone has an inverse relationship (Yu et al., 1981). It has been suggested that hagfish growth hormone may be responsible for the induction and secretion of sex hormones (Yu et al., 1981).

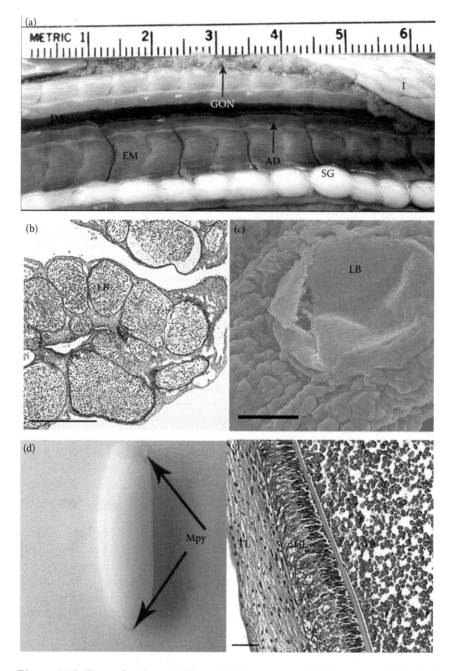

Figure 1.13 (**See color insert.**) Reproductive organs. (a) Gross morphological details of the gonads (GON) which are attached to the intestine (I) via mesentery. DA, dorsal aorta; AD, archinephric duct;   (*Continued*)

Methods

Tissue perfusion and fixation

Pacific hagfish (*E. stoutii*) (80–155 g) were collected from Barkley Sound and housed in outdoor tanks at the Bamfield Marine Sciences Centre in Bamfield, BC, Canada. Nine hagfish were weighed and anesthetized with MS-222 (2 g/L). Hagfish were perfused for 1 h with a 4% paraformaldehyde and buffered formaldehyde solution at pH 7.4. Three hagfish had the following tissues removed and placed in glass scintillation vials of formalin for light microscopy: dental plate, sensory barbels, skin, pharyngeal muscle, body wall muscle (*M. parietalis*), gill pouches, liver, gall bladder, dorsal aorta, postcardinal vein, ventricle and atrium of branchial heart, eggs (if present), testes/ovaries, archinephric duct, slime glands, and intestine at defined regions along the length from mouth to anus. Samples were kept in formalin overnight in a refrigerator (4°C), washed twice with 70% ethanol (EtOH) for 30 min before transport to the University of Alberta (U of A) for light microscopy studies. An additional three hagfish had the aforementioned tissues removed and prepared for both SEM and transmission electron microscopy (TEM). Tissues were cut in $2 \times 2 \times 2$ mm sections before being placed in 4% paraformaldehyde, 2.5% glutaraldehyde, 0.05 mmol L^{-1} sodium cacodylate buffer (pH 7.4) in glass scintillation vials on ice (4°C). Fixation occurred over a 3-h period before washing in 30%, 50%, and 70% EtOH for 1 h each with a second and final wash in 70% EtOH. SEM samples were transported to the U of A. TEM samples were placed in 1% osmium tetroxide in water for 1 h, immediately rinsed twice in 70% EtOH, left for 1 h, and washed three more times in 70% EtOH at 1-h intervals before transport to the U of A. Three hagfish were also used for description of gross morphology. The animals were euthanized, dissected, and photographed with a Canon EOS 500-D (Japan) under bright light illumination. One hagfish had the DA cannulated with PE50 tubing to perfuse the tissues with Coomassie Blue solution (1% in 500 mmol L^{-1} NaCl) to better visualize glomeruli and the capillary network.

Figure 1.13 (Continued) EM, epaxial muscle; SG, slime gland. (b) Light micrograph of the testes. Numerous lobules (LB) of different sizes hold sperm of different maturity. H&E stain. Scale bar = 500 µm. (c) Scanning electron micrograph of the testes lobule (LB), which contains the maturing sperm. Scale bar = 100 µm. (d) Gross portrait of an egg. At either end are tufted micropyles (Mpy) which are used to attach the eggs to one another. (e) Sagittal section of an egg. The outer protective thecal layer (TC) rests atop the granulosa layer (GL). Within the egg itself are numerous yolk platelets (YP) used for nourishment of a developing embryo. H&E stain. Scale bar = 100 µm.

Light microscopy

Tissues were trimmed and embedded in paraffin prior to microtome sectioning (5 μm sections). Tissues were poststained with hematoxylin and eosin (H&E) to emphasize the following features: nuclei (blue/black), cartilage (pink), eosinophil/zymogen granules and keratin (bright pink/orange), cytoplasm (pink), muscles/thick elastic fibers (deep pink), and collagen (light pink; Morrison et al., 2006). The stained tissues were viewed under a Zeiss Scope.A1 microscope with images captured on an optronic camera.

Scanning electron microscopy

Tissues were trimmed and placed in a dehydration series for 1 h. They were then immersed in various 1,1,1,3,3,3-hexamethyldisilazane (HMDS; Electron Microscopy Sciences, Hatfield, PA, USA): ethanol solutions (75:25, 50:50, 25:75) for 15 min each. Tissues were rinsed twice in pure HMDS for 15 min each and then dried overnight in HMDS. Gold sputter coating was completed using a Hummer 6.2 Sputter Coater (Anatech Ltd.) for 2–3 min. Images were taken on an FEI Company Scanning Electron Microscope (Model: XL30) operating at 20 kV (FEI Company, Oregon).

Transmission electron microscopy

Osmicated samples in 70% EtOH were further dehydrated completely in an EtOH series (70%, 90%, 100% (3×)) and then immersed in a Spurr resin:ethanol mix (1:1) for 3 h before pure Spurr resin immersion overnight. Samples were then cured and embedded in pure Spurr resin (Electron Microscopy Sciences, Hatfield, PA, USA). Sections were made on an ultramicrotome and then poststained with uranyl acetate and lead citrate. Images were captured using a Philips FEI transmission electron microscopy (Model: Morgagni 268) operating at 80 kV and a Gatan Orius CCD camera. All images were transferred to Adobe Photoshop CS6 for resizing, sharpening, and color adjustment. The images were combined into plates and labeled using Microsoft Publisher.

Acknowledgments

We thank Arlene Oatway at the University of Alberta Microscopy Suite and staff at Bamfield Marine Sciences Centre. This work was supported by an NSERC Discovery Grant to G.G.G.

References

Adam, H. 1963. Structure and histochemistry of the alimentary canal. In A. Brodal and R. Fänge (eds.), *The Biology of Myxine*. Universitetsforlaget, Oslo, pp. 256–288.

Anderson, I. G. and G. A. Haslewood. 1967. New evidence for the structure of myxinol. *Biochem J* 104:1061–1063.

Andres, K. H. and M. von Düring. 1993. Cutaneous and subcutaneous sensory receptors of the hagfish *Myxine glutinosa* with special respect to the trigeminal system. *Cell Tissue Res* 274:353–366.

Andrew, W. and C. P. Hickman. 1974. *Histology of the Vertebrates (Illustrated)*. Mosby, St. Louis.

Appy, R. G. and R. C. Anderson. 1981. The parasites of lampreys. In M.W. Hardisty (ed.), *The Biology of Lampreys*, Vol. 3, Academic Press, New York, pp. 1–42.

Baldwin, J. W. Davison and M. E. Forster. 1991. Anaerobic glycolysis in the dental plate retractor muscles of the New Zealand hagfish *Eptatretus cirrhatus* during feeding. *J Exp Biol* 260:295–301.

Blackstad, T. 1963. The skin and slime glands. In A. Brodal and R. Fänge (eds.), *The Biology of Myxine*. Universitetsforlaget, Oslo, pp. 195–230.

Braun, C. B. 1998. Schreiner organs: A new craniate chemosensory modality in hagfishes. *J Comp Neurol* 392:135–163.

Braun, M. H. and S. F. Perry. 2010. Ammonia and urea excretion in the Pacific hagfish *Eptatretus stoutii*: Evidence for the involvement of Rh and UT proteins. *Comp Biochem Phys A* 157:405–415.

Chapman, C. B., D. Jensen and K. Wildenthal. 1963. On circulatory control mechanisms in the Pacific hagfish. *Circ Res* 12:427–440.

Chiu, K.-H., S. Ding, Y.-W. Chen, C.-H. Lee and H.-K. Mok. 2011. A NMR-based metabolomic approach for differentiation of hagfish dental and somatic skeletal muscles. *Fish Physiol Biochem* 37:701–707.

Choe, K. P., S. Edwards, A. I. Morrison-Shetlar, T. Toop and J. B. Claiborne. 1999. Immunolocalization of Na$^+$/K$^{(+)}$-ATPase in mitochondrion-rich cells of the Atlantic hagfish (*Myxine glutinosa*) gill. *Comp Biochem Phys A* 124:161–168.

Choe, K. P., A. I. Morrison-Shetlar, B. P. Wall and J. B. Claiborne. 2002. Immunological detection of Na$^{(+)}$/H$^{(+)}$ exchangers in the gills of a hagfish, *Myxine glutinosa*, an elasmobranch, *Raja erinacea*, and a teleost, *Fundulus heteroclitus*. *Comp Biochem Phys A* 131:375–385.

Clark, A. J. and A. P. Summers. 2007. Morphology and kinematics of feeding in hagfish: Possible functional advantages of jaws. *J Exp Biol* 210:3897–3909.

Clark, A. J. and A. P. Summers. 2012. Ontogenetic scaling of the morphology and biomechanics of the feeding apparatus in the Pacific hagfish *Eptatretus stoutii*. *J Fish Biol* 80:86–99.

Coonfield, B. R. 1940. The pigment in the skin of *Myxine glutinosa*. *Trans Am Microsc Soc* 59:398–403.

Cox, G. K., E. Sandblom and A. P. Farrell. 2010. Cardiac responses to anoxia in the Pacific hagfish, *Eptatretus stoutii*. *J Exp Biol* 213:3692–3698.

Cox, G. K., E. Sandblom, J. G. Richards and A. P. Farrell. 2011. Anoxic survival of the Pacific hagfish (*Eptatretus stoutii*). *J Comp Physiol B* 181:361–371.

Currie, S. and S. L. Edwards. 2010. The curious case of the chemical composition of hagfish tissues-50 years on. *Comp Biochem Phys A* 157:111–115.

Dawson, J. A. 1963. The oral cavity, the "jaws" and the horny teeth of *Myxine glutinosa*. In A. Brodal and R. Fänge (eds.), *The Biology of Myxine.* Universitetsforlaget, Oslo, pp. 231–255.

Downing, S. W., R. H. Spitzer, E. A. Koch and W. L. Salo. 1984. The hagfish slime gland thread cell. I. A unique cellular system for the study of intermediate filaments and intermediate filament–microtubule interactions. *J Cell Biol* 98:653–669.

Downing, S. W., R. H. Spitzer, W. L. Salo, J. S. Downing, L. J. Saidel and E. A. Koch. 1981. Threads in the hagfish slime gland thread cells: Organization, biochemical features, and length. *Science* 212:326–328.

Edwards, S. L., J. B. Claiborne, A. I. Morrison-Shetlar and T. Toop. 2001. Expression of Na$^{(+)}$/H$^{(+)}$ exchanger mRNA in the gills of the Atlantic hagfish (*Myxine glutinosa*) in response to metabolic acidosis. *Comp Biochem Phys A* 130:81–91.

Evans, D. H., P. M. Piermarini and K. P. Choe. 2005. The multifunctional fish gill: Dominant site of gas exchange, osmoregulation, acid–base regulation, and excretion of nitrogenous waste. *Physiol Rev* 85:97–177.

Fänge, R. 1963. Structure and function of the excretory organs. In A. Brodal and R. Fänge (eds.), *The Biology of Myxine.* Universitetsforlaget, Oslo, pp. 516–529.

Fänge, R., G. Bloom and R. Strahan. 1963. The portal vein heart of myxinoids. In A. Brodal and R. Fänge (eds.), *The Biology of Myxine.* Universitetsforlaget, Oslo, pp. 340–351.

Farrell, A. P. 2007. Cardiovascular systems in primitive fishes. In D. McKenzie, A. P. Farrell and C. Brauner (eds.), *Primitive Fishes.* Academic Press, London, pp. 53–120.

Fels, L. M., S. Kastner and H. Stolte. 1998. The hagfish kidney as a model to study renal physiology and toxicology. In J. M. Jørgensen, J. P. Lomholt, R. E. Weber and H. Malte (eds.), *The Biology of Hagfishes.* Chapman & Hall, London, pp. 347–363.

Fernholm, B. 1974. Diurnal variations in the behaviour of the hagfish *Eptatretus burgeri. Mar Biol* 27:351–356.

Fernholm, B. 1975. Ovulation and eggs of the hagfish *Eptatretus burgeri. Acta Zool* 204:199–204.

Forster, M. E. 1990. Confirmation of the low metabolic rate of hagfish. *Comp Biochem Phys A* 96:113–116.

Forster, M. E. 1998. Cardiovascular function in hagfishes. In J. M. Jørgensen, J. P. Lomholt, R. E. Weber and H. Malte (eds.), *The Biology of Hagfishes.* Chapman & Hall, London, pp. 237–258.

Fudge, D. S., N. Levy, S. Chiu and J. M. Gosline. 2005. Composition, morphology and mechanics of hagfish slime. *J Exp Biol* 208:4613–4625.

Glover, C. N., C. Bucking and C. M. Wood. 2011. Adaptations to *in situ* feeding: Novel nutrient acquisition pathways in an ancient vertebrate. *Proc R Soc B Biol Sci* 278:3096–3101.

Gorbman, A. 1990. Sex differentiation in the hagfish *Eptatretus stoutii. Gen Comp Endocrinol* 77:309–323.

Greene, C. W. 1925. Notes on the olfactory and other physiological reactions of the California hagfish. *Science* 61:68–70.

Hardisty, M. W. 1979. *Biology of the Cyclostomes.* Chapman & Hall, London.

Hirsch, E. F., M. Jellinek and T. Cooper. 1964. Innervation of systemic heart of California hagfish. *Circ Res* 14:212–217.

Inui, Y. and A. Gorbman. 1978. Role of the liver in regulation of carbohydrate metabolism in hagfish *Eptatretus stoutii*. *J Comp Biol Phys A* 60:181–183.

Jansen, J. K. S. and P. Andersen. 1963. Anatomy and physiology of the skeletal muscles. In A. Brodal and R. Fänge (eds.), *The Biology of Myxine*. Universitetsforlaget, Oslo, pp. 161–194.

Jensen, D. 1966. The hagfish. *Sci Am* 214:82–90.

Jespersen, A. 1975. Fine structure of spermiogenesis in eastern Pacific species of hagfish (Myxinidae). *Acta Zool* 56:189–198.

Johansen, K. 1963. The cardiovascular system of *Myxine glutinosa*. In A. Brodal and R. Fänge (eds.), *The Biology of Myxine*. Universitetsforlaget, Oslo, pp. 289–316.

Johnson, F. R., R. M. McMinn and R. F. Birchenough. 1962. The ultrastructure of the gall-bladder epithelium of the dog. *J Anat* 96:477–487.

Jordan, E. and C. C. Speidel. 1930. Blood formation in cyclostomes. *Am J Anat* 46:355–391.

Koch, E. A., R. H. Spitzer, R. B. Pithawalla, F. A. Castillos and L. J. Wilson. 1993. The hagfish oocyte at late stages of oogenesis: Structural and metabolic events at the micropylar region. *Tissue Cell* 25:259–273.

Korneliussen, H. and K. Nicolaysen. 1973. Ultrastructure of four types of striated muscle fibers in the Atlantic hagfish (*Myxine glutinosa* L.). *Z Zellforsch* 143:273–290.

Krejsa, R. J., P. Bringas and H. C. Slavkin. 1990. A neontological interpretation of conodont elements based on agnathan cyclostome tooth structure, function, and development. *Lethaia* 23:359–378.

Leppi, T. J. 1968. Morphochemical analysis of mucous cells in the skin and slime glands of hagfishes. *Histochemistry* 15:68–78.

Lim, J., D. S. Fudge, N. Levy and J. M. Gosline. 2006. Hagfish slime ecomechanics: Testing the gill-clogging hypothesis. *J Exp Biol* 209:702–710.

Mallatt, J. O. N. and C. Paulsen. 1986. Gill ultrastructure of the Pacific hagfish *Eptatretus stoutii*. *Am J Anat* 177:243–269.

Malte, H. and J. P. Lomholt. 1998. Ventilation and gas exchange. In J. M. Jørgensen, J. P. Lomholt, R. E. Weber and H. Malte (eds.), *The Biology of Hagfishes*. Chapman & Hall, London, pp. 223–234.

Martini, F. H. 1998. The ecology of hagfishes. In J. M. Jørgensen, J. P. Lomholt, R. E. Weber and H. Malte (eds.), *The Biology of Hagfishes*. Chapman & Hall, London, pp. 57–78.

McInerney, J. E. 1974. Renal sodium reabsorption in the hagfish, *Eptatretus stoutii*. *Comp Biochem Physiol* 49:273–280.

Mellgren, S. I. and J. S. Mathisen. 1966. Oxidative enzymes, glycogen and lipid in striated muscle. *Z Zellforsch* 71:169–188.

Menke, A. L., J. M. Spitsbergen, A. P. M. Wolterbeek and R. A. Woutersen. 2011. Normal anatomy and histology of the adult zebrafish. *Toxicol Pathol* 39:759–775.

Morisawa, S. 1999. Fine structure of micropylar region during late oogenesis in eggs of the hagfish *Eptatretus burgeri* (Agnatha). *Dev Growth Differ* 41:611–618.

Morrison, C. M., K. Fitzsimmons and J. R. J. Wright. 2006. *Atlas of Tilapia Histology*. The World Aquaculture Society, Baton Rouge.

Mugnaini, E. and S. B. Harboe. 1967. The liver of *Myxine gluctinosa*: A true tubular gland. *Z Zellforsch* 78:341–369.

Munz, F. W. and R. W. Morris. 1965. Metabolic rate of the hagfish, *Eptatretus stoutii* (Lockington) 1878. *Comp Biochem Physiol* 16:1–6.

Nozaki, M. 2013. Hypothalamic–pituitary–gonadal endocrine system in the hagfish. *Front Endocrinol (Lausanne)* 4:200.

Nozaki, M., T. Ichikawa, K. Tsuneki and M. Kobayashi. 2000. Seasonal development of gonads of the hagfish, *Eptatretus burgeri*, correlated with their seasonal migration. *Zool Sci* 17:27–40.

Papermaster, B. W., R. M. Condie and R. A. Good. 1962. Immune response in the California hagfish. *Nature* 196:355–357.

Parks, S. K., M. Tresguerres and G. G. Goss. 2008. Theoretical considerations underlying Na$^{(+)}$ uptake mechanisms in freshwater fishes. *Comp Biochem Phys C Toxicol Pharmacol* 148:411–418.

Potter, I. C., U. Welsch, G. M. Wright, Y. Honma and A. Chiba. 1995. Light and electron microscope studies of the dermal capillaries in three species of hagfishes and three species of lamprey. *J Zool* 235:677–688.

Powell, M. L., S. I. Kavanaugh and S. A. Sower. 2004. Seasonal concentrations of reproductive steroids in the gonads of the Atlantic hagfish, *Myxine glutinosa*. *J Exp Zool A Comp Exp Biol* 301:352–360.

Powell, M. L., S. I. Kavanaugh and S. A. Sower. 2005. Current knowledge of hagfish reproduction: Implications for fisheries management. *Integr Comp Biol* 45:158–165.

Rall, D. P. and J. W. Burger. 1967. Some aspects of hepatic and renal excretion in Myxine. *Am J Physiol* 212:354–356.

Read, L. J. 1975. Absence of ureogenic pathways in liver of the hagfish Bdellostoma cirrhatum. *Comp Biochem Physiol B* 51(1):139–141.

Rice, R. H., V. J. Wong and K. E. Pinkerton. 1994. Ultrastructural visualization of cross-linked protein features in epidermal appendages. *J Cell Sci* 107:1985–1992.

Riegel, J. A. and D. Bardack. 1998. An analysis of the function of the glomeruli of the hagfish mesonephric kidney. In J. M. Jørgensen, J. P. Lomholt, R. E. Weber and H. Malte (eds.), *The Biology of Hagfishes*. Chapman & Hall, London, pp. 364–376.

Rogers, A. W. 1983. *Cells and Tissues: An Introduction to Histology and Cell Biology*. Academic Press, New York.

Schultz, A. G., S. C. Guffey, A. M. Clifford and G. G. Goss. 2014. Phosphate absorption across multiple epithelia in the Pacific hagfish (*Eptatretus stoutii*). *Am J Physiol—Reg I* 307:R643–R652.

Slavkin, H. C., E. Graham, M. Zeichner-David and W. Hildemann. 1983. Enamellike antigens in hagfish: Possible evolutionary significance. *Evolution* 37:404–412.

Sperry, D. G. and Wassersug, R. J. 1976. A proposed function for microridges on epithelial cells. *Anat Rec* 185(2):253–257.

Spitzer, R. H. and E. A. Koch. 1998. Hagfish skin and slime glands. In J. M. Jørgensen, J. P. Lomholt, R. E. Weber and H. Malte (eds.), *The Biology of Hagfishes*. Chapman & Hall, London, pp. 109–132.

Terakado, K., M. Ogawa, Y. Hashimoto and H. Matsuzaki. 1975. Ultrastructure of the thread cells in the slime gland of Japanese hagfishes, *Paramyxine atami* and *Eptatretus burgeri*. *Cell Tissue Res* 159:311–323.

Tresguerres, M., S. K. Parks and G. G. Goss. 2006. V-H(+)-ATPase, Na(+)/K(+)-ATPase and NHE2 immunoreactivity in the gill epithelium of the Pacific hagfish (*Epatretus stoutii*). *Comp Biochem Phys A* 145:312–321.

Tsuneki, K. and A. Gorbman. 1977. Ultrastructure of the ovary of the hagfish *Eptatretus stoutii. Acta Zool* 58:27–40.

Umezu, A., H. Kametani, Y. Akai, T. Koike and N. Shiojiri. 2012. Histochemical analyses of hepatic architecture of the hagfish with special attention to periportal biliary structures. *Zool Sci* 29:450–457.

Welsch, U., S. Büchl and R. Erlinger. 1998. The dermis. In J. M. Jørgensen, J. P. Lomholt, R. E. Weber and H. Malte (eds.), *The Biology of Hagfishes.* Chapman & Hall, London, pp. 133–142.

Winegard, T. M. and D. S. Fudge. 2010. Deployment of hagfish slime thread skeins requires the transmission of mixing forces *via* mucin strands. *J Exp Biol* 213:1235–1240.

Winegard, T., J. Herr, C. Mena, B. Lee, I. Dinov, D. Bird, M. Bernards et al. 2014. Coiling and maturation of a high-performance fibre in hagfish slime gland thread cells. *Nat Commun* 5:3534.

Worthington, J. 1905. Contribution to our knowledge of the myxinoids. *Am Nat* 39:625–663.

Yan, H. Y. 2009. A histochemical study on the snout tentacles and snout skin of bristlenose catfish *Ancistrus triradiatus. J Fish Biol* 75:845–861.

Yousen, J. H. 1979. The liver. In M. W. Hardisty (ed.), *The Biology of Lampreys,* Vol. 3. Academic Press, New York, pp. 263–332.

Yu, J. Y. L., W. W. Dickhoff, P. Swanson and A. Gorbman. 1981. Vitellogenesis and its hormonal regulation in the Pacific hagfish, *Eptatretus stoutii* L. *Gen Comp Endocrinol* 43:492–502.

Zintzen, V., C. D. Roberts, M. J. Anderson, A. L. Stewart, C. D. Struthers and E. S. Harvey. 2011. Hagfish predatory behaviour and slime defence mechanism. *Sci Rep* 1:131.

chapter two

Hagfish fisheries research

Scott M. Grant

Contents

Introduction

It has been demonstrated that hagfish populations cannot withstand heavy fishing pressure (Martini et al., 1997a; Honma, 1998; Martini, 1998; AHWG, 2003). Subsequently, there is a clear need not only to understand how hagfish populations respond to a commercial fishery, but also conservation measures that avoid overharvesting. Unfortunately, much of the scientific information required to assess the status of fish stocks is completely lacking for hagfish. Unambiguous estimates of abundance are difficult, if not impossible to obtain, growth rate and longevity are unknown, several aspects of the reproductive cycle are poorly understood, and it is unclear as to what extent populations can undergo a compensatory increase in production at low stock size. For example, for Atlantic hagfish, the time required for a clutch of eggs to ripen may approximate the teleost condition (i.e., 6–8 months) or be considerably longer (Patzner, 1998; Powell et al., 2004) and the latter may be expected to make it more difficult to compensate for heavy fishing pressure. However, evidence for the lack of an obligatory spawning season in Atlantic hagfish populations (Cunningham, 1886; Nansen, 1887) and an ability to adjust the duration of various phases of the reproductive cycle to environmental conditions may positively influence production at reduced stock size. Ultimately, sustainable resource management seeks to match the level of removals to a level the resource can sustain. Challenges to sustainable management of hagfish fisheries include high susceptibility to overexploitation due to life-history traits, catchability, and predominance of females in the catch which have been well documented for Atlantic hagfish (Scott and Scott, 1988; Honma, 1998; Martini, 1998; Patzner, 1998; AHWG, 2003; NEFSC, 2003; Grant, 2006). In addition, evidence of only short migrations (Walvig, 1963, 1967) suggests Atlantic hagfish populations are localized and recruitment-dependent.

In the Newfoundland and Labrador region of Atlantic Canada, interest in newly emerging fisheries for Atlantic hagfish dates back to early- to mid-1990s. Early attempts at harvesting and marketing Atlantic hagfish were from exploratory surveys within Subdivision 3Pn, a North Atlantic Fisheries Organization (NAFO) regulatory area on the southwest coast of insular Newfoundland (Figure 2.1). These exploratory surveys were unsuccessful as Atlantic hagfish occurring in this area exhibited a much smaller body length than those that were being harvested in the newly developed (i.e., 1993) Gulf of Maine Atlantic hagfish fishery. Renewed interest in the Newfoundland region came in 2002 from fisher's ecological knowledge of predation by somewhat larger Atlantic hagfish in the monkfish (*Lophius*

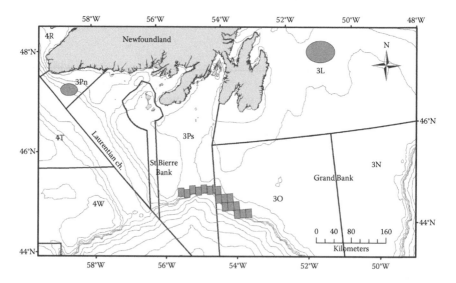

Figure 2.1 **(See color insert.)** Map showing location of Atlantic hagfish fixed grid research survey and commercial fishing areas in NAFO Division 3O and Subdivision 3Ps (shaded squares) and location of exploratory Atlantic hagfish surveys in Division 3L and Subdivision 3Pn (shaded ellipses).

americanus) fishery on the southwest slope of the Grand Bank in NAFO Division 3O. An exploratory survey in 3O revealed not only that Atlantic hagfish were larger than the 3Pn stock, but also that they occurred in commercial quantities (Grant, 2006). However, the apparent boom and bust nature of hagfish fisheries globally (Honma, 1998) and evidence of a downturn in the Gulf of Maine Atlantic hagfish fishery within 9–10 years (AHWG, 2003) provided cause for concern over the long-term economic and biological sustainability of Atlantic hagfish fisheries. Moreover, the outcome of these fisheries emphasize the need for conservation and science at the forefront of new fisheries development and continued monitoring of fishing activities once a fishery is established.

In the mid-2000s, two 5-year limited entry Atlantic hagfish fisheries were initiated in the Newfoundland region, one on the slope of the Grand Bank (NAFO Division 3O) and the other on the slope of St. Pierre Bank (NAFO Subdivision 3Ps) (Figure 2.1). Prior to the autumnal fishing season, baited trap fixed-grid research surveys were carried out annually in areas of interest within 3O and 3Ps to collect Atlantic hagfish samples and assess trends in catch rates, trap selectivity, population size structure, spatial distribution, and life-history traits. Each provisional fishery was monitored and limited spatially to the area encompassed by the fixed-grid survey and by annual allocation (182 metric tons) in an effort to determine the level of commercial fishing the resource can sustain. Areas selected for the surveys (Figure 2.1) were based on fisher's traditional ecological knowledge of Atlantic hagfish scavenging in monkfish gillnet fisheries. The commercial fishing industry participated in all aspects of fishery development for Atlantic hagfish including volunteering commercial fishing vessels as research platforms during the annual fixed-grid surveys. To better understand the life history of Atlantic hagfish dissection, and detailed gonadal analysis was conducted for over 9700 animals collected from 2005 to 2008. Spatial–temporal analyses of maturity were also investigated from samples obtained during exploratory/research surveys and during the 2013 Atlantic hagfish fishery on the slope of the St. Pierre Bank.

Materials and methods

Atlantic hagfish samples were collected from the southwestern slope of the Grand Bank (NAFO Division 3O) and St. Pierre Bank (NAFO Subdivision 3Ps) over four consecutive years from 2005 to 2008 (Figure 2.1). In 2006, samples of Atlantic hagfish were also obtained from an exploratory survey in the Laurentian Channel region of NAFO Subdivision 3Pn (Figure 2.1). All Atlantic hagfish samples were collected in August–September. The Grand Bank and St. Pierre Bank samples were collected during fixed-grid research surveys, with 50 sites randomly sampled annually. Ten fixed-grids comprised of 10′ of latitude by 10′ of longitude were surveyed on the slope of the Grand Bank, while on the slope of the St. Pierre Bank six similar-sized grids and a seventh 2′ × 10′ grid were sampled. The research surveys were carried out with a string of baited 227 L traps that possessed 14.3 and 15.1 mm diameter escape holes and control traps that possessed remove's' 3.2 mm diameter holes. All traps were fitted with four 50-mm diameter one-way entrance funnels that possessed flexible teeth which prevented hagfish from escaping back through the entrance. From each survey, site samples of up to 21 L of Atlantic hagfish were collected from all three trap types, fast frozen at sea, and maintained at −20°C until analysis.

Laboratory analysis of Atlantic hagfish samples collected during the research surveys was carried out to (1) obtain body length comparisons among trap types and depths and (2) obtain a greater understanding of life-history processes. To meet the first objective, individual body length (±1 mm) and weight (±0.1 g) were obtained for Atlantic hagfish collected from all three trap types deployed at a survey site with at least 20 survey sites analyzed per year. In total, 88,490 Atlantic hagfish were examined for length and weight in 2005–2008. To meet the second objective, samples collected from traps with 14.3 and 15.1 mm escape holes were dissected to expose the gonadal components for examination with 5–12 survey sites examined annually. An effort was made not only to examine samples from a broad geographic area and wide range of depths, but also to examine samples from the same heavily fished grids over time to track fishery-induced trends in life-history processes. In total, 9749 Atlantic hagfish were dissected and examined. In addition, more than 1000 Atlantic hagfish samples from control traps were also dissected and examined to provide suitable numbers of small individuals for maturity assessments.

Sex was established by stage of gonadal development (Grant, 2006). Females produce a single clutch of eggs of similar size during a reproductive cycle. The number and average length and diameter (±0.1 mm) of developing eggs in a clutch were recorded. Lengths of degenerating eggs within a clutch were also recorded. Degenerated eggs were no longer ovoid in shape or they were markedly smaller than developing eggs or collapsed and generally exhibited a more proximal location to the ovary than developing eggs (see also Walvig, 1963). When developing eggs enter the final stages of maturation, they begin to form anchor filaments on their apical ends. Eggs in the final stages of maturation were recorded and weighed (±0.1 g). When the eggs are fully developed (i.e., ripe), they are ovulated from their follicle. Subsequently, the postovulatory follicle becomes glandular, then it is gradually resorbed into the ovary. Three stages in the regression of the postovulatory follicle were identified: a glandular corpus luteum stage followed by a medial stage and fully regressed stage. The corpus luteum stage is a large (5–10 mm) white colored glandular structure that hangs under the ovary. As the corpus luteum regresses to the medial stage, it loses its glandular appearance, takes on a more proximal location to the ovary, and becomes smaller (2–5 mm) and amber in color. The fully regressed stage of the follicle is brown in color, further reduced in size (1–2 mm), and fully regressed back into the ovary. When encountered, the number and length (±0.1 mm) of corpus lutea and medial-stage follicles were recorded, while presence or absence was recorded for fully regressed follicles. Once the internal examination was complete, the liver was removed and weighed, and eviscerated (gutted) body weight was also recorded.

Results and discussion

Sex ratios and body length comparisons

The sex ratio of Atlantic hagfish captured in baited traps favored females at 97:1 on the slope of the Grand Bank and 99:1 on the slope of the St. Pierre Bank (Table 2.1). Atlantic hagfish classified as females possessed ovoid eggs ≥2 mm in length and did not show evidence of testicular development. The testis of animals classified as males possessed well-developed lobules and/or fluid-filled vacuoles and did not exhibit ovarian development. It has been suggested that the sexes are geographically separate during certain times of the year (Holmgren, 1946), and Walvig (1963) summarizes cases where male Atlantic hagfish were plentiful in the eastern North Atlantic. Analysis of the sex of 9749 Atlantic hagfish collected over a broad geographic range and at depths of 110–823 m from the slope of Grand and St. Pierre banks during August–September of 2005–2008 failed to identify areas where males were plentiful.

Atlantic hagfish considered to be hermaphrodites were rarely encountered and accounted for <0.2% of the animals examined (Table 2.1). Hermaphroditic animals were highly variable in the stage of gonadal development and degree of overlap or intermingling of gonadal components. For example, two of the hermaphrodites possessed corpus lutea indicative of a recent spawning event and one to three enlarged (5–15 mm diameter) and fluid-filled vacuoles within the testis. Eight of the hermaphrodites possessed medial-stage postovulatory follicles that had not fully regressed into the ovary and a testis in various stages of maturation from well-developed lobules to enlarged, fluid-filled vacuoles. Five additional hermaphrodites possessed large eggs (>20 mm in length) and three of these animals also possessed enlarged, fluid-filled vacuoles within the testis. The eggs in one of these hermaphrodites was in the final stages of maturation (anchor filaments developing) and the testis possessed two enlarged and fluid-filled vacuoles. Overall, these results suggest the occurrence of fully functional hermaphrodites within the Grand and St. Pierre Bank Atlantic hagfish populations but this sexual characteristic is extremely rare.

Table 2.1 Atlantic hagfish body length summary by sex for the Grand Bank (GB) and St. Pierre Bank (SPB)

Area	Sex	No. examined	Total length (mm)				t-test		
			Minimum	Maximum	Mean	±1 SE	df	t-value	p-value
GB	Female	4960	326	661	459.4	0.6	161	1.701	0.091
	Male	158	340	714	449.7	5.7			
	Hermaphrodite	19	402	566	458.8	11.6			
	Sterile	8	543	622	580.4	4.2			
SPB	Female	4553	330	630	447.4	0.6	34	0.399	0.692
	Male	35	325	675	441.9	13.5			
	Hermaphrodite	12	397	519	460.3	12.2			
	Sterile	1	558	558	558.0	—			

Results of t-tests of differences in mean body length between males and females are shown for each area.

Atlantic hagfish considered to be sterile were among the largest animals examined (Table 2.1) not only exhibiting a relatively high body length but also a much higher gutted body weight at length. There was no evidence of gonadal development in these animals which appears to have led to an increase in the allocation of energy to somatic growth rather than reproduction.

There is evidence suggesting that male Atlantic hagfish attain a greater maximum body length than females on the slope of both the Grand and St. Pierre Bank (Table 2.1). However, mean body length did not differ significantly between males and females (Table 2.1).

Distribution and evidence of a spawning migration

Not only does the Atlantic hagfish occur only in the North Atlantic Ocean, it is also the only species of hagfish to occur within the North Atlantic (Martini et al., 1997b; Grant, 2006). The Atlantic hagfish is widely distributed in Arctic seas and southward along both coasts of the North Atlantic (Scott and Scott, 1988). However, little is known about the extent of Atlantic hagfish distribution or abundance within specific regions. In Atlantic Canada, limited available information is attributed to the burrowing and generally immobile behavior of Atlantic hagfish (Martini, 1998), which reduces their vulnerability to bottom trawl surveys, the major source of information on the distribution and abundance of groundfish species.

Both salinity and temperature influence the distribution of all-known species of hagfish, whereas the influence of depth appears to be species-specific (Martini, 1998). On the slope of the Grand and St. Pierre banks, during annual research surveys, Atlantic hagfish were captured in baited traps at depths of 95–951 m and temperatures of 0–9.0°C. However, the highest catches coincided with the highest mean bottom water temperatures (6.8–7.0°C) within depths of 125–300 m (Figure 2.2). Atlantic hagfish appeared to avoid subzero temperatures, which were commonly recorded at shallower depths (95–115 m) in the eastern region of the St. Pierre Bank survey area. Although Atlantic hagfish do not produce antifreeze proteins (G. Fletcher, personal communication, Ocean Sciences Centre, St. John's NL, Canada), they were able to tolerate subzero temperatures (−0.8°C) in captivity over a two-week period but were inactive (S. Grant, personal observation). Tolerance to subzero temperatures may be attributed to the blood plasma of Atlantic hagfish being iso-osmotic to seawater (Fange, 1998), which freezes at approximately −1.9°C. Thus, it appears that Atlantic hagfish can tolerate short-term incursions of subzero water temperatures such as those produced in the Newfoundland region when ocean currents force a subsurface layer of cold water (<0°C) known as the cold intermediate layer (Petrie et al., 1988) to come in contact with the seabed.

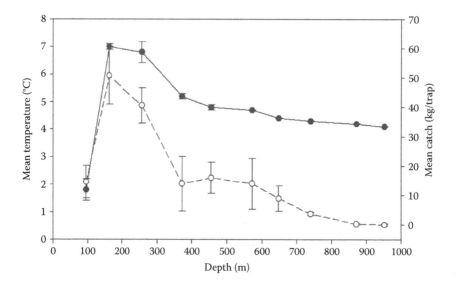

Figure 2.2 Variation in mean (±1 SE) bottom water temperature (solid symbols) and catch rates of Atlantic hagfish in traps with 14.3 mm escape holes (open symbols) during annual research surveys in slope waters of the Grand Bank (NAFO Div. 3O) and St. Pierre Bank (NAFO Subdiv. 3Ps) from 2005 to 2008.

Mean catch rates of Atlantic hagfish in baited trap surveys in continental slope waters of the Grand and St. Pierre banks exhibited a declining trend with an increase in mean depth to 951 m and a coinciding decrease in mean bottom temperature from 7.0°C to 4.1°C (Figure 2.2). Atlantic hagfish appear to avoid temperatures above 10–12°C (Bigelow and Schroeder, 1953; Scott and Scott, 1988), and there is general agreement that they can be held in captivity for extended periods at temperatures as low as 0–4°C (Martini, 1998). Atlantic hagfish were not captured in the northern region of the Grand Bank in NAFO Division 3L (Figure 2.1) during a midsummer exploratory baited trap survey where minimum bottom water temperatures of 0.1–1.9°C were recorded. Subzero bottom water temperatures are commonly recorded on the Grand Bank during the winter months, and the general immobile behavior of Atlantic hagfish (Martini, 1998) is likely to generally limit their distribution and movements in the Newfoundland region to more stable and warmer bottom temperatures that occur in deep water channels between the banks and upper-to-mid-continental slope waters. It is concluded that Atlantic hagfish is a temperature seeker in the Newfoundland and Labrador region, with the highest concentrations occurring within the warmest available bottom waters.

Hagfish distributions are typically described as patchy, and Martini (1998) speculated that low swimming speeds of hagfishes (0.25 m/s) may limit home range and prevent their penetration into areas of high current velocity. Low swimming speeds may in part account for the distribution of Atlantic hagfish to low current velocities that typify muddy bottom sediment deposition zones in continental slope waters of the Grand and St. Pierre banks. Movements of about 2.5 km over a 4.5-year period by Atlantic hagfish tagged within a fjord of the eastern North Atlantic (Walvig, 1967) have been reported, but it is unclear whether such movements represent migrations. Spawning migrations to deeper waters have been documented for the Japanese hagfish, *Eptatretus burgeri* (Kobayashi et al., 1972), and capture of large numbers of small juvenile Atlantic hagfish within a specific depth zone during baited trap surveys of the Grand and St. Pierre banks provide the first evidence of a similar spawning migration by Atlantic hagfish in the western North Atlantic (Figure 2.3).

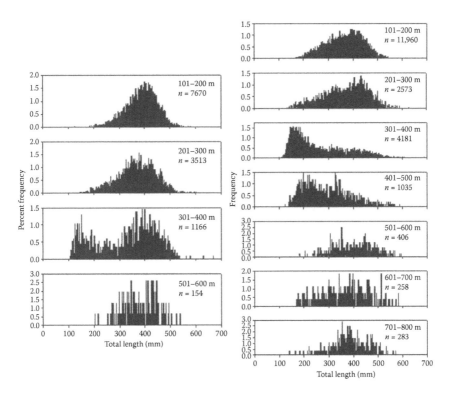

Figure 2.3 Length frequency distributions by 100-m depth interval for Atlantic hagfish captured in control traps during annual research surveys of the St. Pierre Bank (panels on the left) and the Grand Bank (panels on the right). The depth interval and number of hagfish examined (*n*) are illustrated in each panel.

Figure 2.3 illustrates length frequency distributions for Atlantic hagfish captured in control traps and summed across all years for baited trap surveys of both the Grand and St. Pierre banks. At both locations, there is a clear increase within the 301–400 m depth interval in the number of small juveniles within the 100–200 mm length class. On the slope of the Grand Bank, the distribution of small juveniles also extended into the 401–500 m depth interval. The pulse of small juveniles was evident in all samples from these depth intervals; however, it is unclear as to whether small juveniles are widely distributed or form dense aggregations within these depth intervals.

A comparison of the weight of the smallest Atlantic hagfish captured in control traps and the weight of eggs in the final stages of maturation (i.e., anchor filaments present) in females captured during the baited trap research surveys provide strong evidence that the smallest juveniles are aged 0+. Analysis of Atlantic hagfish from the Grand and St. Pierre banks included an assessment of the weight and length (excluding tufts) of eggs in the final stages of maturation for a total of 202 females. The average weight of an egg was 1.4 g (range, 0.7–2.6 g) and the average length was 24.1 mm (range, 20.1–31.6). In addition, a total of 12 Atlantic hagfish captured during the surveys were ≤100 mm in length (range, 90–100 mm) and exhibited a range in weight of 0.9–2.0 g (mean, 1.6 g). On the basis of these findings, it is concluded that in continental slope waters of the Grand and St. Pierre banks, recently hatched Atlantic hagfish are up to 100 mm in length and possibly longer. It is notable that comparable weights among eggs and the smallest Atlantic hagfish captured lend support to the supposition that there is no free-swimming larval stage in Atlantic hagfish (Bigelow and Schroeder, 1953). The smallest recorded length for Atlantic hagfish is about 60–70 mm (Bigelow and Schroeder, 1953; Scott and Scott, 1988). Patzner (1998) indicates that the length of mature eggs of most hagfish species, excluding the anchor filaments, varies from 14 to 25 mm. Thus, small egg size can account for the disparity in size among the smallest recorded lengths of Atlantic hagfish and those captured on the slope of the Grand and St. Pierre Bank. Although unclear for Atlantic hagfish, a survival advantage may be expected for recently hatched individuals that are larger in size.

An unequal sex ratio of far more females than males in baited traps is common in Atlantic hagfish (Walvig, 1963; Martini et al., 1997b; Patzner, 1998; Grant, 2006) and females dominated the catch in baited traps on the slope of the Grand and St. Pierre banks (Table 2.1). In addition, it has been well established that the capture in baited pots of females and males in the final stages of gonadal maturation in preparation for spawning is rare (Patzner, 1998). Indeed, only 202 of the 9513 females examined from the Grand and St. Pierre banks possessed eggs in the final stages of maturation (presence of anchor filaments) and only a single female possessed a clutch of fully developed eggs that were ovulated within the body cavity in preparation for spawning (Figure 2.4). In comparison, of the 189 males examined, three exhibited a fully mature testis. It is notable that the single fully ripened female and all three ripe males were captured within the 301–400 m depth interval—the same depth interval where large numbers of small juveniles were captured (Figure 2.3).

Figure 2.4 illustrates that an entire clutch of eggs was ovulated at about the same time and that the function of the anchor filaments in Atlantic hagfish is to keep the eggs in close contact by forming a chain. Available evidence suggests sperm counts in ripe males are very low (Patzner, 1998). Thus, ovulating all eggs at one time and linking the eggs in a chain would help maximize fertilization. It is notable that eggs at the ends of the chain have anchor filaments at their free ends that can become entangled within the mesovarium. Indeed, remnants of entangled eggs in the anterior and posterior region of the mesovarium were discovered in 0.4–1.3% of repeat spawning mature females examined from samples collected on the slope of the Grand and St. Pierre banks from 2005 to 2008. At the population level, this would lead to a minor reduction in the actual number of eggs (i.e., fecundity) extruded during spawning.

Reasons for the unequal sex ratio and scarcity of ripe male and female Atlantic hagfish in baited traps are unclear. Walvig (1963, 1967) summarizes cases where male Atlantic hagfish were plentiful in the eastern North Atlantic and did not exclude a spawning migration into other areas to explain why ripe animals are not captured in baited traps. Cessation of

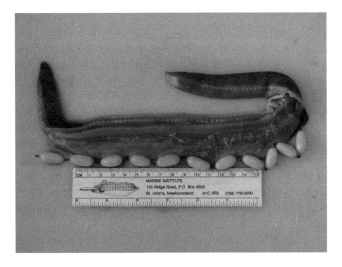

Figure 2.4 **(See color insert.)** Photograph of dissection of female Atlantic hagfish illustrating a chain of mature eggs ovulated from their follicles in preparation for spawning.

feeding prior to spawning and a concomitant migration to deeper waters may account for the scarcity of fully ripe Atlantic hagfish in baited traps. All things considered, the capture of high numbers of very small juveniles and Atlantic hagfish with fully ripe gonads within the same depth zone on the slope of the Grand and St. Pierre banks provides the first evidence of a spawning migration for this species. It is unclear why Atlantic hagfish would migrate down the slope of the banks to spawn in relatively cold bottom waters (Figure 2.2). Perhaps cooler and seasonally stable bottom water temperatures at greater depths maximize larval development and survival. Further, male Atlantic hagfish may be concentrated at these depths, threat of predation on egg, larval, and young juvenile stages may be reduced, or some yet to be identified physical features at these depths may be important not only to the spawning process, but also for the survival of early life stages. The discovery of high concentrations of small juveniles within a specific depth zone emphasizes the importance of putting biological research at the forefront of hagfish fisheries development and presents a unique opportunity to gain a better understanding of the long soughtafter method of reproduction in Atlantic hagfish. Challenges include working at great depths and distance from shore.

Analysis of Atlantic hagfish catch rates and body length frequencies on the slope of the Grand and St. Pierre banks indicates that commercial-sized individuals are concentrated at shallower and somewhat warmer temperatures than very small juveniles. These Atlantic hagfish fisheries target the 100–300 m depth zone, which avoids impacting the nursery and apparent spawning grounds that occur at greater depths.

Length at maturity: Conservation harvesting

From a conservation perspective, it is important to identify the maturity schedule or length at maturity for new emerging fishery species. A lack of understanding of this aspect of a species' life history and its consideration in gear design early in the development of a fishery can lead to overharvesting of juveniles and subsequently contribute to collapse of a fishery. This is particularly relevant to Atlantic hagfish, as juveniles are highly susceptible to capture in baited traps (Grant, 2006). In addition, discard survival may be low, given the low tolerance of hagfishes to abrupt changes in temperature and salinity (Adam and Strahan, 1963) that can occur when traps are hauled from deep water sets.

Grant (2006) presented the first length at maturity curve for female Atlantic hagfish that were collected from continental slope waters of the Grand Bank (NAFO Division 3O). Grant (2006) added a 10% correction factor to the length measurements because the specimens were preserved in 10% formalin. Preserved lengths at first, 50%, and 100% maturity were 354, 378, and 440 mm, respectively. The period of time hagfish remain in preservative is known to significantly affect total length and 10% shrinkage was reported to be common (McMillan and Wisner, 1984), but the actual amount of shrinkage for Atlantic hagfish and whether shrinkage varies with body length is unknown. A more recent analysis of length at maturity of not only nonpreserved female Atlantic hagfish collected from the slope of the Grand Bank, but also several more specimens (828 vs. 250) revealed a total length of 335 mm at first attainment of sexual maturity, 50% were found to mature at 394 mm, and 100% became mature at 460 mm (Figure 2.5). A length at maturity curve derived for 856 nonpreserved Atlantic hagfish occurring on the slope of the neighboring St. Pierre Bank (NAFO Subdivision 3Ps) revealed a similar length at first (337 mm), 50% (391 mm), and 100% (440 mm) sexual maturity (Figure 2.5).

Although recently refuted (Martini et al., 1998; Grant, 2006), Wisner and McMillan (1995) proposed elevating Atlantic hagfish in the western north Atlantic to distinct species status (i.e., *Myxine limosa*) based on a number of factors that appeared to differ from populations in the eastern North Atlantic. This included an increase in maximum body length and an increase in length at first attainment of sexual maturity for western North Atlantic populations. However, an analysis of body length and maturity of a third population of Atlantic hagfish occurring in the Newfoundland region revealed that such differences can occur over small geographic scales (Figures 2.5 and 2.6). High predation by Atlantic hagfish in fixed gear fisheries in the Laurentian Channel of NAFO Subdivision 3Pn led fishermen to investigate their commercial potential during the early 1990s. Although the 3Pn fishing grounds are only 300–500 km west of the well-established Atlantic hagfish fisheries located on the slope of the Grand and St. Pierre banks (Figure 2.1), 3Pn hagfish were deemed

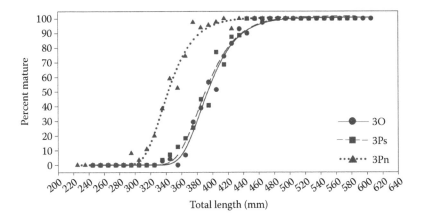

Figure 2.5 Logistic maturity curves for female Atlantic hagfish captured on the slope of the Grand Bank (NAFO Div. 3O) and the St. Pierre Bank (NAFO Subdiv. 3Ps) and within the Laurentian Channel (NAFO Subdiv. 3Pn) in autumn 2006.

too small to meet market demands (TriNav, 1996). A more recent analysis of Atlantic hagfish captured in 227-L traps with 14.3 mm escape holes confirmed that the mean body length of the population occurring in 3Pn was 40–42 mm shorter than those captured on the slope of the Grand and St. Pierre banks (Figure 2.6). Further, an analysis of the maturity curve derived from dissections of 753 female Atlantic hagfish collected from the 3Pn population revealed a substantial negative shift when compared with maturity curves for both the Grand and St. Pierre banks (Figure 2.5). Length at first attainment and 50% maturity were 46–51 mm less than the Grand and St. Pierre Bank populations. In fact, female body length at first attainment of sexual maturity in the 3Pn population fell within the range

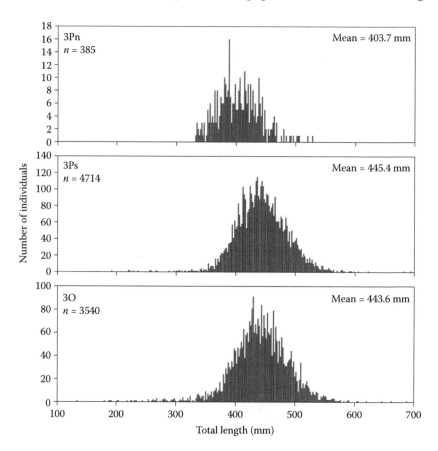

Figure 2.6 Body length distributions for Atlantic hagfish captured in 227-L baited traps with 14.3-mm escape holes within the Laurentian Channel (NAFO Subdiv. 3Pn) and on the slope of the St. Pierre Bank (NAFO Subdiv. 3Ps) and the Grand Bank (NAFO Div. 3O) during autumn 2006. Mean body length and number of animals examined (*n*) are also illustrated.

reported for the eastern North Atlantic (i.e., 254–330 mm; Cunningham, 1886), whereas the Grand and St. Pierre Bank populations were only marginally higher than the maximum reported for the eastern North Atlantic.

Studies show that differences in size at maturity at small spatial scales are not uncommon in less-mobile bottom-dwelling species and are assumed to be linked to local environmental conditions (see Gendron, 1992). For example, intense predation pressure can select for smaller sizes at maturity and reduce maximum size compared with populations subjected to reduced predation (Stearns and Crandall, 1981). Reasons for the small spatial-scale differences in mean body length and length at maturity of Atlantic hagfish occurring in the Newfoundland region are unknown but can be attributed to clines that can manifest in time or space (Pianka, 2000).

Early studies of the selective properties with regard to body length of Atlantic hagfish captured in baited 227-L barrel-type traps with 12.7, 13.5, and 14.3 mm (1/2″, 17/32″, and 9/16″) escape holes led Grant (2006) to recommend the Grand Bank fishermen be encouraged to use traps with 14.3 mm escape holes or larger. Given the high vulnerability of juveniles to capture and uncertainties with regard to juvenile postrelease survival, gear selectivity studies were continued in the Newfoundland region with the addition of 227-L barrel-type traps with 15.1 mm (19/32″) escape holes. Between trap comparisons of the percentage of the catch below both the minimum reported weight for human consumption (80 g) and length at 50% maturity over a four-year (2005–2008) time series indicated that traps with 15.1 mm escape holes captured significantly fewer (i.e., $p < 0.05$) of these undersized hagfish than did traps with 14.3 mm escape holes. Atlantic hagfish below the length at 50% maturity accounted for 10–12% of the catch in traps with 14.3 mm escape holes compared with 5–6% in traps with 15.1 mm escape holes. Atlantic hagfish below 80 g accounted for 34–41% of the catch in traps with 14.3 mm escape holes compared with 20–24% in traps with 15.1 mm escape holes.

Both large and small Atlantic hagfish are highly susceptible to capture in baited traps; however, selectivity studies carried out in the Newfoundland region indicate that catches of undersized hagfish can be significantly reduced with a seemingly minor increase (0.8 mm) in escape hole size (Figure 2.7). This is attributed to the elongated tubular body form of hagfish and resulting knife-edge selectivity of the large barrel-type hagfish trap. Given the difficulty and uncertainties surrounding the estimates of abundance and productivity of hagfish, it was concluded during an assessment of the Grand and St. Pierre Bank Atlantic hagfish stocks that management should be based on fishing effort, size selectivity, and escapement potential of hagfish traps (DFO, 2009). It was further concluded that a conservation objective would be to allow escapement to occur consistent with a length at 50% maturity. Given the knife-edge selectivity

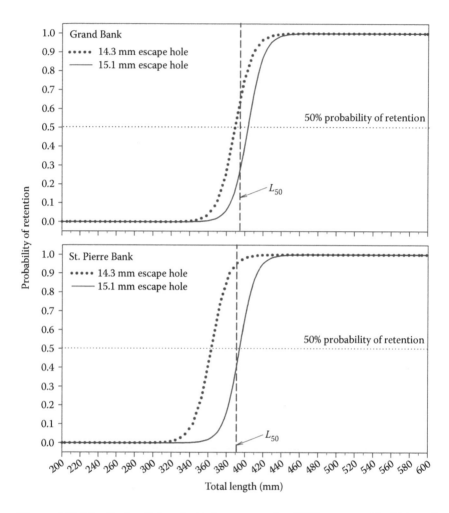

***Figure* 2.7** Atlantic hagfish selectivity curves for 227-L traps with 14.3 and 15.1 mm escape holes for the Grand Bank and the St. Pierre Bank. Female length at 50% maturity (L_{50}) and 50% probability of retention are also illustrated.

curves for the hagfish traps used and the 50% probability of retention for traps with 14.3 and 15.1 mm escape holes encompassing the length at 50% maturity of both Atlantic hagfish stocks (Figure 2.7), it was considered that the greatest escapement of juveniles may be achieved with a 40/60 trap fleet split. Specifically, 40% of the fishing effort is carried out using traps with 14.3 mm escape holes and 60% of the effort is with traps with 15.1 mm escape holes. This remains a condition of license in Newfoundland's Atlantic hagfish fisheries. Size-specific variation in the stage of the reproductive cycle at capture could account for the observed differences in trap selectivity among the Grand Bank and St. Pierre Bank populations (Figure 2.7). Specifically, a female with large eggs in the later stages of the reproductive cycle has a relatively larger girth and is therefore more likely to be retained than a similar-length female with small eggs.

Trap size and soak time are also factors that can influence the escapement of juvenile hagfish. When traps are hauled too soon, juveniles may not have a chance to escape. To facilitate escapement, traps should be soaked for at least 12–24 h. With regard to trap size, when exploratory Atlantic hagfish fisheries in 3Pn were carried out with small 21-L bucket traps with a single one-way entrance funnel and 14.3 mm escape holes, the traps often filled to capacity with juveniles that produced copious amounts of slime. Further, it was not uncommon to haul back buckets full of rotting Atlantic hagfish in overnight sets. It would appear that, owing to their larger numbers, juveniles rapidly filled the small bucket traps to capacity, which led to excessive sliming and suffocation. This problem was alleviated by using larger 227-L traps with 14.3 mm escape holes.

Is spawning a synchronized event in Atlantic hagfish?

Much of the reproductive biology of Atlantic hagfish is poorly understood. Timing of the initiation of gametogenesis is unknown, and the duration of the reproductive cycle is unclear as they may reproduce every one (Powell et al., 2004) or more years (Patzner, 1998); however, low fecundity has been well established (Patzner, 1998; Grant, 2006). The capture in the eastern North Atlantic of female Atlantic hagfish with eggs in all stages of development as well as females with large postovulatory follicles in autumn, winter, and summer led Cunningham (1886) and Nansen (1887) to conclude that spawning is not limited to any particular season. Further, after finding no evidence of a dominant stage in the female reproductive cycle of Atlantic hagfish during three different years, Patzner (1982) concluded that there was no evidence of a synchronized reproductive cycle in the eastern North Atlantic. However, seasonal concentrations of reproductive steroids in the gonads of Atlantic hagfish provided evidence to suggest the existence of a seasonal reproductive cycle in the western North Atlantic (Powell et al., 2004). Studies by Grant (2006) also provided evidence that spawning may be synchronized in the western North Atlantic and suggested that there may be two to three synchronized spawning events per year on the slope of the Grand Bank. Grant (2006) based his conclusions on findings of a bimodal egg length frequency distribution and several females with corpus lutea in a single point-in-time sample collected on the slope of the Grand Bank.

Grant (2006) emphasized the need to revisit past methodologies used to elucidate the nature of the reproductive cycle in Atlantic hagfish. For example, periodic sampling that follows an egg-length mode within a population through time may help us to better understand not only whether synchronized spawning events occur, but also the duration of the egg maturation phase of the reproductive cycle. However, recent findings from dissection and analysis of female Atlantic hagfish collected from the Grand and St. Pierre banks from 2005 to 2008 suggest that egg length alone is not an adequate measure for tracking egg development during the final stage of the egg maturation phase. Specifically, it was discovered that both the maximum length and diameter of the ovoid egg were important factors influencing whether eggs entered the final stage of maturation. Analysis of 202 females that possessed eggs with anchor filaments developing revealed that a minimum length of 20 mm and a minimum diameter of 8 mm had to be attained before anchor filaments began to develop. These results indicate that volume of the ellipsoid egg is a better indicator of eggs entering the final stage of maturation. When egg-length and egg-volume distributions were compared for the same point-in-time samples collected from 2005 to 2008, there was no evidence of a bimodal distribution for egg volume (e.g., Figure 2.8). Further, when the more representative minimum volume of an egg at the onset of anchor filament development is used instead of minimum length, considerably fewer females were found to possess a clutch of eggs entering the final

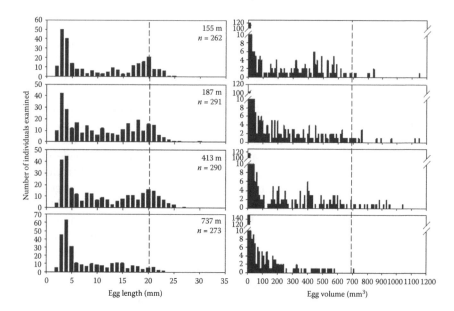

Figure 2.8 Egg length distributions and corresponding egg volume distributions for Atlantic hagfish captured at four broadly spaced research survey sites on the slope of the Grand Bank (NAFO Div. 3O) in autumn 2007. Dashed vertical line represents minimum egg length and minimum egg volume at onset of anchor filament development.

stages of maturation (Figure 2.8). These new results provide no evidence to suggest synchronization of the timing of the final stages of egg maturation in the reproductive cycle of Atlantic hagfish and hence the absence of synchronized spawning. Further, a lack of clear modality in the egg volume distributions would appear to preclude tracking egg development through periodic sampling.

The previous analysis failed to find evidence of synchronization of the final stage of the egg maturation phase in Atlantic hagfish captured in baited traps. However, the capture in baited traps of Atlantic hagfish in final stages of the reproductive cycle is rare, which, as was previously outlined, has led to the speculation that they cease feeding prior to spawning. It is notable however that Patzner (1998) indicates that the Japanese hagfish (*E. burgeri*), which migrates to deeper water to spawn, can be taken in baited traps on the spawning grounds just before they become completely mature. Nevertheless, cessation of feeding for a prolonged period of time prior to spawning would limit the ability to identify synchronized spawning events from analysis of single point-in-time samples of Atlantic hagfish captured in baited traps.

A high proportion of females examined from point-in-time samples collected on the slope of the Grand and St. Pierre banks possessed small eggs, providing evidence of a recent and possibly synchronized initiation of gametogenesis (Figure 2.8). In fact, all 32 samples of Atlantic hagfish examined from the slope of the Grand and St. Pierre banks during 2005–2008 exhibited similar egg length and egg volume distributions with high proportions of females with small eggs. In each year, samples were collected during the baited trap research survey, which took place in August–September, and sample analysis of life-history processes covered a broad geographic area and range of depths (110–823 m). Temperature and day length represent physical cues that are known to initiate gametogenesis in invertebrates and fish. In addition, studies suggest that gametogenesis in deep-water (650 m) habitats can be correlated with the downward flux of detritus following peaks in surface production (Tyler et al., 2007). Overall, seasonal timing of the Atlantic hagfish sample collections and a ubiquitous high percentage of females with small eggs in the egg length and egg volume distributions from relatively shallow and deep water sites on the slope of the Grand and St. Pierre banks in all years sampled suggests not only seasonal gametogenic synchrony among a large segment of the population, but also a common cue. I speculate that a downward flux of detritus following the spring bloom, which has been demonstrated to cover the Grand Banks by the end of April (Wu et al., 2007), provides the cue by providing supplementary nutrition to the benthic fauna, including adult Atlantic hagfish. Although Atlantic hagfish egg length and egg volume distributions provide evidence to suggest a seasonal aspect to the initiation of gametogenesis, these same distributions also suggest that egg development during the reproductive cycle is highly variable (Figure 2.8).

Can fishing pressure on Atlantic hagfish stocks induce a compensatory increase in production?

Similar to intense predation pressure, commercial fisheries remove both large numbers of fish and larger individuals from the population, which can reduce competition, resulting in faster growth and favor individuals that reproduce at a younger age. Reduced competition can also lead to increased fecundity. Further, in the case of Atlantic hagfish, where spawning is not limited to any particular season, changes in the level of competition for resources may influence the duration of the egg maturation phase or the resting phase of the reproductive cycle. Once the eggs of Atlantic hagfish mature and are ovulated from their follicles, the female enters a resting phase marked by the enlargement of the post-ovulatory follicle to form the glandular corpus luteum, which eventually decreases in size as it regresses back into the ovary. Duration of the resting phase can be gaged by the size of the postovulatory follicle and size of developing eggs. For example, females that possess fully regressed follicles and small eggs have had a relatively long resting period, while females that possess medial-stage follicles and medium-to-large eggs have had a shorter resting period.

Baited trap fisheries for Atlantic hagfish were initiated on the slope of the Grand Bank in 2004 and on the slope of the St. Pierre Bank in 2005. The Grand Bank fishery ended after 5 years (i.e., in 2008) owing to a lack of interest on the part of industry, whereas the St. Pierre Bank fishery has continued to this day. Annual prefishery research surveys were carried out for both fisheries from 2005 to 2008, and commercial fishery data were also collected annually. In the Grand Bank fishery, a total of 527 metric tons (mt) of Atlantic hagfish were landed from 2004 to 2008, whereas in the St. Pierre Bank fishery, a total of 1190 mt of Atlantic hagfish were landed from 2005 to 2012. There was, however, no Atlantic hagfish fishery on the slope of the St. Pierre Bank in 2009 owing to the 2008 downturn in the global economy. At an average body weight of 90–100 g for Atlantic hagfish captured in the Grand Bank and St. Pierre Bank commercial fisheries, these landings represent the removal of large numbers of animals. In addition, monitoring of the body length of Atlantic hagfish captured in baited trap surveys from 2005 to 2008 showed a gradual decrease in the percentage of the larger individuals, specifically those greater than the length at 100% maturity (Figure 2.9).

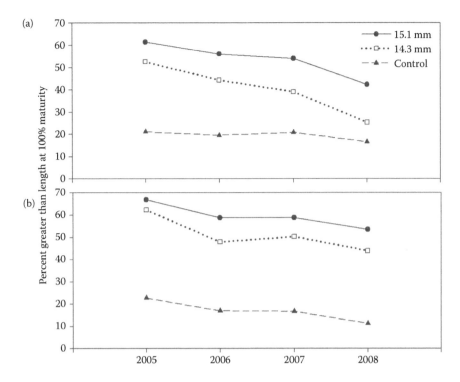

Figure 2.9 Percentage of Atlantic hagfish greater than the length at 100% maturity in research survey traps with 15.1 and 14.3 mm escape holes and control traps fished on the slope of the St. Pierre Bank (a) and the Grand Bank (b) from 2005 to 2008.

Although low fecundity has been well established for Atlantic hagfish, the ability to obtain accurate estimates of fecundity in order to detect any spatial–temporal trends can be a challenge. This is due to the fact that the number of developing eggs present in the ovaries of Atlantic hagfish at the beginning of the reproductive cycle is not a reliable indicator of the number of eggs extruded during spawning (Lyngnes, 1936; Grant, 2006). Lyngnes (1936) indicated that as eggs in Atlantic hagfish from the eastern North Atlantic grow, the total number decreases by about half as a result of resorption. Lyngnes (1936) also found that the degeneration percentage is fairly similar and high among eggs that are 2–17 mm long and then decreases among larger eggs. To avoid overestimating fecundity for a Norwegian population of Atlantic hagfish, Patzner (1998) only considered females with eggs >10 mm in length. Grant (2006) found that the degeneration percentage was considerably lower in Grand Bank Atlantic hagfish, where degenerated eggs were found only in the early stage of the reproductive cycle among females with developing eggs that were less than 12 mm in length. The recent analysis of stage of the reproductive cycle of 9513 female Atlantic hagfish collected from the Grand and St. Pierre banks revealed 7346 females possessed developing eggs >3.0 mm in length and 26% (1906) of these females possessed degenerated eggs. The largest degenerated egg discovered was a single collapsed egg that exhibited a length of 15 mm. On the basis of these findings, analysis of fecundity and relative fecundity (number of eggs/100 g eviserated body weight) for the Grand and St. Pierre Bank Atlantic hagfish populations was based on females that possess eggs that were >15 mm in length.

Over the period 2005–2008, Atlantic hagfish fecundity ranged from 7 to 42 eggs for the Grand Bank and from 6 to 37 for the St. Pierre Bank (Table 2.2). Mean fecundity ranged from 17.3 to 21.9 on the slope of the Grand Bank and from 16.2 to 21.5 on the slope of the St. Pierre Bank. For all 32 samples of Atlantic hagfish examined, $\log_{10}$(fecundity) was found to exhibit a weak to moderately strong positive relationship to $\log_{10}$(body length), with regression correlation coefficients (i.e., R^2) ranging from 0.072 to 0.453 (Table 2.2). It was discovered that $\log_{10}$(relative fecundity) also exhibited a positive relationship to $\log_{10}$(body length), and regression correlation coefficients were found to improve markedly ranging from 0.382 to 0.767 (Table 2.2). On the basis of these findings, relative fecundity–body length relationships were compared among years for two of the most heavily fished regions within the commercial fishing grounds on the slope of the St. Pierre Bank, specifically research survey Grid 2 and Grid 7 (Figure 2.10). Life-history processes were also examined from year to year for these two grids. Grid 2 accounted for 16% and Grid 7 for only 6% of the designated fishing grounds on the slope of the St. Pierre Bank; yet, commercial landings of Atlantic hagfish from these grids accounted for 18% and 23%, respectively, of the total landings (i.e., 694 mt) from 2005 to 2008. Atlantic hagfish relative fecundity at length curves for Grid 7 show a clear increase in elevation from 2005 to 2008, while in Grid 2 curve elevation increased after the first year and remained high thereafter. It is notable that relative fecundity at length curves for Grid 2 has a higher elevation than those illustrated for Grid 7, which could suggest that the former represents an upper limit possibly imposed by space limitations within the body cavity. This may explain why there was no change in elevation from 2006 to 2008.

Analysis of 202 females from both the Grand and St. Pierre Bank Atlantic hagfish populations revealed that there was no relationship between the length of ripe eggs (i.e., those that possessed developing anchor filaments) and body length, indicating that ripe eggs of both the smallest and largest females are similar in size. Therefore, gradual loss of the largest female Atlantic hagfish (Figure 2.9) from a population should not negatively influence juvenile size at hatch and subsequent survival.

In addition to fecundity, many other aspects of the female's life history can be used to monitor the reproductive potential of Atlantic hagfish. For

Table 2.2 Atlantic hagfish fecundity and relative fecundity summary (i.e., mean, minimum, and maximum) for the slope of the Grand Bank and St. Pierre Bank by year and depth (m)

| Location | Year | Depth | No. | Fecundity | | | | Relative fecundity | | | |
				Mean	Min	Max	R^2	Mean	Min	Max	R^2
Grand Bank	2005	133	38	18.0	7	27	0.453	17.8	5.2	38.4	0.700
		237	87	19.7	10	36	0.373	22.5	8.3	63.5	0.609
		182	28	21.9	12	32	0.355	27.5	13.8	57.7	0.767
	2006	258	89	17.6	8	29	0.164	16.5	5.6	31.6	0.498
		251	149	18.6	9	34	0.188	21.2	7.7	51.3	0.538
		488	100	20.2	10	36	0.249	23.8	8.6	68.4	0.516
		823	195	19.7	11	37	0.294	21.8	7.7	51.9	0.586
		210	70	17.5	10	41	0.355	15.6	5.1	66.7	0.651
		251	74	19.6	10	31	0.179	19.6	6.7	42.5	0.490
	2007	187	99	18.9	11	37	0.233	18.5	7.5	59.6	0.532
		413	97	19.0	8	29	0.364	21.9	5.7	48.1	0.557
		737	35	19.7	10	32	0.167	21.6	8.0	44.2	0.408
		155	90	19.3	11	31	0.151	19.6	8.8	37.4	0.397
	2008	164	105	17.3	11	27	0.180	14.9	6.2	29.5	0.533
		329	97	20.0	11	32	0.296	21.5	7.8	38.0	0.564
		750	42	20.8	12	35	0.314	21.4	8.4	46.9	0.645
		110	99	20.3	12	42	0.361	21.8	8.9	71.4	0.621
		230	86	17.2	8	33	0.210	16.7	5.8	47.4	0.604

(Continued)

Table 2.2 (Continued) Atlantic hagfish fecundity and relative fecundity summary (i.e., mean, minimum, and maximum) for the slope of the Grand Bank and St. Pierre Bank by year and depth (m)

Location	Year	Depth	No.	Fecundity				Relative fecundity			
				Mean	Min	Max	R^2	Mean	Min	Max	R^2
St. Pierre Bank	2005	186	113	16.2	7	32	0.232	13.8	4.6	28.4	0.423
	2006	202	97	17.5	9	34	0.146	16.6	6.9	57.8	0.550
		219	108	19.9	10	32	0.227	21.8	8.6	45.9	0.597
		219	107	21.5	10	37	0.235	24.8	7.9	56.8	0.472
		144	67	17.8	9	31	0.262	17.0	5.8	35.0	0.496
		163	63	18.3	10	29	0.260	18.2	4.6	35.8	0.585
		218	74	18.3	10	33	0.317	17.8	8.2	42.0	0.593
	2007	285	126	19.2	6	30	0.335	21.0	4.8	55.2	0.550
		326	90	20.0	10	33	0.198	21.7	8.2	57.9	0.515
		130	43	16.0	8	27	0.072	14.3	6.2	37.5	0.382
		224	72	18.0	8	34	0.261	18.6	7.2	38.3	0.565
	2008	141	32	19.5	12	31	0.386	19.0	8.5	48.8	0.637
		168	88	20.2	10	33	0.262	20.9	6.1	44.6	0.547
		243	63	18.7	10	29	0.271	16.9	6.3	32.8	0.529

Also shown are the number of animals examined and regression correlation coefficients (R^2) for $\log_{10}$(fecundity) and $\log_{10}$(relative fecundity) versus $\log_{10}$(body length) relationships.

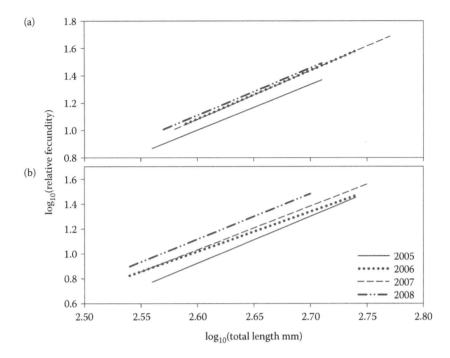

Figure 2.10 Atlantic hagfish relative fecundity–body length relationships for St. Pierre Bank research survey Grid 2 (a) and Grid 7 (b) over the period 2005–2008. Note vertical axis are not equal.

example, changes in the length at maturity curve, percentage of sexually mature females that have initiated the reproductive cycle, duration of the egg maturation or resting phase of the reproductive cycle, and percentage of females with corpus lutea can all signal environmentally mediated changes in the reproductive potential or production. For example, lack of eggs in advanced stages of development in female Atlantic hagfish with postovulatory follicles led Martini et al. (1997b) to conclude that the reproductive cycle must last a significant period of time in the Gulf of Maine. Martini et al. (1997b) also commented on the small number of gravid (<1%) and postovulatory (<5%) females. These results contrasted those obtained by Grant (2006) for the Grand Bank, where postovulatory females were commonly encountered and several of the mature females were not only found to be gravid, but also possessed postovulatory follicles. On the basis of the information available, Grant (2006) concluded that the Grand Bank Atlantic hagfish population had a greater overall reproduction potential than the Gulf of Maine.

Analysis of the gonadal components of female Atlantic hagfish collected from the St. Pierre Bank research survey Grids 2 and 7 from 2005 to 2008 indicates that in all years a high percentage (72–80%) of mature females possessed developing eggs (Figure 2.11). Percentage of females that possessed corpus lutea ranged from 10% to 20% (Figure 2.11). Overall, these results are indicative of a high reproductive potential in all years, with 84–95% of mature females either developing a clutch of eggs or resting from a recent spawning event. There was no evidence to suggest fishery removals had an effect on the percentage of gravid females or percentage of those that had recently spawned (i.e., possessed corpus lutea) in Grid 2 or Grid 7. However, after 2 years of commercial fishing, the percentage of mature females that possessed both developing eggs and medial-stage follicles within their ovaries doubled for each research survey grid and continued to increase in both grids to 2008 (Figure 2.11). In addition, several females with medial-stage follicles were found to possess larger developing eggs in 2007 and 2008 (Figure 2.12). In 2005 and 2006, the largest co-occurring developing eggs and medial-stage follicles were 7.0 and 3.8 mm, respectively. In 2007, two females possessed large (4.5–5.0 mm) medial-stage follicles and developing eggs that were 15.1 and 17.0 mm in length and in 2008 a female that possessed a 5.0 mm follicle possessed eggs that were 19.0 mm in length. Overall, these results provide compelling evidence of a significant reduction in the duration of the resting phase of the reproductive cycle in a large segment of the population.

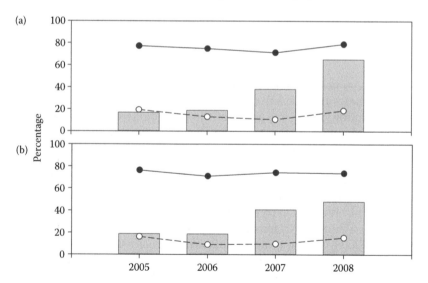

Figure 2.11 Percentage of sexually mature female Atlantic hagfish that possessed developing eggs (solid symbols), percentage of females with both developing eggs and medial-stage follicles (vertical bars), and percentage of females with corpus lutea (open symbols) within the St. Pierre Bank research survey Grid 2 (a) and Grid 7 (b).

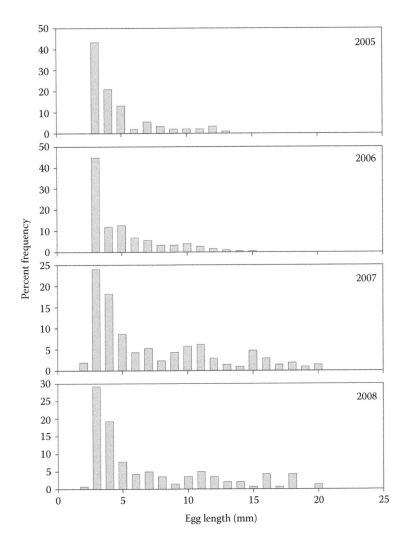

Figure 2.12 Egg length frequency distributions for female Atlantic hagfish that possessed both medial-stage postovulatory follicles and developing eggs in the St. Pierre Bank research survey Grid 7 from 2005 to 2008.

Samples of Atlantic hagfish recently collected during the 2013 autumnal commercial fishery on the slope of the St. Pierre Bank were used to update the length at maturity curve (Figure 2.13). When compared with 2006, the 2013 maturity curve was found to exhibit a negative shift with a 16 mm reduction in body length at 50% maturity and a 10 mm reduction in body length at 100% maturity. There was however no appreciable change in body length at first attainment of sexual maturity.

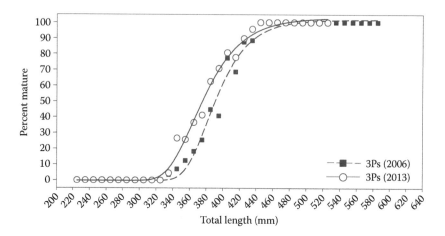

Figure 2.13 Logistic maturity curves for female Atlantic hagfish captured on the slope of the St. Pierre Bank in 2006 and 2013.

In summary, three life-history traits examined for Atlantic hagfish were found to demonstrate an increase in production that coincided with fishery removals of high numbers of individuals and a reduction in the percentage of larger individuals in the population. These results provide evidence that sexually mature Atlantic hagfish are able to rapidly adjust their reproductive potential to changing environmental conditions. In this case, high fishery removals are assumed to have reduced competition for resources, resulting in an increase in relative fecundity at length within 1 year and a decrease in the resting phase of the reproductive cycle within 2 years of intensive fishing pressure. Reduced competition can also result in faster growth among juveniles, which would account for the shift in the maturity curve to a reduction in body length at 50% and 100% maturity within eight years of fishing.

Increased production is consistent with what might be expected in the development of a fishery on a virgin stock and increases in repro-ductive potential by Atlantic hagfish were found to occur rapidly in the Newfoundland region. However, it is unclear at this time whether the increases are sufficient to compensate for fishery removals. Allowing juveniles to reach maturity and maintaining a large enough spawning biomass to replace fishery removals is the key to the development of sustainable fisheries. Unfortunately, with no way to determine abun-dance and productivity for Atlantic hagfish, it is not possible to deter-mine a sustainable removal level. Fortunately, conservation measures that allow escapement of juveniles (i.e., 40/60 trap split, see above) were put in place at the onset (i.e., in 2005) of both the St. Pierre and Grand Bank Atlantic hagfish fisheries. However, only through continued monitoring of the fishery and monitoring of biological parameters that show sensitivity to fishing pressure can it be determined whether production can keep up with removals (i.e., annual quota). Warning signs in the fishery include decreased catch rates, reduction in average individual size of landed Atlantic hagfish (i.e., harvesting juveniles), and geographic shifts in fishing effort or request to expand the fishery to unexploited areas.

In the Grand and St. Pierre Bank Atlantic hagfish populations, it was discovered that eggs must attain a minimum length and diameter before they entered the final stages of maturation and the size of ripe eggs was not related to body length. These results suggest that size of the body cavity will place an upper limit on relative fecundity at length and the ovoid shape of Atlantic hagfish eggs will also cause space limitations. This may explain the lack of an increase in elevation in the relative fecundity body length curves for the St. Pierre Bank research survey Grid 2. If physical limits to fecundity are related to the size of the body cavity, they may be overcome by a decrease in size of ripe eggs. Patzner (1998) reported that ripe egg lengths vary from 14 to 25 mm in most hagfish species; however, it is unclear whether variability in egg length is an environmentally mediated adaptive strategy or whether it would result in an increase in production. For example, a concomitant reduction in juvenile size at hatch with a reduction in the size of ripe eggs may be expected to negatively influence survival.

Lack of an obligatory spawning season in Atlantic hagfish and ability to vary the duration of the resting phase of the reproductive cycle are life-history strategies that appear to have evolved to adjust the reproductive potential to local environmental conditions. Even prior to the development of a fishery, a segment of the St. Pierre Bank population possessed both medial-stage follicles and eggs in varying stages of development, which is indicative of a reduction in the duration of the resting phase. These results suggest that it is the ability to adjust various phases of the reproductive cycle to environmental conditions and individual variability among members of the population that lead to a lack of synchronized spawning. It is unclear whether variability in egg development among females with medial-stage follicles is related to individual variability in postspawning energy reserves, timing of the initiation of gametogenesis, food availability throughout the egg maturation phase, or a combination of these factors. Egg length and egg volume distributions for the Grand Bank and the St. Pierre Bank Atlantic hagfish populations provide evidence to suggest synchronization of gametogenesis. If the downward flux of nutrient-rich detritus following the spring bloom is the cue, then this life-history process would appear to be seasonally timed to maximize the reproductive potential. Under this scenario, an adjustment in the initiation of gametogenesis that is out of phase with the downward flux of nutrients through a reduction in the duration of the resting period would appear counterproductive. However, this need not be the case if females that initiate the gametogenic cycle out of phase with the downward flux of nutrients have higher energy reserves. A fishery-induced reduction in competition and a subsequent increase in food availability could lead to an increase in postspawning energy reserves for a large segment of the population, enabling females to initiate gametogenesis earlier than females with lower energy reserves. Over the long term, a decrease in the resting phase of the reproductive cycle for an increasingly larger segment of the population will increase productivity and thereby help compensate for fishery removals.

Duration of the egg maturation phase of the reproductive cycle may also be expected to be altered by environmental conditions; however, it is not currently possible to determine egg maturation time nor is there a suitable proxy. However, results from the analysis of life-history traits of Atlantic hagfish during development of a fishery may shed some light on the subject of the duration of the egg maturation phase. In the St. Pierre Bank stock, relative fecundity at length was found to increase within one year of fishing, while 2 years of fishing was required before the proxy for a decrease in duration of the resting phase was detected. Because more eggs are produced at the beginning of the reproductive cycle then are spawned, an environmentally mediated change in relative fecundity at length could occur quickly during the current reproductive cycle simply by reducing the number of eggs that degenerate. However, duration of the resting phase cannot be adjusted until the egg maturation, ovulation, and spawning phases of the reproductive cycle are complete. On the basis of a 2-year detection interval for an increase in the percentage of females with both medial-stage follicles and developing eggs, it is conceivable that for a large segment of the population the reproductive cycle lasts more than a year at the onset of the fishery. The findings presented in this chapter provide strong evidence of a fishery-induced reduction in the duration of the reproductive cycle and although the extent of the reduction is unknown, the reproductive cycle could very well be less than a year, particularly in females that possess both large medial-stage follicles and large eggs in their ovaries.

Acknowledgments

This study was funded by the Newfoundland and Labrador Department of Fisheries and Aquaculture and Canadian Centre for Fisheries Innovation. Captain's Andrew Daley and Steve Careen and crews of the CFV *Royal Mariner* and CFV *Covenant II* are acknowledged for their support and cooperation during this study. I thank Wade Hiscock, Georgina Bishop, Rennie Sullivan, Wayne DeGruchy, Chris Keats, Diana Pike, Ryan Pugh,

Darrell Mullowney, Chris Batten, and Luke Hayes for their field and laboratory assistance. We appreciate Paul Brett for sketching Figure 2.1.

References

Adam, H. and R. Strahan. 1963. Notes on the habitat, aquarium maintenance, and experimental use of hagfishes. In A. Brodal and R. Fange (eds.), *The Biology of Myxine*. Universitetsforlaget, Oslo, pp. 33–41.

AHWG Report. 2003. Review of Atlantic hagfish biological and fishery information with assessment and research considerations. A report by the Atlantic Hagfish Working Group for the New England Fishery Management Council, May 30, 2003, 88 pp.

Bigelow, H. B. and W. C. Schroeder. 1953. Fishes of the Gulf of Maine. *US Fish Wildl Serv Fish Bull* 74:viii + 577.

Cunningham, J. T. 1886. On the structure and development of the reproductive elements in *Myxine glutinosa* L. *Q J Micrbiol Sci* 27:49–76.

DFO. 2009. Assessment of NAFO Division 3O and Subdivision 3Ps Atlantic hagfish (*Myxine glutinosa*). DFO Can. Sci. Advis. Sec. Sci. Advis. Rep. 2009/042.

Fange, R. 1998. Hagfish blood cells and their formation, pp. 287–299. In J. M. Jorgensen, J. P. Lomholt, R.E. Weber and H. Malte (eds.), *The Biology of Hagfishes*. Chapman & Hall, London, xix + 578 pp.

Gendron, L. 1992. Determination of the size at sexual maturity of the waved whelk *Buccinum undatum* Linnaeus, 1758, in the Gulf of St. Lawrence, as a basis for the establishment of a minimum catchable size. *J Shellf Res* 11:1–7.

Grant, S. M. 2006. An exploratory fishing survey and biological resource assessment of Atlantic hagfish (*Myxine glutinosa*) occurring on the southwest slope of the Newfoundland Grand Bank. *J Northw Atl Fish Soc* 36:1–20.

Holmgren, N. 1946. On two embryos of *Myxine glutinosa*. *Acta Zool* 27:1–90.

Honma, Y. 1998. Asian hagfishes and their fisheries biology, pp. 45–56. In Jorgensen, J. M., J.P. Lomholt, R.E. Weber and H. Malte (eds.), *The Biology of Hagfishes*. Chapman & Hall, London, xix + 578 pp.

Kobayashi, H., T. Ichikawa, H. Suzuki and M. Sekimoto. 1972. Seasonal migration of hagfish *Eptatretus burgeri*. *Jpn J Ichthyol* 19:191–194.

Lyngnes, R. 1936. Rückbildung der ovulierten und nicht ovulierten Follikel im Ovarium der *Myxine glutinosa* L. *Skrifter norske Videnskaps-Academi i Oslo l. matematisk-naturvidenskapelige Klasse* 4:1–116.

Martini, F. H. 1998. The ecology of hagfishes. In J. M. Jorgensen, J.P. Lomholt, R.E. Weber and H. Malte (eds.), *The Biology of hagfishes*. Chapman and Hall, London, pp. 46–77.

Martini, F. H., J. B. Heiser and M. P. Lesser. 1997b. A population profile for the Atlantic hagfish, *Myxine glutinosa* (L.), in the Gulf of Maine. Part I: Morphometrics and reproductive state. *Fish Bull* 95:311–320.

Martini, F. H., M. P. Lesser and J. B. Heiser. 1997a. Ecology of the hagfish, *Myxine glutinosa* L., in the Gulf of Maine. Part 2: Potential impact on benthic communities and commercial fisheries. *J Exp Mar Biol Ecol* 214:97–106.

Martini, F. H., M. B. Lesser and J. B. Heiser. 1998. A population profile for hagfish, *Myxine glutinosa*, in the Gulf of Maine. Part 2: Morphological variation in populations of *Myxine* in the North Atlantic Ocean. *Fish Bull* 96:516–524.

McMillan, C.B. and R.L. Wisner. 1984. Three new species of seven gilled hag-fishes (Myxinidae, Eptatretus) from the Pacific Ocean. *Proc Calif Acad Sci* 43:249–267.

Nansen, F. 1887. A protandric hermaphrodite (*Myxine glutinosa* L.) amongst the vertebrates. *Bergen Mus Aarsber* 7:1–34.

NEFSC. 2003. Atlantic hagfish. In Report of the 37th Northeast Regional Stock Assessment Workshop (37th SAW): Stock Assessment Review Committee (SARC) Consensus Summary of Assessments. Northeast Fisheries Science Centre Reference Document 03-16, pp. 518–597.

Patzner, R. A. 1982. Die Reproduktion der Myxinoiden. Ein Vergleich von *Myxine glutinosa* und *Eptatretus burgeri. Zoologischer Anzeiger (Jena)* 208:132–44.

Patzner, R. A. 1998. Gonads and reproduction in hagfishes. In J.M. Jorgensen, J.P. Lomholt, R. E. Weber and H. Malte (eds.), *The Biology of Hagfishes*. Chapman & Hall, London, pp. 378–379.

Petrie, B., S. A. Akenhead, S. A. Lazier and J. Loder 1988. The cold intermediate layer on the Labrador and northeast Newfoundland shelves, 1978–86. *Northwest Atlantic Fish Org Sci Council Study* 12:57–69.

Pianka, E. R. 2000. *Evolutionary Ecology*, 6th ed. Pearson Education, 512 pp.

Powell, M. L., S. I. Kavanaugh and S. A. Sower. 2004. Seasonal concentrations of reproductive steroids in the gonads of the Atlantic hagfish, *Myxine glutinosa*. *J Exp Zool* 301A:352–360.

Scott, W. B. and M.G. Scott. 1988. Atlantic fishes of Canada. *Can Bull Fish Aquat Sci* 219:731.

Stearns, S. C. and R. E. Crandall. 1981. Quantitative predictions of delayed maturity. *Evolution* 35:455–463.

TriNav. 1996. A report on the project to investigate the potential of developing an Atlantic hagfish fishery on the southwest coast of Newfoundland Final report (Phase I and Phase II). Report submitted to the Canada/Newfoundland Cooperation Agreement for Fishing Industry Development. 21 pp. + App.

Tyler, P., C. M. Young, E. Dolan, S. M. Arellano, S. D. Brooke and M. Baker. 2007. Gametogenic periodicity in the chemosynthetic cold mussel *Bathymodiolus childressi. Mar Biol* 150(5):829–840.

Walvig, F. 1963. The gonads and the formation of the sex cells. In A. Brodal and R. Fange (eds.), *The Biology of Myxine*. Universitetsforlaget, Oslo, pp. 530–580.

Walvig, F. 1967. Experimental marking of hagfish (*Myxine glutinosa* L.). *Norw J Zool* 15:35–39.

Wisner, R. L. and C. B. McMillan. 1995. Review of new world hagfishes of the genus *Myxine* (Agnatha, Myxinidae) with descriptions of nine new species. *Fish Bull* 93:530–550.

Wu, Y., I. K. Peterson, C. C. L. Tang, T. Platt, S. Sathyendranath and C. Fuentes-Yaco. 2007. The impact of sea ice on the initiation of the spring bloom on the Newfoundland and Labrador shelves. *J Plankton Res* 29:509–514.

chapter three

Fossil hagfishes, fossil cyclostomes, and the lost world of "ostracoderms"

Philippe Janvier and Robert S. Sansom

Contents

Introduction

Fossils of myxinoids and other cyclostomes have wide ramifications for our understanding of vertebrate evolution. Not only can they provide a timescale of events (e.g., calibrating molecular clocks, timing divergences, and deep splits), but they can also bring light to bear on the nature of evolutionary transitions. Have hagfish undergone some form of "degeneracy"? Do cyclostomes exhibit adaptations or radiations that might be associated with the complex sequence of genome changes? What are the relationships of cyclostomes to Early Paleozoic armored jawless vertebrates, or "ostracoderms"? The current phylogenetic interpretation of cyclostomes as monophyletic (Delsuc et al., 2006; Peterson et al., 2008; Heimberg et al., 2010) further underscores the need for a clearer picture of the fossil record. Although the gap has narrowed, the morphology of extant cyclostomes is not currently in accordance with interpretations of cyclostome monophyly as supported by extensive molecular evidence (Heimberg et al., 2010, but see reservations about the use of micro-RNA data expressed by Thomson et al., 2014; Dunn, 2014). This discord only serves to emphasize the necessity for closer scrutiny of the cyclostome fossil record; the long evolutionary history of these groups (potentially the entire Phanerozoic) means that the morphology of extant forms might not be informative about deep splits. Fossils are needed to reconstruct the morphology of common ancestors and to overcome possible long-branch attraction. This problem is not trivial in the case of cyclostomes, however, as living hagfishes and lampreys are completely devoid of a mineralized skeleton. It has been long debated whether this condition was primitive or derived, notably in the context of their presumed "degeneracy," and their presumed relationships to "ostracoderms" (Stensiö, 1927). The absence of a mineralized skeleton also makes it highly improbable that they can be preserved as fossils.

The fossil record of cyclostomes was desperately barren until the discovery of the first fossil lamprey, *Mayomyzon pieckoensis*, from the ca. 310 million year (Myr)-old Late Carboniferous fossil site of Mazon Creek, Illinois (Bardack and Zangerl, 1968, 1971). The exceptional preservation of the soft tissues in this fossil was made possible by its rapid burial in estuarine sediment and subsequent formation of ironstone concretion around the decaying body (Baird et al., 1986). This discovery came as a surprise because the wisdom received at that time was that lampreys and hagfishes had diverged much later, possibly in Early Mesozoic times, from some unknown cyclostome ancestor. Nevertheless, some paleontologists who adhered to Stensiö's hypothesis of an independent derivation of hagfishes and lampreys (Stensiö, 1927, 1939) from different "ostracoderms" groups through progressive loss of their mineralized skeleton considered this finding as a possible corroboration of this view. Later on, paleontologists working of such fossil sites that yield exceptionally preserved soft-bodied organisms (currently referred to as "Konservat-Lagestätten") paid more attention to possible fossil cyclostomes, which resulted in the discovery of some additional Paleozoic (Devonian and Carboniferous) and even Mesozoic lampreys (Janvier and Lund, 1983; Lund and Janvier, 1986; Janvier et al., 2004; Gess et al., 2006; Chang et al., 2006, 2014), as well as the first presumed fossil hagfish, *Myxinikela siroca* (Figure 3.1a), from the same locality and age as *Mayomyzon* (Bardack, 1991, 1998). In addition, other peculiar soft-bodied jawless vertebrates turned up in coeval, Carboniferous Lagerstätten, some of which have been interpreted as possible lampreys or hagfishes (Bardack and Richardson, 1977; Poplin et al., 2001; Germain et al., 2014).

Here we review those taxa that have been suggested as having hagfish or cyclostome affinities and the difficulties associated with their interpretation.

A framework for fossil cyclostome interpretation

The remains of soft-bodied organisms can become preserved as fossils under exceptional circumstances, but they require particular caution

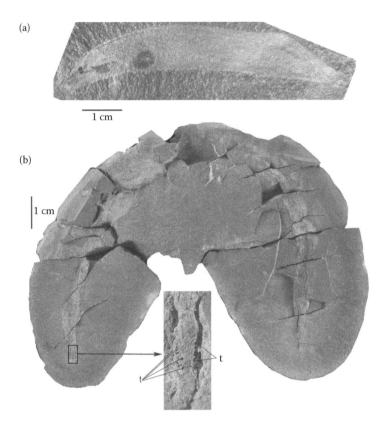

(a)

1 cm

(b)

1 cm

t

t

Figure 3.1 The two presumed fossil hagfishes. (a) *M. siroka* Bardack 1991. Holotype (NEIU MCP 126, Upper Carboniferous of Mazon Creek, Illinois, USA). (b) *Myxineidus gononorum* (Adapted from Poplin, C., D. Sotty and P. Janvier. 2001. *CR Acad Sci IIA* 332:345–350.), Upper Carboniferous of Montceau-les Mines, France. Specimen MNHN-F-SOT098687a (holotype), preserved in a nodule, and details of the natural filling of the oral cavity in ventral view, showing the impressions of the series of nonmineralized "teeth" (t).

when they are interpreted. Morphology of soft-bodied fossils is not an accurate facsimile of the original anatomy of the organism. Instead, it has been subjected to the complex processes of loss and change that occur during decay and fossilization; as such, it can be difficult to identify partially decaying anatomy or distinguish phylogenetically absent structures from those that were taphonomically lost (i.e., decayed away; Donoghue and Purnell, 2009). Fortunately, we have a wealth of data regarding the sequences of change and loss that occur during decay of chordates, which can aid fossil interpretation (Sansom et al., 2010a, 2011, 2013a). The sequences of change and loss of hagfish morphology

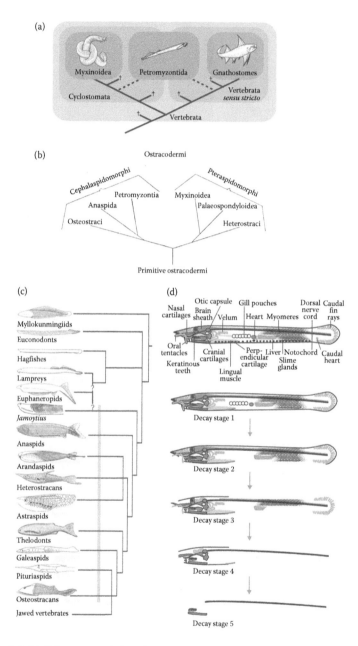

Figure 3.2 (a) Schematic of the alternative interpretations of vertebrate phylogeny regarding cyclostome monophyly and paraphyly, with representative fossil stem groups indicated with a cross. (Adapted from Sansom, R. S., S. E. Gabbott and M. A. Purnell. 2011. *Proc R Soc London B* 278:1150–1157.) (b) Facsimile of

(Continued)

have been revealed in detail (Figure 3.2d; Sansom et al., 2011, 2013a). Regarding their general changes, hagfish decay much faster than their chordate relatives, such that the later stages comprise nothing but the notochord, keratinous teeth, and some of the hard cartilages (Sansom et al., 2011). The decay of hagfish characters also fits the general pattern of character decay observed among chordates, whereby synapomorphic characters (i.e., those characteristic of narrower clades such as Cyclostomata or Myxinoidea) are lost before plesiomorphic characters (characteristic of Chordata and Vertebrata). A consequence of this is "stem-ward slippage," whereby organisms lose synapomorphic characters prior to preservation and the resulting fossil will be interpreted as more primitive than the original organism's true phylogenetic position (Sansom et al., 2010a, 2011), which can be avoided through reference to taphonomic frameworks (Sansom et al., 2011, 2013a). Hagfishes buck this trend in some regards owing to recalcitrance of their characteristic "hard" cartilages (Figure 3.2d; Sansom et al., 2011). Furthermore, the decay of the head cartilages follows repeated patterns related to gene expression—vertebrate-specific type 2a1 collagen is present only in the rapidly decaying soft cartilages, not the more recalcitrant hard cartilages. As such, "soft" cartilages are less likely to be preserved and it will be harder to identify tissues that would indicate gene and genome duplication events, which the multiple versions of collagen are interpreted to have resulted from (Donoghue and Purnell, 2005; Zhang and Cohn, 2008). While this pattern is very marked in hagfish, decay of lamprey collagen and cartilages is more nuanced (Sansom et al., 2011).

Fossil hagfishes

Only two soft-bodied fossils, *Myxinikela siroka* (Figure 3.1a) and *Myxineidus gononorum* (Figure 3.1b), have been interpreted as hagfishes in the literature, and both date back to the Late Carboniferous

---

*Figure 3.2 (**Continued**)* Stensiö's (1927, Figure 103) diagram "showing the interrelationships of the Ostracodermi," and explaining his views on the relationships of living cyclostomes with different groups of "ostracoderms." (c) The current phylogenetic tree of the major living and fossil vertebrates showing the respective relationships of some soft-bodied (or essentially soft bodied) fossils currently regarded as vertebrates. Pattern of relationships (right) and reconstruction of *Jamoytius* essentially based on Sansom et al. (2010). Reconstructions of other fishes based on Janvier (1996a, 2008). The vertical grey bar in the tree indicates the taxa traditionally referred to as "ostracoderms." (d). Hagfish anatomy with cartoons indicating the losses of anatomy during decay experiments. (Adapted from Sansom, R. S., S. E. Gabbott and M. A. Purnell. 2011. *Proc R Soc London B* 278:1150–1157.)

(ca. 310 Myr) from two Konservat-Lagerstätten: Mazon Creek, Illinois, USA, and Montceau-les-Mines, Saône-et-Loire, France, respectively (Bardack, 1991; Poplin et al., 2001). *Myxinikela* is a small, laterally compressed fish preserved as an imprint in the bedding plane of an ironstone concretion (Figure 3.1a). It is currently only known from a single specimen. Despite its rather stout body shape, anterior position of the gill pouches, and lanceolate tail are different from the classical eel-like shape of modern hagfishes, it has been interpreted as a hagfish essentially on the basis of the trace of an elongated prenasal sinus and pairs of nasal and labial tentacles. Other organ imprints in this specimen, notably of possible blood vessels and gill structures, also match relatively well hagfish anatomy (Bardack, 1998). However, *Myxinikela* surprisingly shows no clear indication of keratinous "teeth," although the sediment of this particular fossil site is known for generally preserving impressions of nonmineralized but hard structures, as in the presumed fossil lamprey *Pipiscius* (Bardack and Richardson, 1977), and that the keratinous hagfish teeth have proved to be remarkably decay resistant (Sansom et al., 2011, 2013a).

M. gononorum (Figure 3.1b) is known by at least two specimens, also preserved in an ironstone concretion. It differs from *Myxinikela* by its elongated body, which rather recalls that of modern hagfishes. Contrary to *Myxinikela*, the specimens show virtually no trace of internal structure, apart from the natural, internal filling of the anterior part of the digestive tract, which bears two, chevron-shaped series of impressions interpreted as impressions of nonmineralized denticles (Figure 3.1b). The latter strongly resemble the keratinous "teeth" of hagfishes in terms of both their arrangement and position at the anterior end of the oral region (Poplin et al., 2001; Germain et al., 2014). However, no other anatomical character clearly indicates that *Myxineidus* is a hagfish. Without any other characteristic feature, it is necessary to ask whether *Myxineidus* could represent the remains of an annelid. Loss of segmentation and other nonjaw features is taphonomically viable for annelids (Briggs and Kear, 1993), but Poplin et al. (2001) found that the arrangement of *Myxineidus* apparatus appears to be more compatible with hagfish denticles than that of scolecodonts (annelid "jaws"). Examination of one of the specimens by means of propagation phase-contrast synchrotron radiation computed microtomography confirmed this arrangement of the "teeth" but failed to yield further information about other internal structures. However, it showed a halo of less absorbing matter around the specimen, which is curiously suggestive of the shape of a lamprey, with a large, rounded oral disc positioned anteriorly (Germain et al., 2014). This image is probably an artifact owing to variations in the distribution of remaining organic matter, and a similar halo is frequently seen around other fossils preserved in concretions from the same locality.

Gilpichthys greeni (Bardack and Richardson 1977), another soft-bodied jawless vertebrate from the Carboniferous Mazon Creek Lagerstätte, has been tentatively referred to hagfishes by Janvier (1981, 1996a, b) on the basis of the peculiar series of spike-shaped impressions located in its oral region, which are somewhat suggestive of series of keratinous "teeth." However, Bardack and Richardson (1977) have interpreted these structures as "muscle blocks," and Bardack (1998) rejected Janvier's interpretation. Nevertheless, given their topological position (in the middle of the head) and preservation style (3D relief), interpretation of denticles seems more parsimonious. Similar to *Myxineidus*, *Gilpichthys* also yields few additional diagnostic characters.

The phylogenetic position of *Myxinikela* and *Myxineidus*, if they actually are hagfishes, currently remains uncertain. *Myxineidus* is perhaps the most closely comparable to modern hagfishes because of its elongate body shape and the close resemblance of its "teeth" to those of modern forms, while the stouter body shape of *Myxinikela*, its well-developed presumed eyes, and the uncertainty as to the presence of "teeth" may make it appear less similar to modern forms, or possibly more plesiomorphic. *Myxinikela* has been subject to cladistic analysis but was only securely resolved as a myxinoid with character reweighting (Gess et al., 2006) or enforcing cyclostome monophyly (Sansom et al., 2010b). *Myxineidus*, on the other hand, is unlikely to yield enough characters for formal cladistic analysis on the basis of existing material. Instead, interpretations of its affinity are reliant entirely on the purported homologies of the denticle-like structures.

Other possible stem and crown-group fossil cyclostomes

Current morphology-based phylogenetic analyses that include both extant and fossil taxa express a clear distinction between crown, stem, and total groups (Figure 3.2a). A crown group is a monophyletic group (or clade) that includes the last common ancestor of all the extant species considered, these extant species, and their respective fossil relatives (e.g., all the extant hagfish species and, if any, at least one fossil species that could be proved to be more closely related to, say, *Myxine* than to *Eptatretus*). A stem group is a fossil species or a clade that is more closely related to the crown group than to another one (e.g., a fossil that would share only some unique hagfish characters with the crown group; Figure 3.2a). Stem groups often consist of an array of taxa that share an increasingly larger number of derived characters with the crown group, and are thus a grade.

The total group is a clade that includes the crown group and the stem group and is defined by at least one derived character that is uniquely shared by all its members.

The relationships of the presumed fossil hagfishes and lampreys to the living representatives remain largely undecided and these fossils are generally considered as merely stem hagfishes or stem lampreys because of their early age (Janvier, 2008). Only *Mayomyzon* was regarded by Sansom et al. (2011) as a possible crown-group lamprey. Among fossil lampreys, *Hardistiella* was regarded by Janvier and Lund (1983) and Janvier et al. (2004) as a stem lamprey, essentially because of the presumed persistence of an anal fin and the apparent lack of an oral disc, although this could be due to incomplete preservation of the body margin and taphonomic loss respectively. One specimen of *Hardistiella* displays imprints of which might suggest larval cartilages (Lund and Janvier, 1986), but it could not be ruled out that they represent pigmentation surrounding branchial pouches and openings (Sansom et al., 2010b). The Late Devonian (Famennian) *Priscomyzon*, which possesses a large oral disc, was also considered by Gess et al. (2006) as a possible stem lamprey. The Late Carboniferous *Pipiscius*, coeval with *Mayomyzon* and from the same locality, was not considered as a lamprey by Bardack and Richardson (1977), but was regarded as such by Janvier (1981, 1996a, b), who suggested that it might have been more derived than the latter because of the arrangement of the presumed keratinous plates of its oral disc that somewhat recall those of the living *Ichthyomyzon*. The oral disc of *Pipiscius* has also been compared to the oral "plates" of the enigmatic Cambrian vetulicolians (Shu et al., 1999a), which are regarded as either stem deuterostomes or unusual arthropods (Aldridge et al., 2007), but none of the *Pipiscius* specimens show indications of the characteristic large segments seen in vetulicolians. Finally the Cretaceous lamprey *Mesomyzon* (Chang et al., 2006, 2014), from the Jehol Lagerstätte, China, is almost identical to modern lampreys in overall aspect, but its relationships to any particular modern lamprey taxon are unresolved.

One of the difficulties with respect to the presumed fossil lampreys is that they seem to preserve different character complexes. For example, *Mayomyzon* yields eyes and internal feeding and branchial apparatus, whereas *Pipiscius* yields principally the oral apparatus in relief, *Mesomyzon* preserves the muscles of the trunk, and *Priscomyzon* preserves branchial apparatus in dorsal perspective and some evidence of oral "teeth." The different styles of preservation and the limited overlap in preserved character complexes make it difficult to compare the extinct taxa, or between extinct and extant taxa; it also leaves open the possibility that some of the taxa could be congeneric or even conspecific.

Priscomyzon displays characteristics that are sufficient to ascertain the presence of lampreys as early as the Late Famennian; that is, ca. 360

Myr ago. Consequently, the divergence between hagfishes and lampreys (assuming that they are sister clades) must be older, and this is in agreement with molecular clock dates for the hagfish-lamprey divergence, which is estimated at about 505 Myr (Blair and Blair Edges, 2005; Kuraku and Kuratani, 2006; Peterson et al., 2008). However, a number of Paleozoic taxa have been once regarded as close relatives of either hagfishes or lampreys, or even possible stem cyclostomes, yet with generally tenuous arguments (Janvier, 2015). This is not unexpected, as the closer an organism is to the divergence point, the harder it is to recognize as belonging to either group (i.e., myxinoid or petromyzontid). This is generally true for all fossil taxa, but complicated further in the case of cyclostomes not just by the limitations of soft tissue preservation, and also by the perceived "degeneracy" of hagfish. The former makes it hard to recognize morphology while the later makes it hard to know what morphology to expect.

Palaeospondylus gunni (Traquair, 1890), from the Middle Devonian of Scotland, still remains the most enigmatic early vertebrate, although known by hundreds of specimens. It possesses an axial skeleton composed of vertebrae, a caudal fin with radials and fin supports, possible paired "girdles" or appendages, and its skull consists of a number of peculiar endoskeletal elements that cannot be clearly homologized with the classical components of the vertebrate skull, be it that of a cyclostomes or that of a gnathostome (Sollas and Sollas, 1904; Moy-Thomas, 1940). All the skeletal elements of *Palaeospondylus* are exclusively made up by a spongious calcified matter, which somewhat compares to that of the endoskeleton of *Euphanerops* (see below) and is therefore interpreted as calcified cartilage. However, its resemblance to embryonic osteichthyan cartilage has led Johanson et al. (2010) to consider *Palaeospondylus* as some kind of osteichthyan that failed to develop endochondral and dermal bone. Since its discovery, *Palaeospondylus* has been carefully described, notably by means of grinding sections and chemical preparation (Sollas and Sollas, 1904; Bulman, 1931) and tentatively referred to practically all fish groups: hagfishes, lampreys, placoderms, chimaeroids, sharks, teleosts, lungfishes, or piscine tetrapodomorphs, and generally as a larval form (see review in Moy-Thomas, 1940; Forey and Gardiner, 1981; Janvier, 1996a; Thomson et al., 2003; Joss and Johanson, 2006; Newman and Den Blaauwen, 2008; Johanson et al., 2010). All these interpretations have been more or less convincingly dismissed. However, recent data on hagfish skeletal development (Oisi et al., 2013a, b) seem to confirm the superficial resemblance (already alluded to by Traquair, 1890; Bulman, 1931) between the arrangement of some of the skull elements of *Palaeospondylus*, and that of the cranial cartilages in late stages of hagfish embryos. They also reveal that the absence of vertebral elements in hagfishes may be secondary (Ota et al., 2011, 2013). The extensive vertebral column of *Palaeospondylus* may

thus not preclude close relationships to hagfishes. However, no unambiguous character is uniquely shared by hagfishes and *Palaeospondylus*, which
therefore remains an enigma (Janvier, 2015).

Euphaneropids (Figure 3.2c) are essentially known by *Euphanerops
longaevus* (Woodward, 1900) (probably synonymous with *Endeiolepis
aneri*; Stensiö, 1939), from the Late Devonian (370 Myr) of the Miguasha
Lagerstätte, Canada. *Euphanerops* is one of the best-known soft-bodied
fossil jawless vertebrates, thanks to the presence of calcifications that
appear in its cartilages in large and presumably senescent individuals
(Janvier and Arsenault, 2007). The latter display ovoid spaces, presumably occupied by chondrocytes and surrounded by layers of calcium
phosphate, and aggregated into larger elements, but in a somewhat different manner than in the calcified cartilage of *Palaeospondylus* (Johanson
et al., 2010). Nevertheless, these calcified elements provide evidence for
massive dorsal, caudal, and anal fin radials, possibly elongated paired
fins armed with radials, dorsal, and ventral arcualia, a very large, somewhat lamprey-like branchial basket that reaches as far back as the anal
region, and a circular cartilage in the oral region, which recalls the annular cartilage of lampreys. The morphology of *Euphanerops* is surprising in
many regards. A large branchial basket reaches as far as the anus (Janvier
et al., 2006), behind which there is a possibly paired anal fin (Sansom
et al., 2013b). Because of their strongly hypocercal tail and large anal fins,
Euphanerops and *Endeolepis* have been regarded as "naked anaspids", that
is, late representatives of the "ostracoderm" group Anaspida (Figure 3.2c;
once thought to be closely related to lampreys because of their elongated
body shape and dorsal nasohypophysial opening) that have lost the
dermal skeleton, thereby foreshadowing an evolution toward the lamprey condition (Stensiö, 1939; Arsenault and Janvier, 1991; Janvier, 1996a).
However, there is no evidence that euphaneropids have ever possessed a
dermal skeleton.

Euphanerops has sometimes been allied with *Jamoytius* (Figure 3.2c)
in the Jamoytiiformes. *Jamoytius kerwoodi* (White, 1946) from the Lower
Silurian (ca. 435 Myr) of Scotland is significantly older than all-known
euphaneropids, but is currently regarded as the sister group of the latter, despite its rather different, slenderer body shape and much shorter
branchial region (Sansom et al., 2010b). Like *Euphanerops*, *Jamoytius* seems
to have possessed an annular cartilage and possibly paired, elongated,
ventrolateral fins or fin folds. However, it may have possessed very
thin mineralized scales. *Jamoytius* is often resolved as a jawless stem-
gnathostome (Figure 3.2c) and the most recent analysis of relationships
finds *Jamoytius* and *Euphanerops* as sister taxa on the gnathostome stem-
lineage, owing in part to the probable biomineralization of *Jamoytius*
(Sansom et al., 2010b, Figure 3.2c). This relationship has been recovered
in other analyses (Janvier 1996b; Donoghue et al., 2000), yet *Euphanerops*

has also been interpreted as having lamprey affinities (Gess et al., 2006; Janvier and Arsenault 2007; Figure 3.2c). Either way, it is likely that both taxa are placed very close to the lamprey/cyclostome–gnathostome split (depending on accepted topology). Two other possible jamoytiiform (or euphaneropid) taxa, *Cornovichthys* and *Achanarella*, are known from older, Middle Devonian sites of Scotland, but are poorly preserved (Newman and Trewin 2001; Newman 2002). They have been weakly resolved as stem-gnathostomes and stem-vertebrates, but their affinities are likely a result of decay-induced stem-ward slippage owing to incomplete preservation (Sansom et al., 2010a, b).

Myllokunmingiids (Figure 3.2c; *Myllokunmingia, Haikouichthys, Zhongjianichthys*) are soft-bodied fish-like animals, from the Early Cambrian (532 Myr-old) Chengjiang Lagerstätte of China, currently interpreted as jawless vertebrates, and the earliest vertebrates known to date (Shu et al., 1999b, 2003; Luo et al., 2001; Hou et al., 2002; Zhang and Hou, 2004). Although preserved as imprints, they display a pair of large, presumably anterodorsal eyes, a small dorsal olfactory organ, five or six gill bars with gill-ray-like filaments, chevron-shaped muscle blocks, and a large dorsal fin that extends posteriorly into a lanceolate median fin. The younger (ca. 510 Myr-old) Burgess Shale taxon *Metaspriggina* (Conway Morris and Caron, 2014) is probably very close to myllokunmingiids, but provides more detailed information about the arrangement of the gill bars and eye structure. Despite the small number of characters that can actually be observed on this kind of material, attempted character analyses resolved myllokunmingids as paraphyletic, with *Myllokunmingia* as a stem vertebrate, and *Haikouichthys* as a stem lamprey (Shu et al., 1999a, b). More recent analyses suggest that all myllokunmingiids and probably *Metaspriggina* are stem vertebrates, but appear as a basal polytomy in the vertebrate tree (Sansom et al., 2010b; Conway Morris and Caron, 2014). In the case of *Haikouichthys*, this interpretation is taphonomically consistent—if myllokunmingiids possessed a skull or cranial cartilage, it would be expected to be preserved, given the preservation of other cartilages and more decay-prone characteristic features (Sansom et al., 2011). As such, they possess some key vertebrate synapomophies (e.g., paired sense organs) but lack others (skull), making them a stem-vertebrate in a phylogenetically meaningful sense.

Conodonts are minute tooth- or comb-like elements that are made up by calcium phosphate, like vertebrate odontodes and teeth. They occur in marine sediments from the Cambrian to the Late Triassic (ca. 530–200 Myr ago). Depending on their internal structure, conodonts fall into three groups: protocodondonts, paracododonts, and euconodonts (Figure 3.2c)—the former including elements that are possibly stem chaetognaths and the latter being probably the only monophyletic one. Since the discovery of the first articulated conodont-bearing animal (Briggs et al., 1983), several other such specimens have turned up (Gabbott et al., 1995), but all articulated "conodont animals" known to date are euconodonts. Thanks to these discoveries, it became possible to have a better knowledge of the anatomy of these long enigmatic organisms: a small head with large eyes, a mouth or pharynx containing a large number of probably protractible denticles, an elongated, eel-shaped body with chevron-shaped myomeres, and a small caudal fin armed with radials. This apparently agrees with vertebrate morphology (Aldridge et al., 1993), although the lack of unequivocal evidence for more typically vertebrate structures, such as gill arches or other visceral elements, remains puzzling. The most contentious question was that of the homology of the euconodont denticle tissues with those of the vertebrate odontodes or teeth, which was eagerly defended by, for example, Donoghue (1998) and Donoghue and Aldridge (2001) but opposed by, for example, Blieck et al. (2010), Turner et al. (2010) (see reviews in Aldridge et al., 1993; Janvier, 1995; Aldridge and Purnell, 1996; Schultze, 1996; Aldridge and Donoghue, 1998; Donoghue et al., 2000; Turner et al., 2010; Knell, 2013; Murdoch et al., 2013). The resolution of this particular controversy would have in fact settled that of the affinities of these fossils. This finally happened with the demonstration, by means of high-resolution microtomographic imaging of paraconodonts, that euconodont denticle structure and growth is convergent with vertebrate odontodes, not homologous (Murdoch et al., 2013). Since the publication by Donoghue et al. (2000) on chordate phylogeny, euconodonts were almost constantly considered in the framework of phylogenetic analyses of early vertebrate taxa, and their position as basal-most stem gnathostomes was essentially supported by the presence of the phosphatic conodont elements, assumed to be homologues of odontodes that are present in most total-group gnathostomes, but lacking in all cyclostomes. The only support to that position would thus be a shared ability to produce calcium phosphate, admittedly shared by several other metazoan taxa. However, the old hypothesis that euconodonts might be more closely allied to cyclostomes, and more specifically to hagfishes, sometimes reappears in the literature (Kreijsa et al., 1990; Dzik, 2009; Goudemand et al., 2011). For example, the Carboniferous protocondont-like articulated fossil *Conopiscius* (Briggs and Clarkson, 1987) displays, like the bodies of conodonts, a series of chevron-shaped myomeres (possibly covered with thin scales), but a single pair of hollow, weakly mineralized "teeth" (Dzik, 2009). Kreijsa et al. (1990) and Dzik (2009) suggested the possibility that conodonts were in fact partly or entirely capped with a keratinous tissue, which remains in living cyclostomes. Furthermore, an odd result of the analysis of Jamoytiiformes phylogeny (Sansom et al., 2010b) was the recovery of euconodonts as either stem-petromyzontida or stem-cyclostomes in the case of cyclostome monophyly. This hypothesis, however, was regarded as poorly supported (Aldridge and Donoghue, 1998).

Despite considerable efforts and the use of sophisticated imaging and analytic techniques, the "conodont controversy" remains far from definitively settled. If euconodont hard tissues are interpreted as nonhomologous with those of vertebrates, a natural focus becomes interpretation of the soft tissues. There remain the strikingly chordate- or vertebrate-like features of the euconodont bodies, which does not preclude a position as stem chordates, stem vertebrates, or stem cyclostomes (Sansom et al., 2010a; Janvier, 2013, 2015). Furthermore, a wealth of taphonomic data now exists to provide a framework for interpretation of these soft tissues (Sansom et al., 2010a, 2011, 2013b).

"Ostracoderms": Bony cyclostomes or jawless stem gnathostomes?

Apart from the question of the euconodonts and some still poorly known stem vertebrates, the currently accepted phylogenetic tree of the crown-group vertebrates, that is, living cyclostomes and gnathostomes and their respective fossil relatives, is now relatively well corroborated by both molecule-based and morphological data (Figure 3.2c). In this tree, the stem lineage of gnathostomes comprises a number of extinct jawless fish that possess some combination of presence and absence of jawed vertebrate characteristics, that is, the ability to develop a calcified skeleton made of bone and/or dentinous tissues, perichondrally calcified endoskeleton, externally open endolymphatic ducts, scleral ossifications, paired fins or epicercal tail, and so on. These fossil jawless vertebrates form a paraphyletic array of taxa that have been long gathered into a group called "ostracoderms." The relationships of the cyclostomes to "ostracoderms" now belong to the early history of the researches on vertebrate phylogeny and began in the late nineteenth century when it was realized that "ostracoderms" were effectively jawless, essentially by lack of evidence for jaws (Cope, 1889; see review in Janvier, 2008). From that time on, "ostracoderm" anatomy was progressively interpreted in light of cyclostome anatomy because cyclostomes were the only extant jawless proxies (as if jawlessness, an absence of character, could define a taxon). Stensiö (1927) published a famous diagram of "ostracoderm" relationships that showed hagfishes and lampreys as being related to two different ensembles of "ostracoderms," respectively (Figure 3.2b). Lampreys were presumed to be allied to osteostracans (Figure 3.2c) and anaspids (Figure 3.2c) because of their median dorsal nasohypophysial opening, whereas hagfishes were supposed to be more closely related to *Palaeospondylus* (see above) and heterostracans (Figure 3.2c). However, the poor data on the internal anatomy of heterostracans (not to speak of thelodonts; Figure 3.2c) were meager support for this hypothesis of relationships (save the single pair of common branchial openings that recalled those of myxinid hagfishes). Stensiö's (1968) diagram was often regarded as illustrating the "diphyletic origin of the cyclostomes," although it merely depicted the hypothetical, independent loss of skeletonization in hagfishes and lampreys, respectively. This conception of cyclostome relationships was soon criticized by a number of authors (e.g., Säve-Söderbergh, 1941; Romer, 1945; Moy-Thomas and Miles, 1971; see review in Halstead, 1973), but most of the criticisms bore on the hagfish–heterostracan relationships, whereas the lamprey–osteostracan–anaspid relationships were almost unanimously accepted because the median dorsal nasophypophysial opening was regarded as an obvious shared "specialization" (a derived character in modern terms) (Moy-Thomas and Miles, 1971). Therefore, most of osteostracan internal anatomy was readily interpreted in light of lamprey anatomy, however symplesiomorphic this resemblance could be (the Hennigian notions of apomorphy and plesiomorphy was ignored at that time; Hennig, 1950). The phylogenetic position of heterostracans and thelodonts has been quite erratic until the 1980s; they were often regarded as more closely related to gnathostomes than to other "ostracoderms" and cyclostomes because of their presumably paired (diplorhinous) olfactory organs, thus suggesting paraphyletic jawless vertebrates (Romer, 1945; Halstead, 1973; see review in Janvier and Blieck, 1993). At about the same time, the hypothesis of cyclostome paraphyly (Løvtrup, 1977) was integrated within a new consideration of character distribution among "ostracoderms" and gnathostomes (Janvier, 1978; Janvier and Blieck, 1979). It reinforced the hypothesis of "ostracoderm" paraphyly, but still with a clade corresponding to Stensiö's (1927) "Cephalaspidomorphi" (Figure 3.2b; i.e., including lampreys, anaspids, osteostracans, and possibly galeaspids; Figure 3.2c) as sister group to the gnathostomes, gathered with the latter in the clade "Myopterygii," characterized by muscularized fin radials. In this scheme, heterostracans and hagfishes are sister groups to the myopterygians. In later analyses (Janvier, 1981, 1984), the position of osteostracans and galeaspids (Figure 3.2c) became in turn fluctuating, with lampreys being sister group to anaspids, osteostracans more frequently resolved as sister group to gnathostomes, and thelodonts regarded as paraphyletic, distributed between the divergence of heterostracans and the root of jawed vertebrates (Janvier, 1981).

Vertebrate phylogeny then began to stabilize with advent of the first computerized phylogenetic software. Gagnier (1993) provided a tree where the still paraphyletic cyclostomes diverge before all the "ostracoderms" and jawed vertebrates. However, Janvier (1996a) retained the tree proposed by Forey and Janvier (1993) but admitted that Gagnier's (1993) tree was probably better supported (see also Janvier, 1996b). Finally, Donoghue et al. (2000) published what is practically the current topology, yet with cyclostomes still paraphyletic (Donoghue and Smith, 2001). During the last two decades, cyclostome monophyly has risen in preference, given increasingly strong—at any rate consistent—molecular support (Stock and Whitt, 1992; Delsuc et al., 2006; Heimberg et al., 2010; see also review in Near, 2009), thereby reviving the old, nineteenth-century, phenotype-based classification of the cyclostomes as a "natural group" (Duméril, 1806; Schaeffer and Thomson, 1980; Yalden, 1985).

However, results from the study by Heimberg et al. (2010) based on expression profiles of micro-RNAs surprisingly suggest that the last common ancestor to crown-group vertebrates (i.e., the last common ancestor to cyclostomes and gnathostomes) might have been significantly more complex than living cylostomes. This strangely echoes Stensiö's hypothesis that the cyclostome ancestors possessed an extensive bony skeleton and paired fins, similar to the earliest jawed vertebrates (Janvier, 2010).

Conclusion

Phylogenomics, along with other molecule-based data, have played a major role in the corroboration of cyclostome monophyly and vertebrate phylogeny more generally. Fossils can fill the phenotypic gaps in the vertebrate tree and help in providing a minimum age for characters, clades, and estimates of evolutionary rates. In addition, skeletonized vertebrates, such as "ostracoderms," play this role in documenting the stepwise assembly of the gnathostome body plan. However, the lack of a mineralized skeleton in certain fossil vertebrates practically precludes their use for documenting more stem-ward divergences either within the cyclostomes or the stem vertebrates. Cases of exceptional soft tissue preservation, as well as a better knowledge of the decay and fossilization processes, now allow better understanding of the characters of such nonmineralized (or poorly mineralized) fossils, some of which may reasonably be inserted in the vertebrate tree, however tenuous is the support for their relationships. Fossil lampreys do exist and provide acceptable evidence for their unique characters since ca. 360 Ma ago. In contrast, the two fossil hagfishes recorded to date, both ca. 300 Ma-old, remain poorly informative, if not debated. In fact, for the entire Phanerozoic eon (541 million years), only two fossils specimens (let alone taxa) have been discovered that have been interpreted as fossil hagfish with any degree of confidence (the *Myxinikela* and *Myxineidus* holotypes). This fossil paucity is in accordance with their rapid decay and deep sea environments. Nevertheless, soft-bodied fossils are no longer trivial, and paleontologists now better know where to find them, how to study them, and how to avoid overinterpreting them. There may thus be some hope to discover more stem lampreys and stem hagfishes, and even perhaps demonstrable stem cyclostomes.

References

Aldridge, R. J., D. E. G. Briggs, M. P. Smith, E. N. K. Clarkson and N. D. L. Clark. 1993. The anatomy of conodonts. *Phil Trans R Soc London B* 340:405–421.

Aldridge, R. J. and P. C. J. Donoghue. 1998. Conodonts: A sister-group to hagfish? In: J. M. Jørgensen, J. P. Lomholt, R. E. Weber and H. Malte (eds.), *The Biology of Hagfishes*. Chapman & Hall, London, pp. 16–31.

Aldridge, R. J., X.-G. Hou, D. J. Siveter, D. J. Siveter and S. E. Gabbott. 2007. The systematics and phylogenetic relationships of vetulicolians. *Palaeontology* 50:131–168.

Aldridge, R. J. and M. A. Purnell. 1996. The conodont controversies. *Trends Ecol Evol* 11:463–468.

Arsenault, M. and P. Janvier. 1991. The anaspid-like craniates of the Escuminac formation (Upper Devonian) from Miguasha (Quebec, Canada), with remarks on the anaspid–petromyzontid relationships. In: M. M. Chang, Y.-H. Liu and G.-R. Zhang (eds.), *Early Vertebrates and Related Problems of Evolutionary Biology*. Science Press, Beijing, pp. 19–40.

Baird, G. C., S. D. Sroka, C. W. Shabica and G. J. Kuecher. 1986. Taphonomy of Middle Pennsylvanian Mazon Creek area fossil localities, Northeast Illinois: Significance of exceptional preservation in syngenetic concretions. *Palaios* 1:271–285.

Bardack, D. 1991. First fossil hagfish (Myxinoidea): A record from the Pennsylvanian of Illinois. *Science* 254:701–703.

Bardack, D. 1998. Relationships of living and fossil hagfishes. In: J. M. Jørgensen, J. P. Lomholt, R. E. Weber and H. Malte (eds.), *The Biology of Hagfishes*. Chapman & Hall, London, pp. 16–31.

Bardack, D. and E. S. Richardson, Jr. 1977. New agnathous fishes from the Pennsylvanian of Illinois. *Fieldiana: Geol* 33:489–510.

Bardack, D. and R. Zangerl. 1968. First fossil lamprey: A record from the Pennsylvanian of Illinois. *Science* 162:1265–1267.

Bardack, D. and R. Zangerl. 1971. Lampreys in the fossil record. In: M. W. Hardisty and I. C. Potter (eds.), *The Biology of Lampreys*, Vol. 1. Academic Press, London, pp. 67–84.

Blair, J. S. and S. Blair Edges. 2005. Molecular phylogeny and divergence times of deuterostome animals. *Mol Biol Evol* 22:2275–2284.

Blieck, A., S. Turner, C. J. Burrow, H.-P. Schultze, C. B. Rexroad, P. Bultynck and G. S. Nowlan. 2010. Fossils, histology, and phylogeny: Why conodonts are not vertebrates. *Episodes* 33:234–241.

Briggs, D. E. G. and E. N. K. Clarkson. 1987. An enigmatic chordate from the Lower Carboniferous Granton shrimp-bed of the Edinburgh district, Scotland. *Lethaia* 20:107–115.

Briggs, D. E. G., E. N. K. Clarkson and R. J. Aldridge. 1983. The conodont animal. *Lethaia* 20:1–14.

Briggs, D. E. G. and A. Kear. 1993. Decay and preservation of polychaetes: Taphonomic thresholds in soft-bodied organisms. *Paleobiology* 19:107–135.

Bulman, O. M. B. 1931. Note on *Palaeospondylus gunni* Traquair. *Ann Mag Nat Hist* 8:179–190.

Chang, M. M., F. Wu, D. Miao and J. Zhang. 2014. Discovery of fossil lamprey larva from the lower Cretaceous reveals its three-phased life cycle. *Proc Natl Acad Sci USA* 111(43):15,486–15,490.

Chang, M. M., J. Zhang and D. Miao. 2006. A lamprey from the Cretaceous Jehol biota of China. *Nature* 441:972–974.

Conway Morris, S. and J.-B. Caron. 2014. A primitive fish from the Cambrian of North America. *Nature* 512:419–422.

Cope, E. D. 1889. Synopsis of the families of Vertebrata. *Am Naturalist* 23:1–29.

Delsuc, F., H. Brinkmann, D. Chourrout and H. Philippe. 2006. Tunicate and not cephalochordate are the closest living relatives of vertebrates. *Nature* 486:247–250.

Donoghue, P. C. J. 1998. Growth and patterning in the conodont skeleton. *Phil Trans R Soc London B* 353:633–666.

Donoghue, P. C. J. and R. J. Aldridge. 2001. Origin of a mineralized skeleton. In: P. E. Ahlberg (ed.), *Major Events in Early Vertebrate Evolution. Systematics Association,* Spec. Vol. Ser. 61. Taylor & Francis, London, New York, pp. 85–105.

Donoghue, P. C. J., P. L. Forey and R. J. Aldridge. 2000. Conodont affinity and chordate phylogeny. *Biol Rev* 75:191–251.

Donoghue, P. C. J. and M. A. Purnell. 2005. Genome duplication, extinction and vertebrate evolution. *Trends Ecol Evol* 20:312–319.

Donoghue, P. C. J. and M. A. Purnell. 2009. Distinguishing heat from light in debate over controversial fossils. *BioEssays* 31:178–189.

Donoghue, P. C. J. and P. Smith. 2001. The anatomy of *Turinia pagei* (Powrie), and the phylogenetic status of the Thelodonti. *Trans Roy Soc Edin Earth Sci* 92:15–37.

Duméril, A. M. C. 1806. *Zoologie Analytique, ou Méthode Naturelle de Classification des Animaux.* Didot, Paris.

Dunn, C. W. 2014. Reconsidering the phylogenetic utility of miRNA in animals. *Proc Natl Acad Sci USA* 111(35):12,576–12,577.

Dzik, J. 2009. Conodont affinity of the enigmatic Carboniferous chordate *Conopiscius. Lethaia* 42:31–38.

Forey, P. L. and B. G. Gardiner. 1981. J.A. Moy-Thomas and his association with the British Museum (Natural History). *Bull Br Mus Nat Hist (Geol)* 35:131–144.

Forey, P. L. and P. Janvier. 1993. Agnathans and the origin of jawed vertebrates. *Nature* 361:129–134.

Gabbott, S. E., R. J. Aldridge and J. N. Theron. 1995. A giant conodont with preserved muscle tissue from the Upper Ordovician of South Africa. *Nature* 374:800–803.

Gagnier, P. Y. 1993. *Sacabambaspis janvieri,* Vertébré ordovicien de Bolivie. 2. Analyse phylogénétique. *Ann Paleontol* 79:119–166.

Germain, D., S. Sanchez, P. Janvier and P. Tafforeau. 2014. The presumed hagfish *Myxineidus gononorum* from the Upper Carboniferous of Montceau-les-Mines (Saône-et-Loire, France): New data obtained by means of propagation phase contrast x-ray synchrotron microtomography. *Ann Paleontol* 100:131–135.

Gess, R. W., M. I. Coates and B. S. Rubidge. 2006. A lamprey from the Devonian of South Africa. *Nature* 443:981–984.

Goudemand, N., M. J. Orchard, S. Urdy, H. Bucher and P. Tafforeau. 2011. Synchrotron-aided reconstruction of the conodont feeding apparatus and implications for the mouth of the first vertebrates. *Proc Natl Acad Sci USA* 108(21):8720–8724.

Halstead, L. B. 1973. The heterostracan fishes. *Biol Rev* 48:279–332.

Heimberg, A. M., R. Cowper-Sal-lari, M. Sémon, P. C. J. Donoghue and K. J. Peterson. 2010. MicroRNAs reveal the inter relationships of hag fish, lampreys, and gnathostomes and the nature of the ancestral vertebrate. *Proc Natl Acad Sci USA* 107(45):19,379–19,383.

Hennig, W. 1950. *Grundzüge Einer Theorie der Phylogenetischen Systematik*. Deutscher Zentralverlag, Berlin.

Hou, X., R. J. Aldridge, D. J. Siveter, D.J. Siveter and X. Feng. 2002. New evidence on the anatomy and phylogeny of the earliest vertebrates. *Proc R Soc London B* 269:1865–1869.

Janvier, P. 1978. Les nageoires paires des Ostéostracés et la position systématique des Céphalaspidomorphes. *Ann Paleontol* 64:113–142.

Janvier, P. 1981. The phylogeny of the Craniata, with particular reference to the significance of fossil agnathans. *J Vert Paleontol* 1:121–159.

Janvier, P. 1984. The affinities of the Osteostraci and Galeaspida. *J Vert Paleontol* 3:315–321.

Janvier, P. 1995. Conodonts join the club. *Nature* 374:761–762.

Janvier, P. 1996a. *Early Vertebrates*. Oxford University Press, Oxford.

Janvier, P. 1996b. The dawn of the vertebrates: Characters versus common ascent in the rise of current vertebrate phylogenies. *Palaeontology* 39:259–287.

Janvier, P. 2008. Early jawless vertebrates and cyclostome origins. *Zool Sci* 25:1045–1056.

Janvier, P. 2010. microRNAs revive old views about jawless vertebrate divergence and evolution. *Proc Natl Acad Sci USA* 107:19137–19138.

Janvier, P. 2013. Inside-out turned upside down. *Nature* 502:457–458.

Janvier, P. 2015. Facts and fancies about early fossil chordates and vertebrates. *Nature* 520:483–489.

Janvier, P. and M. Arsenault. 2007. The anatomy of *Euphanerops longaevus* Woodward, 1900, an anaspid-like jawless vertebrate from the Upper Devonian of Miguasha, Quebec, Canada. *Geodiversitas* 29:143–216.

Janvier, P. and A. Blieck. 1979. New data on the internal anatomy of the Heterostraci (Agnatha), with general remarks on the phylogeny of the Craniota. *Zool Scr* 8:287–296.

Janvier, P. and A. Blieck. 1993. L.B. Halstead and the heterostracan controversy. *Mod Geol* 18:89–105.

Janvier, P., S. Desbiens, J. A. Willett and M. Arsenault. 2006. Lamprey-like gills in a gnathostome-related Devonian jawless vertebrate. *Nature* 440:1183–1185.

Janvier, P. and R. Lund. 1983. *Hardistiella montanensis* n. gen., n. sp. (Petromyzontida) from the Lower Carboniferous of Montana, with remarks on the affinities of lampreys. *J Vert Paleontol* 2:407–413.

Janvier, P., R. Lund and E. Grogan. 2004. Further consideration of the earliest known lamprey, *Hardistiella montanensis* Janvier and Lund, 1983, from the Carboniferous of Bear Gulch, Montana, USA. *J Vert Paleontol* 24:742–743.

Johanson, Z., A. Kearsley, J. Den Blaauwen, M. Newman and M. M. Smith. 2010. No bone about it: An enigmatic Devonian fossil reveals a new skeletal framework—A potential loss of gene regulation. *Semin Cell Dev Biol* 21:414–423.

Joss, J. and Z. Johanson. 2006. Is *Palaeospondylus gunni* a fossil larval lungfish? Insights from *Neoceratodus forsteri* development. *J Exp Zool B* 308:163–171.

Knell, S. J. 2013. *The Great Fossil Enigma: The Search for the Conodont Animal*. Indiana University Press, Bloomington.

Kreijsa, R. J., P. Bringas and H. Slavkin. 1990. A neontological interpretation of conodont elements based on agnathan cyclostome tooth structure, function and development. *Lethaia* 23:359–378.

Kuraku, S. and S. Kuratani. 2006. Time scale for cyclostome evolution inferred with a phylogenetic diagnosis of hagfish and lamprey cDNA sequences. *Zool Sci* 23:1053–1064.

Løvtrup, S. 1977. *The Phylogeny of Vertebrata*. Wiley, New York.

Lund, R. and P. Janvier. 1986. A second lamprey from the Lower Carboniferous of Montana (U.S.A.). *Geobios* 19:647–652.

Luo, H., S. Hu and L. Chen. 2001. New early Cambrian chordates from Haikou, Kunming. *Acta Geol Sin* 75:345–348.

Moy-Thomas, J. A. 1940. The Devonian fish *Palaeospondylus gunni* Traquair. *Phil Trans R Soc London B* 230:391–413.

Moy-Thomas, J. A. and R. S. Miles. 1971. *Palaeozoic Fishes*. Chapman & Hall, London.

Murdoch, D. J. E., X.-P. Dong, J. E. Repetski, F. Marone, M. Stampanoni and P. C. J. Donoghue. 2013. The origin of conodonts and of vertebrate mineralized skeletons. *Nature* 502:546–549.

Near, T. J. 2009. Conflict and resolution between phylogenies inferred from molecular and phenotypic data sets for hagfish, lampreys, and gnathostomes. *J Exp Zool B Mol Dev Evol* 312B:749–761.

Newman, M. J. 2002. A new naked jawless vertebrate from the Middle Devonian of Scotland. *Palaeontology* 45:933–941.

Newman, M. J. and J. Den Blaauwen. 2008. New information on the enigmatic Devonian vertebrate *Palaeospondylus gunni*. *Scott J Geol* 44:89–91.

Newman, M. J. and N. Trewin. 2001. A new jawless vertebrate from the Middle Devonian of Scotland. *Palaeontology* 44:43–51.

Ota, K. G., S. Fujimoto, Y. Oisi and S. Kuratani. 2011. Identification of vertebra-like elements and their possible differentiation from sclerotomes in the hagfish. *Nat Commun* 2:373.

Ota, K. G., S. Fujimoto, Y. Oisi and S. Kuratani. 2013. Late development of hagfish vertebral elements. *J Exp Zool B Mol Dev Evol* 320:129–139.

Oisi, Y., K. G. Ota, S. Fujimoto and S. Kuratani. 2013b. Development of the chondrocranium in hagfishes, with special reference to the early evolution of vertebrates. *Zool Sci* 30:944–961.

Oisi, Y., K. G. Ota, S. Kuraku, S. Fujimoto and S. Kuratani. 2013a. Craniofacial development of hagfishes and the evolution of vertebrates. *Nature* 493:175–180.

Peterson, K. J., J. A. Cotton, J. G. Gehling and D. Pisani. 2008. The Ediacaran emergence of bilaterians: Congruence between the genetic and geologic fossil records. *Phil Trans R Soc London B* 363:1435–1443.

Poplin, C., D. Sotty and P. Janvier. 2001. Un Myxinoïde (Craniata, Hyperotreti) dans le Konservat-Lagerstätte Carbonifère supérieur de Montceau-les-Mines (Allier, France). *CR Acad Sci IIA* 332:345–350.

Romer, A. S. 1945. *Vertebrate Paleontology*. University of Chicago Press, Chicago.

Sansom, R. S., K. Freedman, S. E. Gabbott, R. J. Aldridge and M. A. Purnell. 2010b. Taphonomy and affinity of an enigmatic Silurian vertebrate, *Jamoytius kerwoodi* White. *Palaeontology* 53:1393–1409.

Sansom, R. S., S. E. Gabbott and M. A. Purnell. 2010a. Decay of chordate characters causes bias in fossil interpretation. *Nature* 463:797–800.

Sansom, R. S., S. E. Gabbott and M. A. Purnell. 2011. Decay of vertebrate characters in hagfish and lamprey (Cyclostomata) and the implication for the vertebrate fossil record. *Proc R Soc London B* 278:1150–1157.

Sansom, R. S., S. E. Gabbott and M. A. Purnell. 2013a. Atlas of vertebrate decay: A visual and taphonomic guide to fossil interpretation. *Palaeontology* 56:457–474.

Sansom, R. S., S. E. Gabbott and M. A. Purnell. 2013b. Unusual anal fin in a Devonian jawless vertebrate reveals complex origins of paired appendages. *Biol Lett* 9:20130002.

Säve-Söderbergh, G. 1941. Notes on the dermal bones of the head in *Osteolepis macrolepidotus* Ag. and the interpretation of the lateral line system in certain primitive vertebrates. *Zool Bidrag Uppsala* 20:523–541.

Schaeffer, B. and K. S. Thomson. 1980. Reflections on agnathan–gnathostome relationships. In: L. L. Jacobs (ed.), *Aspects of Vertebrate Life*. Museum of Northern Arizona Press, Flagstaff, pp. 19–33.

Schultze, H.-P. 1996. Conodont histology: An indicator of vertebrate relationship? *Mod Geol* 20:275–285.

Shu, D.-G., S. Conway Morris, X.-L. Zhang, L. Chen, Y. Li and J. Han.1999a. A pipiscid-like fossil from the Lower Cambrian of South China. *Nature* 400:746–749.

Shu, D. G., S. Conway Morris, J. Han, Z.-F. Zhang, K. Yasui, P. Janvier, L. Chen et al. 2003. Head and backbone of the early Cambrian vertebrate *Haikouichthys*. *Nature* 421:526–529.

Shu, D. G., H.-L. Luo, S. Conway-Morris, X.-L. Zhang, S.-X. Hu, L. Chen, J. Han, M. Zhu and L.-Z. Chen. 1999b. Lower Cambrian vertebrates from South China. *Nature* 402:42–46.

Sollas, W. J. and I. B. J. Sollas. 1904. An account of the Devonian fish, *Palaeospondylus gunni* Traquair. *Phil Trans R Soc London B* 196:267–294.

Stensiö, E. A. 1927. The Devonian and Downtonian vertebrates of Spitsbergen. 1. Family Cephalaspidae. *Skr. Svalbard Ishav* 12:1–391.

Stensiö, E. 1939. A new anaspid from the Upper Devonian of Scaumenac Bay in Canada, with remarks on the other anaspids. *K Svenska Vetensk Akad Handl* 18:1–25.

Stensiö, E. 1968. The cyclostomes with special reference to the diphyletic origin of the Petromyzontida and the Myxinoidea. In T. Ørvig (ed.), *Current Problems in Lower Vertebrate Phylogeny*. Almqvist and Wiksell, Stockholm, pp. 13–71.

Stock, D. W. and G. S. Whitt. 1992. Evidence from 18S ribosomal RNA sequences that lampreys and hagfishes form a natural group. *Science* 257:787–789.

Thomson, R. C., D. C. Plachetzki, D. Luke Mahler, and B. R. Moore. 2014. A critical appraisal of the use of microRNA data in phylogenetics. *Proc Natl Acad Sci USA* 111:E3659–E3668.

Thomson, K. S., M. Sutton and B. Thomas. 2003. A larval Devonian lungfish. *Nature* 426:833–834.

Traquair, R. H. 1890. On the fossil fishes at Achanarras Quarry. *Ann Mag Nat Hist* 5:479–486.

Turner, S., C. J. Burrow, H.-P. Schultze, A. Blieck, W.-E. Reif, C. B. Rexroad, P. Bultynck and G. C. Nowlan. 2010. False teeth: Conodont–vertebrate phylogenetic relationships revisited. *Geodiversitas* 32:545–594.

White, E. I. 1946. *Jamoytius kerwoodi*, a new chordate from the Silurian of Lanarkshire. *Geol Mag* 83:89–97.

Woodward, A.S. 1900. On a new ostracoderm fish (*Euphanerops longaevus*) from the Upper Devonian of Scaumenac Bay, Quebec, Canada. *Ann Mag Nat Hist* Ser 7 5:416–419.

Yalden, D. W. 1985. Feeding mechanisms as evidence for cyclostome monophyly. *Zool J Linn Soc* 84:291–300.

Zhang, G.-J. and M. J. Cohn. 2008. Genome duplication and the origin of the vertebrate skeleton. *Curr Opin Genet Dev* 18:387–393.

Zhang, X.-G. and X.-G. Hou. 2004. Evidence for a single median fin-fold and tail in the Lower Cambrian vertebrate, *Haikouichthys ercaicunensis*. *J Evol Biol* 17:1162–1166.

chapter four

Hagfish embryology
Staging table and relevance to the evolution and development of vertebrates

Tetsuto Miyashita and Michael I. Coates

Contents

Introduction

Although the biology of hagfishes overall remains highly controversial, authors past and present unanimously agree on the crucial role of hagfish in elucidating the origin and early evolution of vertebrates. Still, an ongoing quest since the late nineteenth century has produced surprisingly little information on one of the most anticipated mysteries about these archaic-looking fish: prenatal development. Hagfish embryos are notoriously difficult to obtain. Consequently, hagfish embryology throughout the twentieth century relied on a small sample of specimens occasionally collected from the field. As multiple projects converged on the same—and suboptimally prepared—materials, erroneous, anecdotal, or incomplete observations reinforced dubious interpretations, which stood uncontested for a long time (e.g., neural crest origins in an epithelial outpocket: Conel, 1942; endodermal origins for the nasohypophyseal canal: Gorbman, 1983).

However, modern efforts to collect and study embryos have overcome early logistical and technological challenges (Ota and Kuratani, 2006, 2007, 2008; Kuratani and Ota, 2008a; Ota et al., 2007). This recent success makes hagfish an attractive comparative taxon to constrain developmental features on a phylogenetic tree of vertebrates (Shimeld and Donoghue, 2012), and hagfish embryology now enjoys renewed enthusiasm and growing interest. In this chapter, we will review the history and current state of hagfish embryology (Table 4.1), describe the development of hagfish embryos and propose a staging scheme (Table 4.2; Figure 4.1), and outline future challenges.

Hagfish: A vertebrate that wasn't?

The image of a hagfish evokes almost visceral curiosity about its systematic affinity. This was apparently the case for Pehr Kalm, who noted a slimy lamprey-like fish during his brief stay at the Norwegian coast and reported it to the attention of Carl Linnaeus (Fänge, 1998). Linnaeus (1758) erected *Myxine glutinosa* for this curious fish and—though noting similarity to lampreys—classified it as "vermes intestinalis" (intestinal worm). To our knowledge, hagfish are the only "vertebrate" that Linnaeus failed to recognize as such in his founding work of animal taxonomy. It had to wait Abildgaard's (1792) classification to properly place hagfish among fish. This early confusion has not entirely been settled, however, as the proposed hagfish–lamprey sister-group relationship (Cyclostomi[*]) awaits corroboration by morphological and physiological data. With a long list of mainly molecular data sets that support a monophyletic Cyclostomi, the support is cumulatively consistent but the signal within individual data sets is weak (Near, 2009). This is expected for naked long branches such as hagfish and lampreys, both of which only contribute soft-tissue

---

[*] Cyclostomi Duméril, 1806, is the original name coined for the group of hagfish and lampreys (Janvier, 2008). The same group is termed Cyclostoma in Haeckel (1866). Cyclostomata came into common usage, perhaps because of its appearance in influential works such as Goodrich's (1930) *Studies in the Structure and Development of Vertebrates* and Jarvik's (1980) *Basic Structure and Evolution of Vertebrates*. This trend is now maintained by high-profile molecular and developmental studies (e.g., Kuratani and Ota, 2008a; Heimberg et al., 2010). To make matters more complicated, Cyclostomata Busk, 1852, is a living order of bryozoans. No rules exist regarding homonyms above family level in the International Code of Zoological Nomenclature (ICZN, 1999).

Table 4.1 Chronology of hagfish embryology research

Year	Milestones	References
1890–1910	Successful early collecting efforts and preliminary accounts of embryos of *E. stoutii*; descriptive developmental anatomy of excretory organs	Price (1896a, b, 1897, 1904); Dean (1898); Doflein (1898)
	Descriptive embryology of *E. stoutii* based on a nearly complete series to the prehatching stage	Dean (1899)
	Development of the brain, pharynx, and oral and nasohypophyseal cavities in *E. stoutii* based on histological observations	von Kupffer (1899, 1900, 1906)
	Descriptive embryology of the mouth, gill pouches, and thyroid gland in *E. stoutii*	Stockard (1906a, b)
1911–1940	Description of the brain development (external and internal) in *E. stoutii* based on the Dean-Conel histological sections	Conel (1929, 1931)
	Publication of the Bashford Dean festschrift part I (*"Archaic Fishes"*)	Gudger and Smith (1933); Conel (1933)
	Description of the chondrocranial development in *E. stoutii* based on the Dean-Conel sections	Neumayr (1938)
1941–1970	Neural crest development in *E. stoutii* based on the Dean-Conel sections suggests (incorrectly) an origin from an epithelial outpocketing between the surface ectoderm and the neural tube	Conel (1942)
	Detailed morphological description of two relatively developed embryos of *M. glutinosa*	Holmgren (1946)
	Pituitary development in *M. glutinosa* recovers no evidence for endostylar arrangement	Fernholm (1969)
1971–2000	Re-examination of the Dean-Conel histological sections for the pituitary development suggests (incorrectly) an endodermal origin of the nasohypophyseal duct in *E. stoutii*	Gorbman (1983); Gorbman and Tamarin (1985)
	Three-dimensional reconstruction of the Dean-Conel histological sections reveals lateral line and other neurogenic cranial placodes in *E. stoutii*	Wicht and Northcutt (1995); Wicht and Tusch (1998)

(Continued)

Table 4.1 (*Continued*) Chronology of hagfish embryology research

Year	Milestones	References
2001–current	K.G. Ota and S. Kuratani began successful efforts to obtain embryos from *E. burgeri* in captivity; descriptive accounts of the newly obtained embryos	Ota and Kuratani (2006, 2008); Kuratani and Ota (2008a)
	Neural crest development in *E. burgeri* is similar to that in other vertebrates; no epithelial outpocketing	Ota et al. (2007)
	Expression of collagen type 2α1 genes in noncartilaginous connective tissues in *E. burgeri*	Ota and Kuratani (2010)
	Presence of sclerotomal cargilaginous elements in the caudal region of *E. burgeri*	Ota et al. (2011, 2013, 2014)
	Transcriptome profiles in *E. burgeri*	Takechi et al. (2011)
	Nasohypophyseal development in *E. burgeri* supports the pan-cyclostome posthypophyseal process; secondary closure of an anlage for oral and nasohypophyseal cavity	Oisi et al. (2013a)
	Description of the chondrocranial development in *E. burgeri* and *P. atamii*	Oisi et al. (2013b)
	Partly dorsoventrally patterned expression of *Dlx* cognates in *E. burgeri*	Fujimoto et al. (2013)
	"Hypobranchial" musculature and nerve configuration in relation to a caudal shift of pharyngeal pouches	Oisi et al. (2015)

Abbreviations for generic names: *E. = Eptatretus; M. = Myxine; P. = Paramyxine.*

characters to the data set and have small chance of fossilization (Løvtrup, 1977, Chapter 3). Therefore, consistent support should not be interpreted as additive reinforcement of the inference.

If cyclostomes form a clade, hagfish fall into the crown-group Vertebrata [defined by the last common ancestor between lampreys and gnathostomes (jawed vertebrates) and all its descendants; Janvier, 1981]. If cyclostomes are paraphyletic (hagfish falling outside the clade [lampreys + gnathostomes]), then hagfish lie outside the Vertebrata. In that case, the clade [hagfish + (lampreys + gnathostomes)] is called the Craniata (Janvier, 1981). These names, however, represent just two of various nouns and adjectives used by early natural historians to group vertebrates—and other chordates in some cases (Nielsen, 2012).

Table 4.2 Proposed embryonic staging of hagfish based on the embryos of
E. stoutii reported by Dean (1899) from Monterey Bay, California, USA

Stage	Morphological Criteria	Figures
1	Fertilization; zygote	
2	Meroblastic cleavage: discoidal; likely asymmetric beyond the eight-cell stage	P—16, 73–78; T—19–23
3	Blastula: asymmetric downgrowth of the cellular cap; syncytial zone between blastomeres and yolk	P—17, 19, 20, 78
4	Early gastrula: neural axis as epiblastic thickening; mesenchymal mesoendoderm; periblasts merging with syncytium; blastoderm elongation (rostral < caudal)	P—18, 21–23; T—24
5	Late gastrula: tail bud reaching blastopore lip; neural fold as a shadow along midline; "primitive streak"	P—24, 25, 79, 80, 91
6	Neurula: head plate into medullary fold; neural fold possibly fused into a tube cranially; enhanced depression of stomodeum (oral and nasohypophyseal cavity); blastoderm extending greater than one-third the length of the egg	P—31–33, 81–82 (?: 24, 26, 79, 83, 91, 92); T—25
7	Late neurula: clear distinction among fore-, mid-, and hindbrain; otic capsule; rhombomeres; mid-ventral vessel as heart; <54 somites	P—26, 83, 92
8	Late neurula; oral and nasohypophyseal cavity secondarily closed; 59 somites; further differentiation of the brain vesicles; optic vesicle; hepatic vein anlage	P—27, 84, 93 (?: 30)
9	Late neurula—early pharyngula: hyomandibualr pouch; otic capsule close to hindbrain; foregut diverticulum	P—28, 29, 99
10	Pharyngula: well-formed eyes; roots for trigeminal (V), facial (VII), and vestibulocochlear (VIII) nerves; five pharyngeal pouches; hyomandibular pouch elongate; 73 somites	P—34, 35, 94, 100, 129
11	Pharyngula: six to eight pharyngeal pouches; thickened pharyngeal arches; efferent pharyngeal arteries; folding of lateral wall of the brain less pronounced; 105 somites	P—36, 39, 85, 88, 95, 97 101–103, 109, 121, 130
12	Pharyngula: embryo as long as yolk; midbrain shortened; eye posterior to forebrain and anterior to hindbrain; 9 to 12 pharyngeal pouches; onset of relative posterior displacement of the pouches; enlarged hyomandibualr pouch incorporated well within foregut; oral cavity; onset of barbells; nasal cavity with folded epithelium; anlage of lingual apparatus	P—37, 38, 40–44, 86, 87, 89, 90, 96, 110, 112

(*Continued*)

Table 4.2 (Continued) Proposed embryonic staging of hagfish based on the embryos of *E. stoutii* reported by Dean (1899) from Monterey Bay, California, USA

Stage	Morphological Criteria	Figures
13	Pharyngula: 13–14 pharyngeal pouches; pouches rolled ventromedially as part of lateral body wall	P—45, 46, 104, 111, 122
14	Late embryo: nasohypophyseal canal anterior to mouth; enlarged hyomandibular pouch below otic capsule; muscular upper lip behind mouth; lower lip convex; main retractor of lingual apparatus; pharyngeal pouches largely enclosed within body wall; somites shifting rostrally; heart directly underneath head	P—12, 13, 47–50, 105, 113, 123, 132
15	Late embryo: head around end of yolk; wall of forebrain thickened; triangular hindbrain; growing barbells clustered around nasohypophyseal aperture and around mouth; lower lip no longer markedly convex; labial margin folding over mouth; nasohypophyseal aperture (secondarily) opening; cartilaginous otic capsule; pharyngeal pouches displaced caudally; roots of spinal nerves (and somites) well within head region; slime glands	P—14, 51, 55–59, 106, 114–117, 124, 134 (?: 52–54, 125, 133)
16	Late embryo: embryo three quarters the way around yolk; gill (pharyngeal) pouches far displaced caudally relative to head; muscle plates well-developed; parachordal and velar cartilage visible; "hypobranchial" musculature; heart and major vessels assume adult positions	P—15, 60, 61, 108, 118, 126, 135
17	Late embryo: well-developed chondrocranium; tendon of lingual apparatus; gill pouches with inner epithelial folding and with blood irrigation	P—62–65, 127
18	Latest embryo: reduced ventricles of brain; oral barbells folded onto mouth; chondrocranium complete; lingual apparatus proportional to adult counterpart; cartilages present in caudal fin	P—66–70, 98, 107, 119, 137
19	Hatchling	P—71, 72, 120, 128

Although Dean (1899) did not formulate a staging scheme, we designate stages and morphological criteria based on the descriptions by Dean (1899) and Oisi et al. (2013a, b, 2015). Figure numbers in Dean's (1899) plates do not refer to developmental stages as sometimes implied in the literature. Although Dean (1899) arranged his embryos in approximate order of development, many morphological criteria overlap. The overlap may be due partly to small sample size and partly to abnormal development in some of the embryos. These uncertainties may be resolved with future studies, and this will lead to a more finely resolved staging scheme. Figures refer to those from Dean (1899), with P standing for plate figures and T for text figures.

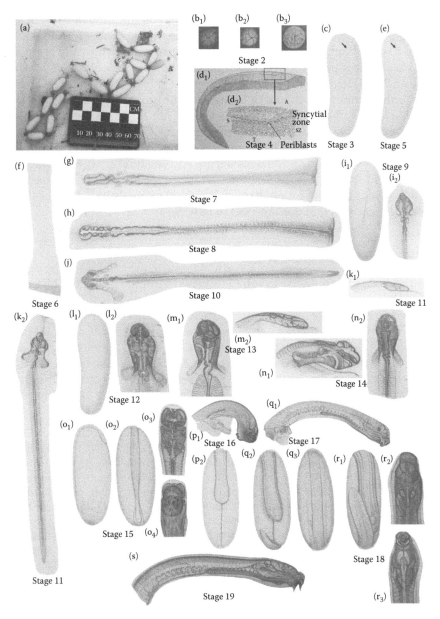

Figure 4.1 Prenatal development of the hagfish *E. stoutii*. (a) Egg capsules approx. 24 h postdeposition in captivity (photograph by T. Miyashita). (b) Embryo at stage 2 at second (b₁), third (b₂), and fourth (b₃) cleavage viewed from the animal pole. (c) Embryo at stage 3 (blastula) indicated by an arrow. (d) Embryo at stage 4 (gastrula) in lateral view (d₁) and in close-up of the syncytial zone near the caudal end (d₂). (e) Embryo at stage 5 (late gastrula) indicated by an arrow. (*Continued*)

This conflict between Vertebrata/Craniata classification has sustained interest in cyclostome relationships. Whether cyclostomes form a clade or not, both hagfish and lampreys represent long branches, and support levels for either phylogenetic hypothesis are low in each published dataset (Near, 2009; Heimberg et al., 2010; Thomson et al., 2014). As support levels are unlikely to improve markedly in the near future, heavily used terms such as the Vertebrata and Craniata should not hinge on the position of one unstable branch. To focus on relationships without confusing terminology, we refer to the Vertebrata under the stem-based (or "total-group") definition (Vertebrata: all taxa closer to, and including, *M. glutinosa* and *Danio rerio* than to *Ciona intestinalis* or *Branchiostoma floridae*) (Miyashita, 2012). Under that definition, hagfish are "vertebrates" regardless of their phylogenetic position with respect to lampreys.

Vertebrate origins and hagfish

Vertebrate animals are clearly distinguished from the rest of the Metazoa by a great number of uniquely shared characters, many of which are linked to the highly organized sensory and motor systems. These vertebrate features include: the skull, tripartite brain, vertebrae, and all other structures that differentiate from neural crest cells and neurogenic placodes. Although potential homologues exist for the brain, neural crest, and neurogenic placodes in several closely related invertebrates (tunicates, cephalochordates, and hemichordates; Wada et al., 1998; Boorman and

---

Figure 4.1 **(Continued)** (f) Embryo at stage 6 (neurula) in dorsal view. (g) Embryo at stage 7 (late neurula) in dorsal view, dissected out from the egg capsule, with head pointing to the left. (h) Embryo at stage 8 (late neurula) as in (g). (i) Embryo at stage 9 (pharyngula) within egg capsule (i_1) and showing details of the head (i_2) in dorsal view. (j) Embryo at stage 10 (pharyngula) as in (g). (k) Embryo at stage 11 (pharyngula) in lateral view (k_1) and dissected out from the egg capsule with head pointing upward (k_2). (l) Embryo at stage 12 (pharyngula) in egg capsule (l_1) and showing details of the head (l_2) in dorsal view. (m) Embryo at stage 13 (pharyngula) showing details of the head in dorsal view (m_1) and in lateral view (m_2). (n) Embryo at stage 14 (late embryo) showing details of the head in lateral view (n_1) and in dorsal view (n_2). (o) Embryo at stage 15 (late embryo) in egg capsule in lateral (o_1) and dorsal (o_2) views and showing details of the head in dorsal (o_3) and ventral (o_4) views. (p) Embryo at stage 16 (late embryo) showing details of the head in lateral view (p_1) and in egg capsule in ventral view (p_2). (q) Embryo at stage 17 (late embryo) showing details of the head in lateral view (q_1) and in egg capsule in ventral (q_2) and dorsal (q_3) views. (r) Embryo at stage 18 (near hatching) in egg capsule in ventral view (r_1) and showing details of the head from dorsal (r_2) to ventral (r_3) views. (s) Hatchling in right lateral view. (b–s) Reproduced from Dean (1899); stage numbers as per this chapter. Illustrations not to scale.

Shimeld, 2002; Christiaen et al., 2002; Mazet et al., 2005; Abitua et al., 2012; Pani et al., 2012), the rich repertoire of features that derive from neural crest and neurogenic placodes are undoubtedly unique to vertebrates.

Hagfish make a unique contribution to estimates of origins and primitive conditions for vertebrate characters irrespective of their alternative phylogenetic positions. Lampreys have a long history of providing insights on this topic, but their lengthy larval stage and metamorphic phase present formidable challenges to linking early developmental processes and adult structures. Lampreys are also problematic because gene expression patterns (e.g., cognates of the otherwise conservative *Hox6* or *Bagpipe*) appear to vary considerably between species (Miyashita, 2015). Comparison with hagfish provides a unique test—albeit limited because of the extreme branch lengths—of whether particular aspects of lamprey development represent retained ancestral states or lineage-specific specializations.

The neural crest gives rise to migratory cells that delaminate from the neural plate border upon folding of the plate into a neural tube (Hall, 2009). Neural crest cells play numerous crucial roles during vertebrate development by differentiating into cartilages, bones, teeth, peripheral neurons, glia, melanocytes, and others, and by regulating differentiations and patterning of nonneural crest structures such as mesodermally derived connective tissues (Matsuoka et al., 2005; Rinon et al., 2007). Neural crest is so essential to vertebrate development that the evolutionary origin of vertebrates has long been tied to the emergence of neural crest (Gans and Northcutt, 1983; Gans, 1993; Northcutt, 2005). Early observations suggest that the neural crest of hagfish is specified in a unique way (Conel, 1942).

Neurogenic placodes are ectodermal thickenings that develop into sensory capsules or ganglia (Baker, 2005). The placodes appear to be conserved in both number and position across vertebrates (Schlosser, 2005). However, sensory structures that develop from the placodes often provide controversial morphological characters. Examples in hagfish include nonimage-forming eyes (image-forming in lampreys and other vertebrates), a single semicircular canal (two in other jawless vertebrates; three in gnathostomes), and lateral line restricted to the facial region below the eye. Whether these conditions represent primitive or degenerate states remains unclear because of limited comparative data from outgroups.

Vertebrae might be considered the sine qua non of vertebrates. Despite this basic understanding, it remains uncertain whether all vertebrates ancestrally shared vertebrae, or evolved vertebrae independently. For example, hagfish have no axial skeleton except for the notochord, but develop cartilaginous nodules around the dorsal aorta within the caudal fin that some have suggested may qualify as vertebrae (Ayers and Jackson, 1900; Ota et al., 2011).

The skull is another skeletal component that characterizes vertebrates (De Beer, 1937; Hanken and Hall, 1993). Both hagfish and lampreys have a chondrocranium—a cartilaginous skull. In gnathostomes, the chondrocranium consists of broadly conserved components that can be identified across lineages. However, homologies are more difficult to establish between the crania of hagfish, lampreys, and gnathostomes. Not surprisingly, skull morphology is tightly linked to overall craniofacial development. Here, hagfish may show a pattern distinct from other vertebrates. For example, the adenohypophysis develops in the epithelium under the forebrain, which appears to be externally closed (Gorbman, 1983).

In the following sections, we first summarize the development of hagfish and then review current evidence that bears on these issues. Finally, we use parsimony to test different scenarios of developmental evolution in the earliest vertebrates and to highlight areas for future investigation.

Development of hagfish

Fertilization

Ripe female hagfish carry a variable number of egg capsules (Figure 4.1a; 13–67 in *Eptatretus burgeri*; 20–35 in *E. cirrhatus*; 12–45 in *E. stoutii*; 6–32 in *Myxine glutinosa*; Fernholm, 1975; Patzner, 1998; Martini and Beulig, 2013). The gonads increase in size from summer to fall (Kobayashi et al., 1972; Tsuneki et al., 1983; Yuuki et al., 2003), consistent with seasonal changes in the level of gonadotropin (Nozaki et al., 2000; Powell et al., 2004, 2005; Kavanaugh et al., 2005). The spawning season likely ranges over October to early December in *E. burgeri* and is probably similar in other Northern Hemispheric species (Ota and Kuratani, 2006; Ota et al., 2007). Sex ratio varies from species to species. There is no controlled comparison between species, and the capture method or species-specific ecology can bias the result. Still, males tend to be rare in the populations of *E. stoutii* and *M. glutinosa*, whereas the ratio is nearly equal in those of *E. burgeri* and *E. cirrhatus* (Tsuneki et al., 1983; Martini et al., 1997; Patzner, 1998; Martini and Beulig, 2013). Mature males are even rarer: only a small fraction (5%) of males had developing testes in a New Zealand population of *E. cirrhatus* (Martini and Beulig, 2013); not one male in a sample of more than 2000 contained a mature testis in *E. burgeri* and *M. glutinosa* (Patzner, 1998); motile sperm was recovered from only one male out of 1000 *M. glutinosa* (Jaspersen, 1975).

Reproductive behavior has never been observed in hagfish. The fact that all hagfish lack external sexual organs (Conel, 1933) appears to rule out chondrichthyan-like internal fertilization, although internal fertilization can happen in some fish without conspicuous external organs (e.g., coelacanths; Griffith and Thomson, 1973; Smith and Rand, 1975). The rarity of mature males and the scarcity of motile sperm imply that external fertilization requires conditions that allow an extremely small amount of sperm to enter an egg capsule. Three hypotheses have been proposed: eggs are deposited and sperms are released (a) in space between rocks (Dean, 1899); (b) in sediment or a burrow (Holmgren, 1946; Gorbman, 1997); or (c) in slime (Gorbman, 1997). No direct evidence exists for any of these hypotheses. However, anoxic conditions in sediment may be unfavorable to embryonic development (Gorbman, 1997). Suitable rock shelters may be limited in availability. When at least two females of *E. stoutii* spawned in captivity, the egg capsules (later confirmed to be unfertilized) were recovered in a residue of slime, food debris, and shed tooth plates (Figure 4.1a; T.M., personal observation). The eggs might have been deposited in slime, or a drifting mass of slime might have nondiscriminatorily gathered debris on the tank floor.

There is little doubt that fertilization occurs, as the egg capsule has a small pore for sperm entry at the animal pole. However, it would be worth testing whether asexual reproduction takes place in natural populations of hagfish as well. If unfertilized eggs can undergo normal development, it could explain both the rarity of sexually mature males in field samples and limited sperm production.

Early development: Zygote to neurula

Little is known about the timing and duration of development in hagfish, except that somites appear approximately five months after deposition of eggs of *E. burgeri* in captivity (Ota et al., 2007). However, both the timing and duration likely vary with environmental conditions because an embryo obtained from a culture in a natural setting was more advanced in stage than those reared in the laboratory (Ota and Kuratani, 2008). Morphologically, numerous embryos (*E. stoutii*) collected from Monterey Bay, California, USA in the late nineteenth century provide a fairly complete picture of prenatal development (Figure 4.1). Most of these embryos are preserved as histological sections (the Dean-Conel collections) at the Museum of Comparative Zoology, Harvard University. The following description, and our proposed staging table (Table 4.2), is primarily based on Dean's (1899) account unless otherwise cited. Although Dean's (1899) figure numbers were previously adopted as substitutes for developmental stages (Ota et al., 2007, 2011; Oisi et al., 2013a, b, 2015), these numbers do not represent stages. On the other hand, Neumayr's (1938) scheme only concerns chondrocranial development. As it only consists of four stages, his scheme is unsuitable for whole embryos.

Cleavages (stage 2) are meroblastic and discoidal (Figure 4.1b), in contrast to lampreys, which have holoblastic cleavage. At fourth cleavage, marginal cells show retarded division while central blastomeres continue to divide normally. The cellular cap of the blastoderm grows down over the yolk asymmetrically (stage 3, Figure 4.1c), and a syncytial zone forms at the boundary with the yolk. The blastocoel has not been identified. Near gastrulation, the embryo consists of a several-cell-thick layer of epiblasts and an equally thick layer of loosely associated mesoendoderm underneath. Although Dean (1899) identified a slight indentation as the possible site of invagination, the loose association of the cells suggests that the mesoendoderm rather forms via ingression. There is no evidence for or against other types of cell movement such as involution, delamination, or epiboly at this stage (Keller and Davidson, 2004). In stage 4, spindle-shaped cells (periblasts) lie beneath the mesoendoderm, although their boundary with the underlying syncytium is obscured by some cells embedded in the syncytial zone (ventral wall of the archenteron) (Figure 4.1d). The tail bud extends to reach the blastopore lip ventrally. The neural plate begins folding (stage 5) as recognized by shadow along the midline (Figure 4.1e). At this stage, the embryo superficially resembles an amniote primitive streak, but almost no information is available on cell movement along the streak. All this time, the blastoderm continues to expand to occupy a greater portion over the length of the yolk. The head plate becomes a medullary fold, and the head outline becomes distinct with the thickening of the mesoblasts. At stage 6, the neural fold begins to close into a tube cranially (Figure 4.1f).

The distinction between gastrula (stage 5) and neurula (stage 6) is an arbitrary one (Table 4.2) partly because of the small sample size and partly because of a temporal overlap between processes that characterize both stages. The embryos around these stages have the folded medullary head plate rostrally and retain the "primitive streak" caudally, and the blastoderm appears to be still rapidly expanding longitudinally. To make matters more complex, three of the four embryos used to describe these two stages exhibit some inconsistencies such as complete closure of the blastopore or separation of the medullary fold caudally (e.g., Dean, 1899: pl. 17, figs 31–33). It is possible that one or all of these embryos are abnormal. Another nonmutually exclusive explanation is temporal overlap and intraspecific variation in the sequence of developmental events between gastrulation and neurulation. This problem can only be resolved with careful study of new embryos.

The large absolute size relative to other fish egg capsules—and the tough membrane of the yolk-filled egg capsule—accounts for many features of early hagfish embryogenesis, as there is little space between the yolk surface and the membrane. Superficial resemblances suggest that amniote early development (Schoenwolf, 1997) may be the best analogue for hagfish, and this interesting comparison is worth exploring in the future. However, some of Dean's (1899) histological observations may be an artifact of his sectioning procedure. In particular, compression can cause the periblasts to merge with the syncytial zone and may explain the difficulty in identifying the archenteron around gastrulation. Clearly, these early observations need to be repeated with new material.

Late development: Pharyngula to hatchiling

At late neurula to early pharyngula stages (stages 7–10) (four to five months after egg deposition; Kuratani and Ota, 2008a), embryos become externally visible (Figures 4.1g–j). The appearance of pharyngeal pouches is preceded by the elaboration of brain vesicles and by the development of the heart and somites (Table 4.2). By late neurula stage (stages 7–8), the tripartite brain organization (fore-, mid-, and hindbrain) is morphologically recognizable (Figures 4.1g, h). The forebrain gives rise to five or six outpocketings into the rostral ventricular wall and, more caudally, the optic vesicle. The lateral wall of the midbrain region is folded asymmetrically and to varying degrees among embryos of the same stage (stages 8–11). This folding becomes increasingly pronounced in later stages (Figure 4.1j). However, this morphology must be interpreted with caution because the folding may partly represent an artifact of fixation methods. The otic capsule assumes a position close to the hindbrain (stage 9; Figure 4.1i). Rhombomeres are recognizable around the onset of somitogenesis (stage 7), and the number and morphology of rhombomeres are consistent with those of other vertebrates (Ota and Kuratani, 2008; Oisi et al., 2013a). The heart forms from the mid-ventral vessel draining from the transverse vein rostral to the head, which collects from the developing vitelline vessels (stage 7). The hepatic vein anlage forms earlier than other major vessels (stage 8). The foregut develops a diverticulum (stage 9; Figure 4.1i), which likely forms a caudal-most part of the nasohypophyseal canal (nasopharyngeal duct) in adult.

As the pharyngeal pouches sequentially form, the head has a spade-shaped outline (Figures 4.1j, k). Meanwhile, somitogenesis continues caudally. In the pharyngula with five pharyngeal pouches (stage 10; Figure 4.1j), the hyomandibular pouch is elongate, whereas other pouches are slit like. The roots for the trigeminal (V), facial (VII), and vestibulocochlear (VIII) nerves are visible in the same embryo. The pharyngeal arches (the area between the pouches) now have efferent arteries and thicken from the stage of six pharyngeal pouches onward, preluding a main stage of migration of the mesodermal mesenchyme and neural crest ectomesenchyme. In the embryo with seven to eight pharyngeal pouches (stage 11; Figure 4.1k), the paraxial mesoderm visibly extends rostrally (without any trace of segmentation) past the otic capsule.

A number of significant developmental events occur in the embryos with 9 to 12 pharyngeal pouches (stage 12; Figure 4.1l). Although the limited sample size makes the precise order of events difficult to identify, the brain and sensory capsules establish postnatal positions. The midbrain is shortened rostrocaudally relative to the forebrain and hindbrain, and its folding is less pronounced. The eye sits between the forebrain and hindbrain in lateral view. The nasal capsule develops with the folded olfactory epithelium (stage 12; Figure 4.1l). The barbells (tentacles) start to form. Among the pharyngeal pouches, the hyomandibular pouch remains lateral to the oral cavity and becomes incorporated in the enlarged oral cavity, whereas the other pouches become slit like and displaced caudally relative to the head (stage 12 onward; Figures 4.1l–q). The lingual apparatus also begins to form. The embryos take on a more three-dimensional topography so that the laterally spread body wall with the pouches is progressively rolled into the body wall ventromedially (Figure 4.1l, m). The process continues across the stages with 13 or 14 pharyngeal pouches (stages 13 and 14), which is consistent with the number of adult gill pouches; Figures 4.1m, n).

Continued displacement of the pharyngeal pouches during the prenatal development accompanies trunks of the branchiomeric nerves that innervate them (glossopharyngeal and vagus; cranial nerve IX, X) on the medial side of the lateral body wall (stage 14; Oisi et al., 2015). Meanwhile, the rostral shift of the myomeres causes the roots of the spinal nerves to migrate into the head region (Oisi et al., 2015). The net result of these movements is a hagfish-specific pattern in which the suprapharyngeal spinal nerves extend ventrally to innervate the ventral trunk muscles ("hypobranchial" musculature) rather than circumvent the pharyngeal region caudally as in other vertebrates (Müller, 1834). Also with the displacement of the pouches, the development of the "hypobranchial" musculature is delayed until stage 16 (Figure 4.1p), extending ventrally rather than taking a circumpharyngeal path (Oisi et al., 2015).

Embryos at latest stages elaborate feeding and ventilation structures (stages 15–18; Figures 4.1o–r). Muscular lip structures are clearly differentiated, as are the protractors and retractors of the lingual apparatus (stages 16, 17; Figures 4.1p, q). Related to the growth of the lingual apparatus, the pharyngeal pouches continue to shift caudally relative to the rest of the head (Figures 4.1p, q). The exception is the hyomandibular pouch, which remains as the caudal wall of the oral cavity and enlarges to allow extension of the velum, a musculoskeletal pump for respiratory flow (Strahan, 1958). Embryos wrap more than two-thirds the way around the yolk when major vessels configure into their adult positions (stage 16; Figure 4.1p). Eventually, the head is on the ventral side of the yolk, nearly diagonal to the animal pole, where the tail overlaps the head (stage 18; Figure 4.1r). The chondrocranium and the cartilages of the caudal fin are complete before hatching. The barbells assume the adult morphology (stage 18).

Embryos likely hatch with the yolk still attached to the belly. Dean's hatchlings still carry a yolk sac (pl. 20, fig. 72 in Dean, 1899), and so do those in a publicly posted film of hatching in *Paramyxine atami* (Mimori, 2010). The cap of the egg capsule on the animal pole pops open, and a hatchling emerges from the head. The semitransparent hatchlings show some coordinated undulation of the body but appear incapable of swimming in the water column. As the embryos of *P. atami* were collected near the hatching stage (R. Mimori, personal communication, 2011), this hatching behavior was likely to be stress induced, perhaps by a temperature difference in a tank.

Evolutionary implications of hagfish development

In this section, we review major recent contributions to hagfish embryology from members of the Evolutionary Morphology Group at RIKEN (Kinya Ota, Yasuhiro Oisi, Shigeru Kuratani, and their colleagues). After decades where no new embryos were found, they made a breakthrough in 2005 by successfully obtaining embryos of *E. burgeri*, mainly from an artificial spawning tank (Ota and Kuratani, 2006, 2008; Ota et al., 2007). Since then, they revived interest in hagfish embryology with a steady production of high-profile papers (Ota et al., 2007, 2011, 2013, 2014; Kuratani and Ota, 2008b; Ota and Kuratani, 2010; Fujimoto et al., 2013; Oisi et al., 2013a, b, 2015). Their narrative consistently accepts hagfish as a cyclostome and depicts hagfish development as consistent with that phylogenetic position and as representing a generally plesiomorphic state of vertebrate development.

Despite the claim that vertebrate-like development of hagfish supports a monophyletic Cyclostomi (Ota et al., 2007, 2011, 2013, 2014; Kuratani and Ota, 2008a, b; Ota and Kuratani, 2010; Oisi et al., 2013a, b), the data presented in these papers alone do not necessarily reject paraphyletic cyclostomes on the basis of parsimony. Neither do they provide sufficient evidence for a plesiomorphic state of cyclostome development with respect to gnathostomes.

Neural crest

Until recently, the single most problematic feature of hagfish development had been the formation of neural crest. Histological cross sections of Dean's embryos show bilateral deformity in the lateral wall of the neural tube just below the surface ectoderm (asterisks in Figure 4.2c). This "outpocketing" continues cranially to the anlage of eye (Conel, 1942). Other vertebrates do not form such epithelial pockets. Instead, delaminated migratory neural crest cells occur in this region (Hall, 2009). The observation suggested that hagfish neural-crest cells originate as an epithelial pocket (Conel, 1942). If true, such outpocketing might represent a primitive state of neural crest origins with respect to other vertebrates.

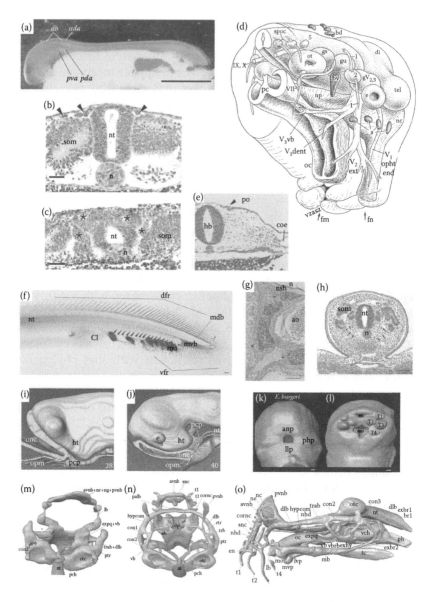

Figure 4.2 A gallery of recent advances in hagfish embryology. (a) An embryo of
E. burgeri (stage 14) in left lateral view. (b) Delaminating neural crest cells indicated by
arrowheads in a transverse section of an embryo of *E. burgeri* at trunk level (stained
with hematoxylin and eosin). (c) Fixation and sectioning of an embryo of *E. stoutiiin
ovo* resulted in artifacts indicated by asterisks (*) that mimic epithelial outpocketing
described earlier for embryos of *E. stoutii* by Conel (1941) (stained with hematoxylin and eosin). (d) Three-dimensional schematic reconstruction of an embryo of *E.
stoutii* (stage 14–15) based on serial transverse sections in oblique right   (*Continued*)

Figure 4.2 (**Continued**) lateral view (from a posteroventral angle). Three sets of lateral-line placodes (a, bd, bv) are stippled dark. (e) Transverse section of an embryo of *E. burgeri* at the level of the posterior hindbrain, showing the ectodermal thickening of a postotic placode (stained with hematoxylin and eosin). (f) Caudal region of an adult of *E. burgeri* in left lateral view, cleared and stained with alcian blue to reveal cartilaginous nodules (arrowheads) around the dorsal aorta. Scale = 1 cm. (g) Transverse section of the caudal region, showing the dorsal, lateral, and midventral cartilaginous nodules around the dorsal aorta. Scale = 1 mm. (h) Transverse section of the caudal region of a mid-pharyngula of *E. burgeri* (stage 12–13?) stained with hematoxylin and eosin, showing sclerotomal mesenchyme (arrows). Scale bar for H–L = 100 μm. (i) Three-dimensional reconstruction of an embryo of *E. burgeri* (stage 9) showing the externally open oronasohypophyseal cavity and the nasohypophyseal placode. (j) Three-dimensional reconstruction of an embryo of *E. burgeri* (stage 12) showing the secondarily closed oronasohypophyseal cavity. (k) Rostral region of an embryo of *E. burgeri* (stage 14) in ventral view. (l) Rostral region of an embryo of *E. burgeri* (stage 15) in ventral view. (m) Three-dimensional reconstruction of the chondrocranium of an embryo of *E. burgeri* (stage 15, early) in dorsal view. (n) Three-dimensional reconstruction of the chondrocranium of an embryo of *E. burgeri* (stage 15, late) in dorsal view. (o) Three-dimensional reconstruction of the chondrocranium of an embryo of *E. burgeri* (stage 16, late) in left lateral view. (a) From Ota and Kuratani (2008); (b, c) From Ota et al. (2007); (d) From Wicht and Northcutt (1995); (e) From Kuratani and Ota (2008b); (f–h) From Ota et al. (2011); (i–l) From Oisi et al. (2013a); (m–o) From Oisi et al. (2013b). Abbreviations and labels: 1, nerve lateralis a; 2, ganglion for nerve lateralis a; 3, nerve lateralis b; 4, distal ganglion of nerve lateralis b; V_{1opht}, ophthalmic branch of trigeminal nerve; V_{2ext}, external branch of trigeminal nerve; V_{3dent}, dental branch of trigeminal nerve; V_{3vb}, velobuccal branch of trigeminal nerve; VII, facial nerve; IX, glossopharyngeal nerve; X, vagus nerve; a, preocular skin grooves; anp, anterior nasal process; ao, dorsal aorta; ah, adenohypophysis; avnb, anterior vertical nasal bar; bd, dorsal set of postocular skin grooves; br1/2, internal branchial arch 1/2; bv, ventral set of postocular skin grooves; cl, cloaca; coe, coelom; con1–3, rostral commissure of dorsal longitudinal bar; cornc, cornual cartilage; db, dorsal branch of unpaired dorsal aorta; dfr, dorsal fin ray; di, diencephalon; dlb, dorsal longitudinal bar; dp, dental plate; e, eye; en, external naris; end, nasohypophyseal canal; exbr1/2, extrabranchiale 1/2; exhy, extrahyal; expq, extra palatoquadrate; fm, future mouth opening; fn, future nasal opening; ge, glossopharyngeal epibranchial placode; gu, utricular ganglion; hb, hindbrain; ht, hypothalamus; hypcom, hypophyseal commissure; lb, lateral bar; llp, lower lip; lvp, rostrolateral basal plate; mb, rostromedial basal plate; mdb, median dorsal bar; mo, mouth; mu, mucous (slime) gland; mvb, median ventral bar; n, notochord; nc, nasal cavity; ne, nasal epithelium; ng, nasal duct cartilage; nhd, nasohypophyseal duct; nhp, nasohypophyseal placode; np, nasopharyngeal duct; nsh, notochordal sheath; nt, neural tube; oc, oral cavity; onc, oronasohypophyseal cavity; opm, oropharyngeal membrane; ot, otic vesicle; otc, otic capsule; palb, palatine bar; pc, pharyngeal cavity; pcp, prechordal plate; pch, parachordal cartilage; pda, paired dorsal aorta; ph, pharynx; php, posthypophyseal process; po, postotic placode; ptr, posterior trabecula; pva, paired ventral aorta; pvnb, posterior vertical nasal bar; rtr, rostral trabecula; snc, subnasal cartilage; som, somites; spoc, spinooccipital nerve; t1–4, tentacles (barbells); tc, tongue cartilage; tel, telencephalon; trab, trabecula; uda, unpaired dorsal aorta; vb, velar bar; vbrb, ventral branchial bar; vch, velar chamber; vfr, ventral fin ray.

Several embryos of *E. burgeri* provided an opportunity to test the original observation of outpocketing (Ota et al., 2007). Although Dean's (1899) embryos were fixed within the egg capsule (*in toto*), Ota et al. (2007) fixed some embryos *in toto* and others after removal from the egg capsule. When embryos were removed before fixation, no outpocketing was observed, even though neural crest ectomesenchyme was present (arrowheads in Figure 4.2b). Those fixed *in toto* had significant distortion and showed apparent "outpocketing" similar to Conel's (1942) description (asterisks in Figure 4.2c). Therefore, the purported outpocketing of the neural tube in hagfish appears to be an artifact of differential tissue shrinkage in confined space (Ota et al., 2007). Gene expression patterns (*Pax6, Pax3/7, Soxea,* and *Sox9*) suggest that neural crest cells in *E. burgeri* delaminate in proximity to the dorsal neural tube in craniocaudal sequence and migrate in a segmental manner between the somites just as in any other vertebrate (Ota et al., 2007).

As Ota et al. (2007) acknowledge, delaminating neural crest does not discriminate cyclostomes as a clade or grade, because all vertebrate taxa under comparison share that one state (Character 1; Figure 4.3). It takes one step at the last common ancestor between hagfish and gnathostomes to acquire neural crest in either scenario. However, Kuratani and Ota (2008a) explicitly use neural crest delamination in hagfish as support for a monophyletic Cyclostomi. It is no less parsimonious to map this character onto the tree with cyclostome paraphyly. If hagfish and lampreys shared a feature of neural crest development to the exclusion of gnathostomes, it still would not eliminate cyclostome paraphyly on the basis of parsimony. This is a general conundrum in any character comparison among these three taxa. Unless the gnathostome state is rooted, it is equally parsimonious for the hagfish–lamprey similarity to be a symplesiomorphy (shared primitive state) within vertebrates or a synapomorphy (shared derived state) for cyclostomes. However, the gnathostome state often cannot be rooted because developmental features are not readily preserved in fossils of stem gnathostomes.

Neurogenic placodes and lateral line

An ongoing interest in the development of neurogenic placodes in hagfish arose from a question of character polarity in adult morphology. Adult hagfish have: (a) eyes without a lens or extraocular muscles, (b) inner ears with one semicircular canal, and (c) facial skin grooves without intraepidermal canals or neuromasts. Do they represent a primitive or degenerated state of the eyes, inner ears, and lateral line within vertebrates, respectively (e.g., Fernholm and Holmberg, 1975; Fernholm, 1985; Braun and Northcutt, 1997)? These sensory structures and ganglia that innervate them develop from (or are induced by) neurogenic placodes.

	Invertebrate chordates	Hagfishes	Lampreys	Gnathostomes (total group)
1. Delaminating neural crest	Absent (0)	Present (1)	Present (1)	Present (1)
2. Lateral line	Absent (2a: 0; 2b: -)	Facial skin grooves - Ep (2a: 1; 2b: 0); Absent - My (2a: 0; 2b: -)	Present (2a, b: 1)	Present (2a, b: 1)
3. Neurogenic placodes	Absent (3a: 0; 3b: -)	Present (3a: 1); Horseshoe pattern (3b: 0)	Present (3a: 1) Unknown (3b: ?)	Present (3a: 1) Individual (3b: ?)
4. Skeletogenic slerotome	Absent (0)	Present (1) - Ep Unknown (?) - My	Present (1)	Present (1)
5. Mode of axial skeleton	Inapplicable (-)	Haemal arches? - Ep (5a: 0; 5b: 0; 5c: 1) Absent (5a-c: 0) - My	Neural arch? (5a: 0; 5b: 1; 5c: 0)	Full set - Eu, Cr (5a-c: 1)
6. Craniofacial patterning	Inapplicable (-)	Posthypophyseal process (0)	Posthypophyseal process (0)	Posthyp. Pr.? (0?) - Ost Maxillary process (1) - Cr

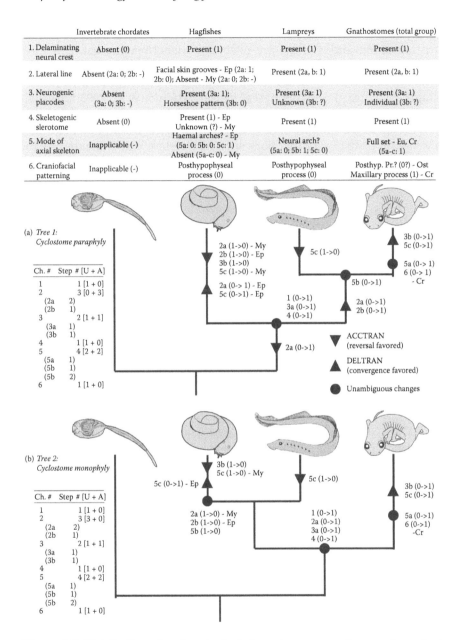

Figure 4.3 Distribution of embryological characters among three major extant lineages of vertebrates (hagfish, lampreys, and gnathostomes) with tunicates (invertebrate chordates) as an outgroup, under two possible topologies: (a) cyclostome paraphyly and (b) cyclostome monophyly. Predicted character changes are equally parsimonious and compatible between cyclostome paraphyly and cyclostome monophyly. Therefore, embryological evidence favors neither   *(Continued)*

The placodes occur lateral to the neural-plate border. In approximate rostral–caudal order, these are olfactory, adenohypophyseal, lens, profundal, trigeminal, otic, lateral line, and epibranchial in the gnathostome head. In cyclostomes, the olfactory and adenohypophyseal placodes are a single structure (nasohypophyseal placode) (Uchida et al., 2003; Oisi et al., 2013a). Although the evolutionary origin of each placode remains contentious, they all develop from the pan-placodal primordium that delineates the neural plate border in a horseshoe-like pattern and is marked by the expression of *Six1/2*, *Eya*, and others (Schlosser, 2005; Schlosser et al., 2014; Saint-Jeannet and Moody, 2014).

The facial skin grooves—a possible homologue of the lateral line—in *Eptatretus* spp. have generated extensive debate. It began with an early observation of the intraepidermal canal and neuromast-like structures in the epidermal grooves on the face of adult hagfish (Ayers and Worthington, 1907). Although both the canals and neuromast-like structures turned out to be histological artifacts (Fernholm, 1985; Wicht and Northcutt, 1995),

---

Figure 4.3 (*Continued*) phylogenetic hypothesis over the other. Characters with multiple components are denoted by letters (Character 2: a, lateral line absent [0] or present [1]; b, lateral line restricted to facial skin grooves [0] or extended to a whole body with functional chemo- and electro-receptors [1]; Character 3: a, absence [0] or presence [1] of neurogenic placodes; b, cranial neurogenic placodes arranged in a horseshoe pattern [0] or individually specified [1]; Character 5: a, arcualia forming around notochord, absent [0] or present [1]; b, neural arches forming around spinal cord, absent [0] or present [1]; c, hemal arches forming around dorsal aorta, absent [0] or present [1]). Characters are inapplicable (–) if the taxon lacks nonindependent, prerequisite characters. Characters scored inapplicable in the outgroup cannot be constrained, and the plesiomorphic state is unknown for the Vertebrata node. The outgroup cannot be coded for the following characters: Character 2b, the morphology of lateral line (the lateral line is absent); Character 3b, the pattern of neurogenic placodes (they do not have vertebrate-like neurogenic placodes); or Character 6, craniofacial patterning (they do not have a vertebrate-like head). Variation within the lineage is considered in coding the characters either if plesiomorphic states are equivocal (hagfish) or if stem taxa express symplesiomorphic conditions (gnathostomes). Total-group gnathostomes include jawless stem taxa such as *Euphanerops* and osteostracans. Unambiguous character changes were mapped first. Ambiguous character changes were optimized under ACCTRAN (accelerated transformation; reversals preferred) and DELTRAN (delayed transformation; convergence preferred). Numbers of steps were counted as U + A, that is, unambiguous changes + ACCTRAN- or DELTRAN-optimized ambiguous changes (number of ambiguous changes is equal between ACCTRAN and DELTRAN). Abbreviations: Cr, crown gnathostomes (Chondrichthyes + Osteicht-hyes); Ep, *Eptatretus*; Eu, *Euphanerops* (jawless stem gnathosotme); My, *Myxine*; Ost, osteostracans (jawless stem gnathostome).

the grooves are innervated by cranial nerves that appear to correspond to the lateral line nerves in other vertebrates (Worthington, 1905; Kishida et al., 1987). In embryos, early observations contradict each other whether the grooves develop from a lateral line placode. Dean (1899) did not find any morphology linked to primordial lateral line, whereas von Kupffer (1900) described three sets of placodes in that region and their trigeminal, facial, and vagal innervations. However, the latter observation is inconsistent with two nerves innervating the grooves in adults (Worthington, 1905). To test whether these grooves represent a lateral line, Wicht and Northcutt (1995) reconstructed three-dimensional models based on serial sections of the embryos of *E. stoutii* described by Dean (1899) and Conel (1931, 1942). Shigeru Kuratani and his colleagues followed up on this recently, with additional morphological description at earlier stages of development and basic gene expression profiles (Kuratani and Ota, 2008b; Ota and Kuratani, 2008; Oisi et al., 2013a).

At an early pharyngula stage of *E. burgeri*, the surface ectoderm along the neural plate border thickens in a horseshoe pattern as predicted by the pan-placode primordium model (Figure 4.2e; Schlosser, 2005; Kuratani and Ota, 2008b; Ota and Kuratani, 2008). Among individual placodes, the otic placode expresses *Sox9* (Ota and Kuratani, 2008). Histological sections of the older pharyngula of *E. stoutii* have revealed a series of placodes that correspond to the lens, epibranchial, and lateral line, as well as placodal derivatives, including the ganglia of the profundus and trigeminus (Wicht and Northcutt, 1995; Wicht and Tusch, 1998). The lateral line placodes in *E. stoutii* consist of three sets: (1) infraocular; (2) postocular, ventral; and (3) postocular, dorsal (arranged transversely); and all of these correspond topographically with the skin grooves in adults (a, bd, and bv in Figure 4.2d; Wicht and Northcutt, 1995). They receive innervation from two sets of the lateral line nerves: (1) a ganglion beside the trigeminal ganglia complex (to the infraocular skin grooves); and (2) a ganglion above the otic capsule (to the postocular grooves) (Wicht and Northcutt, 1995).

Curiously, *Myxine* spp. lack these skin grooves, even though *M. glutinosa* has the nerve that corresponds to the one that innervates the infraocular skin grooves in *E. stoutii* (Lindström, 1949). However, the "lateral line" placodes have never been observed in three described embryos of *M. glutinosa* (Holmgren, 1946; Fernholm, 1969). Lampreys, on the other hand, have a functional lateral line at both the larval and adult stages (Bodznick and Northcutt, 1981; Gelman et al., 2007, 2009), and various stem gnathostome lineages have sensory canals (Janvier, 1996). Parsimoniously, this variation suggests that the facial skin grooves in hagfish represent a degenerate lateral line, which underwent reduction in the hagfish lineage. The lateral line therefore likely arose once in the common ancestor of all living vertebrate lineages. This character distribution is equally parsimonious between cyclostome monophyly and paraphyly (Character 2a;

Figures 4.3a,b). However, on the ground of parsimony, independent origins cannot be rejected either under paraphyletic cyclostomes (Character 2a, DELTRAN; Figure 4.3a). Clearly, mere presence or absence of the lateral line provides little insight as to how this sensory system originated and evolved. This question awaits further comparative data.

The horseshoe pattern of the placode development in hagfish is unique among vertebrates (Character 3, Figure 4.3), although the placodes are difficult to observe at the surface level in lamprey embryos owing to yolk granules in epidermal cells (Kuratani and Ota, 2008b). If the pan-placode primordium model is assumed correct, and if the horseshoe pattern is indeed absent in lampreys (currently unknown), it would take one less step to map this character variation onto a tree with cyclostome paraphyly than onto the one with monophyly. Nevertheless, the current distribution of character states is equally parsimonious on both trees (Figures 4.3a, b).

Vertebrae

The taxon name Vertebrata clearly implies the presence of vertebrae. Although the long-assumed general and basic components of arcualia (such as basidorsals and interdorsals) occur in fossil jawless vertebrates such as *Euphanerops* (Janvier, 2011), hagfish have long been considered to lack any trace of vertebrae. Overlooked in this interpretation was an early anatomical account of *E. stoutii* in which cartilaginous nodules form on the ventral side of the notochord in the postcloacal region (Ayers and Jackson, 1900). These cartilaginous nodules form around the dorsal aorta below the notochord in three sets, irregularly fused to each other between segments: (1) paired dorsal elements; (2) paired lateral elements; and (3) a mid-ventral element (arrowheads in Figures 4.2f, g; Ota et al., 2011). At the pharyngula stages of *E. burgeri*, somites have three distinct regions marked by characteristic expression patterns as in other vertebrates, namely the dorsolateral dermomyotome (*Pax3/7*) and myotome (*MyoD*) and the ventromedial sclerotome (*Pax1/9, Twist*) (Ota et al., 2011). The dermomyotome and myotome are epithelial structures, but the sclerotomal cells are mesenchymal caudally (arrows in Figure 4.2h). Within each somite, the mesenchyme migrates ventrally in two craniocaudally segregated streams, implying the presence of metameric organization known in gnathostomes (Ota et al., 2011). These *Pax1/9-Twist*-positive cells arguably migrate farther ventrally around the dorsal aorta to form the cartilaginous nodules. At a later stage of the same species (roughly corresponding to stage 15 or 16), the mesenchyme around the dorsal aorta expresses a biglycan/decorin (*BGN/DCN*) gene, which is linked to the noncollagenous extracellular matrix of cartilages in gnathostomes (Ota et al., 2013). On the basis of histological observations and expression data, Ota et al. (2011, 2013, 2014) proposed that these cartilaginous nodules are homologues of vertebrae—degenerate vertebral elements that remained only on the ventral side of the notochord toward the tail, whereas the rest of the vertebral column was lost.

The assessment of homology between these caudal cartilaginous nodules and vertebrae requires a character analysis at multiple levels. Developmentally, Ota and colleagues assumed that the mesenchymal cells expressing *Pax1/9* and *Twist* in pharyngulas are those expressing *BGN/DCN* in late embryos, but alternatives have not been ruled out. For example, Ota et al. (2011, 2013, 2014) clearly compare the cartilaginous nodules in hagfish with pieces of arcualia in other taxa. Although both derive from sclerotomes, so do other axial elements such as ribs and hemal arches in gnathostomes. Further comparison is difficult because none of the reported expression patterns in hagfish are specific to the arcualia in gnathostomes, and because of anatomical inconsistencies. Anatomically, the cartilaginous elements form around the dorsal aorta rather than on the ventral side of the notochord (Figure 4.2g). This configuration suggests that the nodules in hagfish compare more closely with hemal arches than with arcualia. More precisely, the hemal series of cartilages in *Euphanerops* (Janvier and Arsenault, 2007) may provide a useful comparison. In addition, such caudal cartilaginous nodules are absent in *Myxine* (Cole, 1905; Ota et al., 2014); so the ancestral state even in hagfish is uncertain.

On the basis of this information, we propose an alternative hypothesis: sclerotomal specification by *Pax1/9* and *Twist* allowed skeletogenic mesenchyme to be deployed differently among early vertebrate lineages (Character 4; Figure 4.3). In this sense, these caudal nodules cannot be compared specifically with vertebrae (or arcualia). Instead, they represent one variant of sclerotomal differentiation in anatomical positions that correspond to the gnathostome hemal series. The presence of parachordal cartilages across vertebrates indicates that the paraxial mesoderm can de-epithelialize and skeletonize in all vertebrates including hagfish (De Beer, 1937; Ota et al., 2013, 2014). As for the cartilaginous nodules in *Eptatretus*, they develop from a population of the skeletogenic mesenchyme that chondrifies around the dorsal aorta. This genus of hagfish is frequently observed in tightly coiled posture, with tail bent inside the spiral (Miyashita and Palmer, 2014) and with a pumping blood sinus near the tip of the tail (Greene, 1900). Rather than being vestigial on the ventral side of the notochord, these cartilaginous nodules could be a lineage-specific adaptation for structural support of the dorsal aorta. *Myxine*, which lacks such nodules, has never been observed in coiling posture. Whether the caudal nodules in *Eptatretus* are a remnant of the ancestral hemal series or an independently derived feature depends on the phylogenetic position of *Euphanerops*, which appears to have all of the sclerotomal differentiations: neural, vertebral, and hemal (Janvier and Arsenault, 2007).

Treating *Euphanerops* as a stem gnathostome, we consider three different character transformation scenarios: (1) degenerate (a full set of the vertebral and hemal series was present in the first vertebrate and variably lost among lineages); (2) independent (elements of the arcualia, neural arches, and hemal series were derived independently); and (3) stepwise (elements were acquired gradually along the main stem). In both cyclostome paraphyly (Figure 4.3a) and cyclostome monophyly hypotheses (Figure 4.3b), the independent-origins scenario requires the fewest character changes (4 changes in Character 5, ACCTRAN and DELTRAN), even though individual elements may be parsimoniously interpreted as degenerate (Character 5b, c). This scheme holds even if caudal nodules are treated as modified ventral components of the arcualia, in which case the shift in position of the ventral vertebral elements also adds to the number of inferred character changes. The independent-origins scenario is appealing also because it does not require an archetypical set of vertebral and hemal elements in the hypothetical common ancestor—a morphology that has never been observed in a stem vertebrate. The scenario of independent specialization is consistent with an unusual aspect of skeletogenesis in hagfish that two *Col2α1* cognates are expressed differentially between mesodermal connective tissues and in cartilages (Ota and Kuratani, 2010).

Craniofacial patterning and the skull

Craniofacial development of hagfish has two historical foci: the origin of the nasohypophyseal canal (and adenohypophysis) and compositions of the chondrocranium. The adenohypophysis derives from an ectodermal placode in lampreys and gnathostomes. On the other hand, its origin was unclear for hagfish. Initially, von Kupffer (1899) identified the deeply depressed surface ectoderm below the forebrain as containing an adenohypophyseal anlage at the neurula stage of *E. stoutii*. At later developmental stages, however, the nasohypophyseal canal and oral cavity appear to develop without opening externally, and the adenohypophysis arises in this domain (Fernholm, 1969; Gorbman, 1983; Gorbman and Tamarin, 1985; Wicht and Tusch, 1998). In the latter interpretation, the adenohypophysis must arise in the endodermal epithelium, a pattern distinct from that in the rest of vertebrates. A series of embryos of *E. burgeri* has recently revealed that: (1) the presumptive oral and nasohypophyseal cavity arises as a deeply depressed ectodermal epithelium tucked beneath the forebrain at the neurula stage; (2) this cavity secondarily closes externally by the membrane extending from the lateral side (a feature unique to hagfish); and (3) the adenohypophysis develops in the enclosed ectodermal epithelium below the forebrain as in all other vertebrates (onc and opm in Figures 4.2i, j; Oisi et al., 2013a). The adenohypophyseal anlage is the posterior part of a single median placode (nasohypophyseal placode) similar to that in lampreys and has identical expression patterns (cognates of *Six3/6* and *Sox2/3*; *Fgf8* cognate in posterior part; *Shh* and *Nkx2.1* cognate in hypothalamus; Oisi et al., 2013a). The similarity with lampreys was further reinforced by the presence of the posthypophyseal process—an anlage of the muscular upper lip unique to cyclostomes among living vertebrates (php in Figure 4.2k, l).

Another focal topic in hagfish development concerns chondrocranial development. Early comprehensive descriptions of the chondrocranium, based on the embryos of *E. stoutii* (four different stages) and *M. glutinosa* (two different stages) (Neumayr, 1938; Holmgren, 1946), laid the foundation for a more complete series of the chondrocranial development in *E. burgeri* (Figures 4.3m–o; Oisi et al., 2013b). Each of these studies has been driven by the continued demand for well-supported homology hypotheses relating the elements of hagfish chondrocrania to those of lampreys and gnathostomes. Although resultant schemes of homology differ between these individual studies, they share two conclusions: (1) hagfish and lampreys display a greater number of mutual similarities than either shares with gnathostomes; and (2) these similarities—identified mostly on the basis of the ectomesenchyme populations and relative positions within the head—are considered homologues rather than homoplasies.

As in the delaminating neural crest, the narrative is that the purported homologies in the craniofacial patterning are consistent with cyclostome monophyly (Oisi et al., 2013a, b). Alternative interpretations are possible, however (character 6 in Figure 4.3). The chondrocranium is a unique vertebrate character (De Beer, 1937). In the absence of an outgroup with a corresponding chondrocranium, the hagfish–lamprey similarities can represent synapomorphies of the Cyclostomi, but might also represent symplesiomorphies of the Vertebrata. If any one of the several stem gnathostome clades (e.g., osteostracans) shared the cyclostome-like craniofacial patterning such as a single median nasohypophyseal opening (Janvier, 1985, 1996), it is more parsimonious to interpret these similarities as symplesiomorphies. Significantly, the internal anatomy of the galeaspid *Shuyu* suggests a mix of cyclostome- and gnathostome-like morphology (Gai et al., 2011; Oisi et al., 2013a, b; Dupret et al., 2014). Therefore, phylogenetic resolution of the developmental characters awaits a detailed systematic analysis to distinguish synapomorphies and symplesiomorphies.

Concluding remarks

Nearly all papers on hagfish development begin with the contrast of monophyly (primarily with support from molecular data) versus paraphyly (primarily with support from phenotypic data) of cyclostomes and set out to resolve this dichotomy by finding similarities to, or differences from, the rest of vertebrates. Although this review is no exception, and although phylogenetic questions underpin all comparative studies, that narrative has perhaps been overused. Hagfish development does not exhibit unequivocally primitive conditions with respect to all other vertebrates. In this sense, no developmental feature excludes hagfish from vertebrates. The existing evidence, in terms of developmental morphology, is compatible with both monophyletic and paraphyletic cyclostomes because none of the character states can be compared beyond a three-taxon system of hagfish, lampreys, and gnathostomes. The equivocal nature of this evidence—coupled with the need to incorporate stem gnathostomes that can only be placed in the tree on the basis of skeletal and inferred soft tissue morphology from adult specimens—illustrates why the resolution of tree topology will remain elusive, no matter how closely hagfish development parallels that of other vertebrates. It also merits a stem-based definition for the clade Vertebrata, which recognizes hagfish as a vertebrate regardless of whether or not cyclostomes form a clade.

The developmental anatomy of hagfish used to be depicted as primitive with respect to vertebrates, but has recently been reinterpreted as derived with some degenerate features. Some character states in hagfish must represent reversals in order to be consistent with a hypothesis of cyclostome monophyly (e.g., vertebrae; Ota et al., 2011, 2013, 2014). However, simpler alternatives exist for the putative vertebrae in *Eptatretus*: the structures could have arisen independently. Moreover, if the derived states of hagfish are increasingly explained away as degenerate, the net result will be an implicit resurrection of something akin to Owen's (1848) vertebrate archetype theory. This approach is flawed: it both assumes a pattern that has never been observed in stem vertebrates, and provides little insight as to how the character originally evolved.

In addition to their bearing on relationships of early vertebrate lineages, the hagfish clade is a valuable resource for studying the evolution of developmental mechanisms underlying vertebrate characters. For example, independent origins appear to be the most parsimonious explanation for the axial skeletal elements among vertebrates. The implications are: (1) the subdivision of somites by differential gene expressions is a prerequisite to the vertebrate axial skeleton and is subject to comparison with invertebrate chordates; (2) the sclerotomal mesenchyme specified via *Pax1/9* and *Twist* was deployed to develop various axial skeletal elements; and (3) the origin of the vertebrae in gnathostomes likely required an additional developmental program, which may be identified via comparison with hagfish and lampreys. The horseshoe pattern of the placode development in hagfish is another example that can test the existing model of placode origins (Schlosser, 2005).

Hagfish embryology has experienced a profound transformation in the last decade. Old observations have been critically tested, and new interpretations exploiting digital imaging methods have revealed widespread vertebrate features of development (Ota et al., 2007, 2011; Oisi et al., 2013a). Thanks to the effort of K. Ota, S. Kuratani, and their colleagues, the subject is no longer the intractable "final frontier" (or backwater) of vertebrate embryology (Ota and Kuratani, 2006). Coming decades will see an increase in laboratories using hagfish embryos and refinement of comparative approaches to elucidate the still poorly known developmental process of this curious vertebrate. Some unresolved questions that have not been outlined in this chapter include: the patterning of the circulatory system, the mechanism of posterior displacement of the pharyngeal pouches, the molecular profile of neural crest specification and delamination, and the development of the eyes, inner ears, and their peripheral structures. We are optimistic that comparative embryology of hagfish will produce many more insights into the rise of vertebrates.

Acknowledgments

We thank Greg Goss (University of Alberta) and Susan Edwards (Appalachia State University, USA) for invitation to this volume and their patience with our hagfish-like sluggishness in producing this chapter. Karsten Hartel and Andrew Williston (Museum of Comparative Zoology, USA) provided T.M. with access to the Dean-Conel collections. T.M. benefitted from discussion with Ted Allison (University of Alberta), Andrew Clark (College of Charleston, USA), Philippe Janvier (Museum national d'Histoire naturelle, France), Shigeru Kuratani (RIKEN, Japan), the late Joe Nelson (University of Alberta), Yasuhiro Oisi (Max Planck Florida Institute for Neuroscience), Kinya Ota (Academia Sinica, Taiwan), and Adam Summers (Friday Harbor Laboratories, USA). M.I.C. benefitted from ongoing discussions with Robert Ho and Victoria Prince (University of Chicago). T.M. also thanks Hillary Maddin (Carleton University, Canada), Zack Lewis, James Hanken (Harvard University, USA), and Kesia Miyashita (Spencer Environmental Management, Canada) for logistical assistance. We thank an anonymous referee for insightful comments, especially about nomenclatural history.

References

Abildgaard, P. C. 1792. Kurze anatomische Beschreibung des Säugers (*Myxine glutinosa* Linn.). *Schriften der Berlinischen Gesellschaft Naturforschender Freunde* 10:193–200.

Abitua, P. B., E. Wagner, I. A. Navarrete, and M. Levine. 2012. Identification of a rudimentary neural crest in a non-vertebrate chordate. *Nature* 492:104–107.

Ayers, H. and C. Jackson. 1900. Morphology of the Myxinoidei. I. Skeleton and musculature. *Journal of Morphology* 17:185–226.

Ayers, H. and J. Worthington. 1907. The skin end organs of the trigeminus and lateralis nerves of *Bdellostoma dombeyi*. *American Journal of Anatomy* 7:327–336.

Baker, C. 2005. Neural crest and cranial ectodermal placodes. In M. S. Rao and M. Jacobson (eds.), *Developmental Neurobiology*. Kluwer Academic Press, New York, pp. 67–127.

De Beer, G. R. 1937. *The Development of the Vertebrate Skull*. Oxford University Press, London.

Bodznick, D. and R. G. Northcutt. 1981. Electroreception in lampreys: Evidence that the earliest vertebrates were electroreceptive. *Science* 212:465–467.

Boorman, C. J. and S. M. Shimeld. 2002. *Pitx* homeobox genes in *Ciona* and amphioxus show left–right asymmetry is a conserved chordate character and define the ascidian adenohypophysis. *Evolution & Development* 4:354–365.

Braun, C. B. and R. G. Northcutt. 1997. The lateral line system of hagfishes (Craniata: Myxinoidea). *Acta Zoologica* 78:247–268.

Busk, G. 1852. An account of the Polyzoa and sertularian zoophytes collected in the voyage of the Rattlesnake on the coast of Australia and the Louisiade Archipelago. In J. MacGillivray (ed.), *Narrative of the Voyage of H.M.S. Rattlesnake, Commanded by the Late Captain Owen Stanley, During the Years 1846–1850*. Boone, London, pp. 343–402.

Christiaen, L., P. Burighel, W. C. Smith, P. Vernier, F. Bourrat, and J. S. Joly. 2002. *Pitx* genes in Tunicates provide new molecular insight into the evolutionary origin of pituitary. *Gene* 287:107–113.

Cole, F. J. 1905. A monograph on the general morphology of the myxinoid fishes, based on a study of *Myxine*. Part I. The anatomy of the skeleton. *Transactions of the Royal Society of Edinburgh* 41:749–791.

Conel, J. L. 1929. The development of the brain of *Bdellostoma stouti*. I. External growth changes. *Journal of Comparative Neurology* 47:343–403.

Conel, J. L. 1931. The development of the brain of *Bdellostoma stouti*. II. Internal growth changes. *Journal of Comparative Neurology* 52(3):365–499.

Conel, J. L. 1933. The genital system of the Myxinoidea: A study based on notes and drawing of these organs in *Bedellostomata* made by Bashford Dean. In E. W. Gudger (ed.), *The Bashford Dean Memorial Volume, Archaic Fishes*. The American Museum of Natural History, New York, pp. 63–110.

Conel, J. L. 1942. The origin of the neural crest. *Journal of Comparative Neurology* 76:191–215.

Dean, B. 1898. On the development of the Californian hagfish, *Bdellostoma stouti*. Lockington. *Quarterly Journal of Microscopical Science* 40:269–279.

Dean, B. 1899. On the embryology of *Bdellostoma stouti*. A genera account of myxinoid development from the egg and segmentation to hatching. In *Festschrift zum 70ten Geburststag Carl von Kupffer*. Gustav Fischer Verlag, Jena, pp. 220–276.

Doflein, F. 1898. Bericht über eine wissenschaftliche Reise nach Californien. *Sitzungsber Gesellsch Morph Physiol München* 14:105–118.

Duméril, A. M. C. 1806. *Zoologie Analytique, ou Mèthode Naturelle de Classification des Animaux*. Didot, Paris.

Dupret, V., S. Sanchez, D. Goujet, P. Tafforeau, and P. E. Ahlberg. 2014. A primitive placoderm sheds light on the origin of the jawed vertebrate face. *Nature* 507:500–503.

Fänge, R. 1998. Introduction: Early hagfish research. In J. M. Jørgensen, J. P. Lomholt, R. E. Weber, and H. Malte (eds.), *The Biology of Hagfishes*. Chapman & Hall, London, pp. xiii–xix.

Fernholm, B. O. 1969. A third embryo of *Myxine*: Considerations on hypophysial ontogeny and phylogeny. *Acta Zoologica* 50:169–177.

Fernholm, B. 1975. Ovulation and eggs of the hagfish *Eptatretus burgeri. Acta Zoologica* 56:199–204.

Fernholm, B. 1985. The lateral line system of cyclostomes. In E. E. Foreman, A. Gorbman, J. M. Dodd and R. Olsson (Eds.) *Evolutionary Biology of Primitive Fishes.* NATO ASI Series A: Life Sciences. Volume 103. Plenum Press, New York, pp. 113–122.

Fernholm, B. and K. Holmberg. 1975. The eyes in three genera of hagfish (*Eptatretus, Paramyxine* and *Myxine*)—A case of degenerative evolution. *Vision Research* 15:253–259.

Fujimoto, S., Y. Oisi, S. Kuraku, K. G. Ota, and S. Kuratani. 2013. Non-parsimonious evolution of hagfish Dlx genes. *BMC Evolutionary Biology* 13:15.

Gai, Z., P. C. J. Donoghue, M. Zhu, P. Janvier, and M. Stampanoni. 2011. Fossil jawless fish from China foreshadows early jawed vertebrate anatomy. *Nature* 476:324–327.

Gans, C. 1993. Evolutionary origin of the vertebrate skull. In J. Hanken and B. K. Hall (eds.), *The Skull*, vol. 2. The University of Chicago Press, Chicago, pp. 1–35

Gans, C. and R. G. Northcutt. 1983. Neural crest and the origin of vertebrates: A new head. *Science* 220:268–273.

Gelman, S., A. Ayali, E. D. Tytell, and A. H. Cohen. 2007. Larval lampreys possess a functional lateral line system. *Journal of Comparative Physiology A* 193:271–277.

Gelman, S., A. H. Cohen, and E. Sanovich. 2009. Developmental changes in the ultrastructure of the lamprey lateral line nerve during metamorphosis. *Journal of Morphology* 270:815–824.

Goodrich, E. S. 1930. *Studies on the Structure and Development of Vertebrates.* Macmillan, London.

Gorbman, A. 1983. Early development of the hagfish pituitary gland: Evidence for the endodermal origin of the adenohypophysis. *American Zoologist* 23:639–654.

Gorbman, A. 1997. Hagfish development. *Zoological Science* 14:375–390.

Gorbman, A. and A. Tamarin. 1985. Early development of oral, olfactory and adenohypophyseal structures of agnathans and its evolutionary implications. In E. E. Foreman, A. Gorbman, J. M. Dodd and R. Olsson (Eds.) *Evolutionary Biology of Primitive Fishes.* NATO ASI Series A: Life Sciences. Volume 103. Plenum Press, New York, pp. 165–185.

Greene, C. W. 1900. Contributions to the physiology of the California hagfish, *Polistotrema stoutii.* I. The anatomy and physiology of the caudal heart. *American Journal of Physiology* 3:366–382.

Griffith, R. W. and K. S. Thomson. 1973. *Latimeria chalumnae*: Reproduction and conservation. *Nature* 242:617–618.

Gudger, E. W. and B. G. Smith. 1933. The segmentation of the egg of the Myxinoid, *Bdellostoma stouti.* In E. W. Gudger (ed.), *The Bashford Dean Memorial Volume, Archaic Fishes.* The American Museum of Natural History, New York, pp. 43–62.

Haeckel, E. 1866. *Generelle Morphologie der Organismen.* Georg Reimer, Berlin.

Hall, B. K. 2009. *The Neural Crest and Neural Crest Cells in Vertebrate Development and Evolution.* Springer, New York.

Hanken, J. and B. K. Hall. 1993. *The Skull. Volume 1: Development.* The University of Chicago Press, Chicago.

Heimberg, A. M., R. Cowper-Sal, M. Sémon, P. C. J. Donoghue, and K. J. Peterson. 2010. microRNAs reveal the interrelationships of hagfish, lampreys, and gnathostomes and the nature of the ancestral vertebrate. *Proceedings of the National Academy of Sciences* 107:19, 379–19, 383.

Holmgren, N. 1946. On two embryos of *Myxine glutinosa*. *Acta Zoologica* 27:1–90.

ICZN. 1999. *International Code of Zoological Nomenclature*, 4th Edition. The International Trust for Zoological Nomenclature, London.

Janvier, P. 1981. The phylogeny of the Craniata, with particular reference to the significance of fossil "agnathans". *Journal of Vertebrate Paleontology* 1:121–159.

Janvier, P. 1985. *Les Céphalaspides du Spitsberg. Anatomie, phylogénie et systématique des Ostéostracés siluro-dévoniens. Révision des Ostéostracés de la Formation de Wood Bay (Dévonien inférieur du Spitsberg).* Cahiers de Paléontologie, Centre national de la Recherche scientifique, Paris.

Janvier, P. 1996. *Early Vertebrates*. Clarendon Press, Oxford.

Janvier, P. 2008. Early jawless vertebrates and cyclostome origins. *Zoological Science* 25:1045–1056.

Janvier, P. 2011. Comparative anatomy: All vertebrates do have vertebrae. *Current Biology* 21:R661–R663.

Janvier, P. and M. Arsenault. 2007. The anatomy of *Euphanerops longaevus* Woodward, 1900: An anaspid-like jawless vertebrate from the Upper Devonian of Miguasha, Quebec, Canada. *Geodiversitas* 29:143–216.

Jarvik, E. 1980. *Basic Structure and Evolution of Vertebrates*, 2 vols. Academic Press, London.

Jaspersen, A. 1975. Fine structure of spermiogenesis in eastern Pacific species of hagfish (Myxinidae). *Acta Zoologica* 56:189–198.

Kavanaugh, S. I., M. L. Powell, and S. L. Sower. 2005. Seasonal changes of gonadotropin-releasing hormone in the Atlantic hagfish *Myxine glutinosa. General Comparative Endocrinology* 140:136–143.

Keller, R. and L. Davidson 2004. Cell movements of gastrulation. In C. D. Stern (ed.), *Gastrulation: From Cells to Embryo*. Cold Spring Harbor Laboratory Press, Cold Spring Harbor, pp. 291–304.

Kishida, R., R. C. Goris, H. Nishizawa, H. Koyama, T. Kadota, and F. Amemiya. 1987. Primary neurons of the lateral line nerves and their central projections in hagfishes. *Journal of Comparative Neurology* 264:303–310.

Kobayashi, H., H. Suzuki, and M. Sekimoto. 1972. Seasonal migration of the hagfish *Eptatretus burgeri. Japanese Journal of Ichthyology* 19:191–194.

von Kupffer, C. 1899. Zur Kopfentwicklung von *Bdellostoma. Sitzungsberichte der Gesellschaft fur Morpholigie und Physioligie* 15:21–35.

von Kupffer, C. 1900. *Studien zur vergleichenden Entwicklungsgeschichte des Kopfes der Kranioten, Heft 4: Zur Kopfentwicklung von Bdellostoma.* Verlag von J. F. Lehmann, München.

von Kupffer, C. 1906. Die Morphogenie des Centralnervensystems. *Handbuch der vergleichenden und experimentellen Entwicklungslehre der Wirbeltiere* 2(Part 3):1–272.

Kuratani, S. and K. G. Ota. 2008a. Hagfish (Cyclostomata, Vertebrata): Searching for the ancestral developmental plan of vertebrates. *BioEssays* 30:167–172.

Kuratani, S. and K. G. Ota. 2008b. Primitive versus derived traits in the developmental program of the vertebrate head: Views from cyclostome developmental studies. *Journal of Experimental Zoology Part B* 310:294–314.

Lindström, T. 1949. On the cranial nerves of the cyclostomes with special reference to n. trigeminus. *Acta Zoologica* 30:315–458.

Linnaeus, C. 1758. *Systema Naturae*, Xth edition, Vol. 1 (Systema naturae per regna tria naturae, secundum classes, ordines, genera, species, cum characteribus, differentiis, synonymis, locis. Tomus I. Editio decima, reformata). *Holmiae Salvii*, 824 pp.

Løvtrup, S. 1977. *The Phylogeny of Vertebrata*. Wiley, London.

Martini, F. H. and A. Beulig. 2013. Morphometrics and gonadal development of the hagfish *Eptatretus cirrhatus* in New Zealand. *PLoS ONE* 8:e78740.

Martini, F., J. B. Heiser, and M. P. Lesser. 1997. A population profile for Atlantic hagfish, *Myxine glutinosa* (L), in the Gulf of Maine. 1. Morphometrics and reproductive state. *Fishery Bulletin* 95:311–320.

Matsuoka, T., P. E. Ahlberg, N. Kessaris, P. Iannarelli, U. Dennehy, W. D. Richardson, A. P. McMahon, and G. Koentges. 2005. Neural crest origins of the neck and shoulder. *Nature* 436:347–355.

Mazet, F., J. A. Hutt, J. Milloz, J. Millard, A. Graham, and S. M. Shimeld. 2005. Molecular evidence from *Ciona intestinalis* for the evolutionary origin of vertebrate sensory placodes. *Developmental Biology* 282:494–508.

Mimori, R. 2010. New perspectives (cont.): The World's first? Hatching of hagfish [In Japanese]. *Tokyo Zoo Net News*. http://www.tokyo-zoo.net/topic/topics_detail?kind=news&link_num=13592 (obtained August 30, 2014).

Miyashita, T. 2012. Comparative analysis of the anatomy of the Myxinoidea and the ancestry of early vertebrate lineages. Unpublished M.Sc. thesis, University of Alberta.

Miyashita, T. 2015. Fishing for jaws in early vertebrate evolution: A new theory of mandibular confinement. *The Biological Reviews*. doi: 10.1111/brv.12187.

Miyashita, T. and A. R. Palmer. 2014. Handed behavior in hagfish—An ancient vertebrate lineage—And a survey of lateralized behaviors in other invertebrate chordates and elongate vertebrates. *The Biological Bulletin* 226:111–120.

Müller, J. 1834. Vergleichende anatomie der Myxinoiden, der Cyclostomen mit durchbohrtem Gaumen. Osteologie und Myologie. *Abhandlungen der Königlichen Akademie der Wissenschaften zu Berlin* 1834:65–340.

Near, T. J. 2009. Conflict and resolution between phylogenies inferred from molecular and phenotypic data sets for hagfish, lampreys, and gnathostomes. *Journal of Experimental Zoology Part B* 312:749–761.

Neumayr, L. 1938. Die entwicklung des kopskelettes von *Bdellostoma* St. L. *Archivio Italiano di Anatomica e di Embriologia* 40:1–222.

Nielsen, C. 2012. The authorship of higher chordate taxa. *Zoological Scripta* 41:435–436.

Northcutt, R. G. 2005. The new head hypothesis revisited. *Journal of Experimental Zoology Part B* 304:274–297.

Nozaki, M., T. Ichikawa, K. Tsuneki, and H. Kobayashi. 2000. Seasonal development of gonads of the hagfish, *Eptatretus burgeri*, correlated with their seasonal migration. *Zoological Science* 17:225–232.

Oisi, Y., S. Fujimoto, K. G. Ota, and S. Kuratani. 2015. On the peculiar morphology and development of the hypoglossal, glossopharyngeal and vagus nerves and hypobranchial muscles in the hagfish. *Zoological Letters* 1:6.

Oisi, Y., K. G. Ota, S. Fujimoto, and S. Kuratani. 2013b. Development of the chondrocranium in hagfishes, with special reference to the early evolution of vertebrates. *Zoological Science* 30:944–961.

Oisi, Y., K. G. Ota, S. Kuraku, S. Fujimoto, and S. Kuratani. 2013a. Craniofacial development of hagfishes and the evolution of vertebrates. *Nature* 493:175–180.

Ota, K. G., S. Fujimoto, Y. Oisi, and S. Kuratani. 2011. Identification of vertebra-like elements and their possible differentiation from sclerotomes in the hagfish. *Nature Communications* 2:373.

Ota, K. G., S. Fujimoto, Y. Oisi, and S. Kuratani. 2013. Late development of hagfish vertebral elements. *Journal of Experimental Zoology Part B* 320:129–139.

Ota, K. G., S. Kuraku, and S. Kuratani. 2007. Hagfish embryology with reference to the evolution of the neural crest. *Nature* 446:672–675.

Ota, K. G. and S. Kuratani 2006. The history of scientific endeavors towards understanding hagfish embryology. *Zoological Science* 23:403–418.

Ota, K. G. and S. Kuratani. 2007. Cyclostome embryology and early evolutionary history of vertebrates. *Integrative and Comparative Biology* 47:329–337.

Ota, K. G. and S. Kuratani. 2008. Developmental biology of hagfishes, with a report on newly obtained embryos of the Japanese inshore hagfish, *Eptatretus burgeri*. *Zoological Science* 25:999–1011.

Ota, K. G. and S. Kuratani. 2010. Expression pattern of two collagen type 2α1 genes in the Japanese inshore hagfish (*Eptatretus burgeri*) with special reference to the evolution of cartilaginous tissue. *Journal of Experimental Zoology Part B* 314:157–165.

Ota, K. G., Y. Oisi, S. Fujimoto and S. Kuratani. 2014. The origin of developmental mechanisms underlying vertebral elements: Implications from hagfish evo-devo. *Zoology* 117:77–80.

Owen, R. 1848. *On the Archetype and Homologies of the Vertebrate Skeleton.* J. van Voorst, London.

Pani, A. M., E. E. Mullarkey, J. Aronowicz, S. Assimacopoulos, E. A. Grove, and C. J. Lowe. 2012. Ancient deuterostome origins of vertebrate brain signalling centres. *Nature* 483:289–294.

Patzner, R. A. 1998. Gonads and reproduction in hagfishes. In J. M. Jørgensen, J. P. Lomholt, R. E. Weber, and H. Malte (eds.), *The Biology of Hagfishes.* Chapman & Hall, London, pp. 378–395.

Powell, M. L., S. I. Kavanaugh, and S. A. Sower. 2004. Seasonal conditions of reproductive steroids in the gonads of the Atlantic hagfish, *Myxine glutinosa*. *Journal of Experimental Zoology Part A* 301:352–360.

Powell, M. L., S. I. Kavanaugh, and S. A. Sower. 2005. Current knowledge of hagfish reproduction: Implications for fisheries management. *Integrative and Comparative Biology* 45:158–165.

Price, G. 1896a. Some points in the development of a myxinoid (*Bdellostoma stouti* Lockington). *Anatomischer Anzeiger* 12:81–86.

Price, G. 1896b. Zur Ontogenie eines Myxinoiden (*Bdellostoma stouti* Lockington). Sitzungsberichte der Mathematisch—Physikalischen Classe der K. B. Akademie der Wissenschaften zu München Bd 36: Heft 167–174.

Price, G. 1897. Development of the excretory organs of a Myxioid, *Bdellostoma stoutii* Lockington. *Zoologische Jahrbücher. Abteilung für Anatomie und Ontogenie der Tiere* 10:205–226.

Price, G. 1904. A further study of the development of the excretory organs in *Bdellostoma stouti*. *American Journal of Anatomy* 4:117–138.

Rinon, A., S. Lazar, H. Marshall, S. Büchmann-Møller, A. Neufeld, H. Elhanany-Tamir, M. M. Taketo, L. Sommer, R. Krumlauf, and E. Tzahor. 2007. Cranial neural crest cells regulate head muscle patterning and differentiation during vertebrate embryogenesis. *Development* 134:3065–3075.

Saint-Jeannet, J. P. and S. A. Moody. 2014. Establishing the pre-placodal region and breaking it into placodes with distinct identities. *Developmental Biology* 389:13–27.

Schlosser, G. 2005. Evolutionary origins of vertebrate placodes: Insights from developmental studies and from comparisons with other deuterostomes. *Journal of Experimental Zoology Part B* 304:347–399.

Schlosser, G., C. Patthey, and S. M. Shimeld. 2014. The evolutionary history of vertebrate cranial placodes. II. Evolution of ectodermal patterning. *Developmental Biology* 389:98–119.

Schoenwolf, G. C. 1997. Reptiles and birds. In S. F. Gilbert and A. M. Raunio (eds.), *Embryology: Constructing the Organism*. Sinaeur Associates, Sunderland, pp. 437–458.

Shimeld, S. M. and P. C. J. Donoghue. 2012. Evolutionary crossroads in developmental biology: Cyclostomes (lamprey and hagfish). *Development* 139:2091–2099.

Smith, C. L. and C. S. Rand. 1975. *Latimeria*, the living coelacanth, is ovoviviparous. *Science* 190:1105–1106.

Stockard, C. R. 1906a. The development of the mouth and gills in *Bdellostoma stoutii*. *American Journal of Anatomy* 5:481–517.

Stockard, C. R. 1906b. The development of the thyroid gland in *Bdellostoma stoutii*. *Anatomischer Anzeiger* 29:91–99.

Strahan, R. 1958. The velum and the respiratory current of *Myxine*. *Acta Zoologica* 39:227–240.

Takechi, M., M. Takeuchi, K. G. Ota, O. Nishimura, M. Mochii, K. Itomi, N. Adachi et al. 2011. Overview of the transcriptome profiles identified in hagfish, shark, and bichir: Current issues arising from some nonmodel vertebrate taxa. *Journal of Experimental Zoology Part B* 316:526–546.

Thomson, R. C., D. C. Plachetzki, D. L. Mahler, and B. R. Moore. 2014. A critical appraisal of the use of microRNA data in phylogenetics. *Proceedings of the National Academy of Sciences* 111:E3659–E3668.

Tsuneki, K., M. Ouji, and H. Saito. 1983. Seasonal migration and gonadal changes in the hagfish *Eptatretus burgeri*. *Japanese Journal of Ichthyology* 29:429–440.

Uchida, K., Y. Murakami, S. Kuraku, S. Hirano, and S. Kuratani. 2003. Development of the adenohypophysis in the lamprey: Evolution of epigenetic patterning programs in organogenesis. *Journal of Experimental Zoology B* 300:32–47.

Wada, H., H. Saiga, N. Satoh, and P. W. Holland. 1998. Tripartite organization of the ancestral chordate brain and the antiquity of placodes: Insights from ascidian *Pax-2/5/8, Hox* and *Otx* genes. *Development* 125:1113–1122.

Wicht, H. and R. G. Northcutt. 1995. Ontogeny of the head of the Pacific hagfish (*Eptatretus stouti*, Myxinoidea): Development of the lateral line system. *Philosophical Transactions: Biological Sciences* 349:119–134.

Wicht, H. and U. Tusch. 1998. Ontogeny of the head and nervous system of myxinoids. In J. M. Jørgensen, J. P. Lomholt, R. E. Weber, and H. Malte (eds.), *The Biology of Hagfishes*. Chapman & Hall, London, pp. 431–451.

Worthington, J. 1905. Contribution to our knowledge of the myxinoids. *American Naturalist* 39:625–663.

Yuuki, Y., K. Ishida, and S. Yasugi. 2003. The ecology of hagfish *Eptatretus burgeri* and the fisheries actual condition in the Japan Sea off Shimane Prefecture. *Report of the Shimane Prefectural Fisheries Experimental Station* 11:1–6.

chapter five

Photoreception in hagfishes
Insights into the evolution of vision

Shaun P. Collin and Trevor D. Lamb

Contents

Introduction

Hagfishes are the only surviving members of the group of jawless, marine chordates or craniates that diverged from the ancestors of lampreys more than 500 million years ago (mya). Previous studies have revealed that many characteristics about the anatomy and physiology of hagfishes appear "simple" or "primitive" and even resemble nonvertebrate chordates such as amphioxus and sea squirts (Forey and Janvier, 1993; Lacalli, 2004; Janvier, 2007). These features include differences in the vertebral column, the ability to control water content by osmoregulation, and the presence of lymphocytes.

These differences have led many to consider hagfishes to represent degenerate cyclostomes, that is, hagfishes have lost several vertebrate characters that lampreys and jawed vertebrates have retained. Thus, it is possible that the eyes of hagfishes have degenerated as a consequence of the reduced reliance on vision under low light conditions (Conway Morris, 1989; Braun, 1996; Jeffrey and Martasian, 1998). Fernholm and Holmberg (1975) even consider that the variability in the morphology of the eye reflects the levels of degeneracy, with the genus *Eptatretus* representing the most plesiomorphic condition and *Myxine* representing the most derived condition, with *Paramyxine* occupying an intermediate position.

An alternative view is that hagfishes are actually the most primitive vertebrates and constitute the "sister group" of all vertebrates (Northcutt, 1996; Janvier, 2007), although this view is now generally discounted. Instead, a large body of molecular evidence indicates that hagfishes and lampreys are monophyletic (Kuratani et al., 2002; Kuraku and Kuratani, 2006; Heimberg et al., 2010). Furthermore, recent embryological studies of the origin of the neural crest also support monophyly (Ota and Kuratani, 2007; Kuratani and Ota, 2008).

In a recent review, Lamb et al. (2007) proposed that the "eyes" of hagfishes reflect an early stage in the evolution of the vertebrate eye, prior to the acquisition of image-forming capabilities and more sophisticated neural processing and have retained features from an earlier stage in eye evolution. They proposed that the eyes of hagfishes are bilateral expansions of the circadian light-sensitive region of the brain and are therefore comparable to a (bilateral) pineal organ, which hagfishes lack.

At first sight, this hypothesis might seem compatible only with the view that hagfishes are the sister group of lampreys and gnathostomes, but there is another possibility that was suggested by the observation that (unlike lampreys) hagfishes do not undergo metamorphosis and that instead there are remarkable similarities between adult hagfishes and premetamorphic lampreys (ammocoetes). Thus, it is possible that, in developmental terms, the eye of an adult hagfish corresponds broadly to the eye of an ammocoete and likewise to the eye of a very immature gnathostome embryo (Lamb et al., 2007; Lamb, 2013). On this basis, hagfishes, or at least their eye development, might be viewed as neotenous.

In this scenario, the last common ancestor of hagfishes, lampreys, and gnathostomes was capable of developing a camera-style eye during embryogenesis. In the line that evolved to become hagfishes, it is proposed that embryogenesis of the eyes proceeded only as far as a stage corresponding to that reached in larval lampreys. If this is indeed the case, then the eyes of hagfishes should not be considered as degenerate, but may instead be viewed as providing a unique window into an early stage in vertebrate evolution, prior to the advent of imaging vision. This theory has yet to be tested, but potentially it can explain why these nonimage-forming dorsal "eyes," which have been thought by others to be degenerate, have been retained apparently unchanged for hundreds of millions of years (see "The insights into the evolution of the vertebrate eye").

Hagfishes generally inhabit the deeper parts of the oceans (over 1000 m, Adam and Strahan, 1963) and are all benthic, foraging over the soft sediment in low light conditions (Foss, 1968; Tambs-Lyche, 1969). However, some species (*Eptatretus burgeri*) frequent much shallower depths (~10 m), where light levels are much higher (McInerney and Evans, 1970; Fernholm, 1974). There are also differences in burrowing habits between the two main groups of hagfishes, the eptatretids and the myxinids (Worthington, 1905), but whether this is due to differences in the photic environments or morphological differences in their lateral line systems (Braun and Northcutt, 1997) is currently unknown.

The head of the hagfish is without jaws but possesses an impressive range of adaptations for feeding. Typically, hagfishes burrow into their prey, that is, dead or dying fish, and devour the flesh and viscera, stripping the bones using their rasping tongue. Under some circumstances, they are forced to compete for food (Auster and Barber, 2006) but feeding is aided by the ability of the tail to curve back on itself to form a knot that can be moved toward the head to enable the hagfish to anchor itself against a prey object (Jensen, 1966). In addition to this unique method of scavenging and extracting nutrition, recent behavioral observations using remote underwater video cameras deployed off the coast of New Zealand have revealed dramatic images of teleosts and cartilaginous fishes attacking hagfishes but quickly releasing them upon the rapid expulsion of mucous (Zintzen et al., 2011). This video footage (http://www.nature.com/srep/2011/111027/srep00131/extref/srep00131-s1.mov) shows that at least *Eptatretus* spp. have localized control of their mucous glands, which produce an almost instantaneous (<0.4 s) secretion of mucous from a specific part of their body that appears to be distasteful and may clog the gills of would-be predators. Zintzen et al. (2011) also showed that *Neomyxine* spp. actively search out live prey (teleosts) by entering their burrows, apparently somehow immobilizing the fish and extracting their body held firmly by their fast-acting dental plates (http://www.nature.com/srep/2011/111027/srep00131/extref/srep00131-s2.mov).

This chapter extends the review of Locket and Jørgensen (1998) and concentrates on the sensory tissues in hagfishes considered to be photoreceptive and traces this afferent input to the central nervous system. However, it is clear that although anatomical studies of the "eyes" of hagfishes have been undertaken and provide predictive value in structure/function relationships, there is a critical need to undertake further studies to understand photoreception in hagfishes, how light influences behavior in this unique group of jawless fishes, and how their eyes can inform us of the evolution of photoreception.

The "eyes" of hagfishes

The eyecup

The eyes of hagfishes are rudimentary, small (~1.0–1.5 mm in diameter), and conical and completely lack a lens or intra- or extraocular muscles, although a lens placode is present in eptatretid embryos (Price, 1896; Stockard, 1906; Wicht and Northcutt, 1995). The unpigmented eyes are buried beneath a patch of translucent skin in eptatretids, whereas they lie buried under muscle and have no vitreous body in myxinids (Fernholm and Holmberg, 1975; Locket and Jørgensen, 1998; Figure 5.1). The eyecup in *Myxine* has no lumen, the margins meeting at a fibrous plug, whereas *Eptatretus* species have a vitreous cavity, with scattered collagen fibrils, some forming dense aggregates (Locket and Jørgensen, 1998; Figure 5.2). The opacity of the skin overlying the eye varies between the two groups of hagfishes; *E. burgeri* possesses a relatively clear, unpigmented "window,"

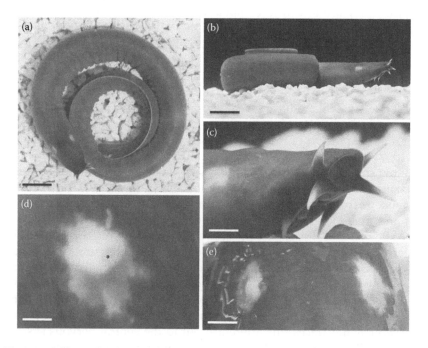

Figure 5.1 (**See color insert.**) Series of photographs of both the Pacific hagfish, *E. stoutii* (a through c), and the New Zealand hagfish, *E. cirrhatus* (d and e). Note the bilateral position of the opaque patches of unpigmented skin overlying the "eyes." Note in *E. cirrhatus* that the skin overlying the eyes is relatively clear, revealing a circular gap in the surrounding adipose tissue (aperture signified by the asterisk in d). Scale bars, 2 cm (a and b), 5 mm (c), 0.5 mm (d), and 1.5 mm (e). Photographs by S. E. Temple and S. P. Collin (a through c) and S. P. Collin and T. D. Lamb (d and e).

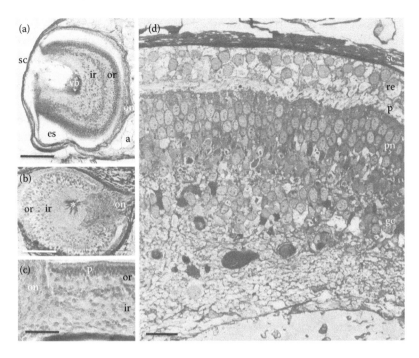

Figure 5.2 The eyes and retina of hagfishes. (a) Sagittal section through the eye of *E. stoutii.* A scleral covering (sc) or primitive cornea surrounds the eye, which is embedded in adipose tissue (a). An extracellular space (es) envelops the retina, which can be divided into inner (ir) and outer (or) regions. (b) Saggital section through the eye of *M. glutinosa.* Note the difference in the presence (remnant in a) and absence (b) of a vitreous body (vb). Note the central fibrous layer surrounding a connective tissue strand (asterisk). (With kind permission from Springer Science+Business Media: Z Zellforsch Mikrosk Anat, The hagfish retina: fine structure of retinal cells in *Myxine glutinosa,* L., with special reference to receptor and epithelial cells, 111, 1970, 519–538, K. Holmberg.) (c) Transverse section (10 μm paraffin) through the retina of *E. stoutii,* where axons leave the eye via the optic nerve (on). (Kindly provided by R. G. Northcutt.) (d) Transverse section (1 μm resin) through the retina of *E. stoutii.* gc, presumed ganglion cell layer; p, layer of photoreceptor outer segments; pn, layer of photoreceptor nuclei; re, retinal epithelium; sc, sclera. Scale bars, 0.2 mm (a), 0.1 mm (b), 40 μm (c), and 15 μm (d).

whereas the overlying skin of *Myxine glutinosa* is not lighter than the pigmented skin in other parts of the body. Fernholm and Holmberg (1975) measured the transparency of the skin overlying the eye and found that 68% and 61% of the incident light passed through the dermis in *E. burgeri* and *M. glutinosa,* respectively, in contrast to 33% in *Paramyxine atami.* Interestingly, the region underlying the skin (and surrounding the eye) is filled with blood in *Eptatretus stoutii* (Wicht and Northcutt, 1990; S. P. Collin, personal observation).

Although the eyecup is surrounded by adipose tissue, this tissue does not extend over the front of the eyecup, where a window (aperture) into the eyecup exists, thereby allowing light to enter via the overlying and unpigmented patch of (often translucent) epithelium (Figure 5.1). Hagfishes (i.e., *Paramyxine* and *Myxine* spp.) are considered to lack an iris and lens (Holmberg, 1977) and therefore do not possess the optical apparatus to focus incident light onto the back of the eye. The eyes of hagfishes also do not possess intraocular eye muscles, a ciliary ganglion, or an Edinger–Westphal nucleus. However, a protective epithelial layer supported by collagen lamellae contiguous with the sclera covers the anterior part of the eyecup in *E. stoutii* (Fernholm and Holmberg, 1975; S. P. Collin and H. B. Collin, unpublished data). In *Eptatretus cirrhatus*, this rudimentary "cornea" may be optically transparent and the overlying patch of unpigmented skin is translucent, thereby allowing light to fall on all parts of the conical eyecup from a region above and to the sides of the head.

The retina

The retina of hagfishes is typically avascular (with the exception of *E. stoutii*, which contains blood vessels; Holmberg, 1971) and differentiated, but it appears to lack the typical layering (three neural layers and two plexiform layers) of jawed vertebrates. The retina is bounded by distinct inner and outer limiting membranes, but the retinal layering of myxinids appears even less organized than eptatretids. Allen (1905) and Walls (1942) report that, in some cases, the embryonic fissure persists in adult hagfishes. The neuroretina and the epithelium are continuous at the margins and do not extend to form a ciliary body or iris (Locket and Jørgensen, 1998).

The inner retina

Characterization of the neuron types and detailed examination of the stratification of the hagfish retina have not yet been performed, although four layers have been previously described, that is, an inner synaptic layer, an inner layer of cell bodies, an outer synaptic layer, and an outer receptor cell layer (Fernholm and Holmberg, 1975; Holmberg, 1977; Figure 5.2). Retrograde labeling with neurobiotin from the optic nerve in *E. burgeri* has revealed that the cell bodies of the inner layer are retinal ganglion cells, which lie in two ill-defined sublayers (proximal and distal) within the inner retina (Sun et al., 2014) and may correspond to the two cell types of the inner retinal layer described by Holmberg (1970).

The outer retina and retinal epithelium

The layers of the outer retina are difficult to discriminate, but could be considered to include the photoreceptors (and their synaptic terminals) and the retinal epithelium (which is unpigmented in all species examined). The photoreceptors of hagfishes occur at much lower packing density than in gnathostome retinas, and in some studies, it has been reported that their outer segments project into an extracellular space, but this could be artifactual due to difficulties in preservation and the fact that hagfishes are osmoconformers and have an osmotic concentration that is similar to seawater. The photoreceptors are ciliary in nature, with an obvious outer segment and an extensive cylindrical region containing a nucleus and synaptic terminal, but with a poorly delineated inner segment region (Figure 5.3). In *M. glutinosa*, the outer segments are arranged in membranous whorls rather than with the distinct arrangement of orderly discs as seen in the outer segment discs in *Polistotrema* (*Eptatretus*) *stoutii* (Holmberg, 1971) and in the cones and rods of gnathostomes (Holmberg, 1970). The connecting cilium has the classical 9 + 0 filament structure and usually protrudes into the center of the outer segment, although the more familiar lateralized location has also been described (Holmberg, 1970; Figure 5.3). On morphological criteria, only a single rod-like photoreceptor has been identified in *M. glutinosa* (Locket and Jorgensen, 1998), although Vigh-Teichmann et al. (1984) found two classes of outer segments in this species that could be distinguished on immunocytochemical, but not ultrastructural, criteria. The outer segment discs presumably containing the photopigment appear to be completely enclosed in a plasma membrane (Holmberg, 1970, 1971; Locket and Jørgensen, 1998), suggesting that the photoreceptors may be rod-like, although their whorl-like appearance has been compared with the pineal receptors of teleost fishes and anuran amphibians (Holmberg, 1970).

The visual pigment (opsin) genes contained within the retinal photoreceptors of hagfishes are currently unknown, although one of the two photoreceptor types identified by Vigh-Teichmann et al. (1984) was immunoreactive to bovine rhodopsin. In a biochemical study of the levels of vitamin A, Steven (1955) found only vitamin A_1, suggesting that the photoreceptor cells may be utilizing a *retinal*-based chromophore. The photoreceptor terminals form spherical synaptic bodies (but not in *M. glutinosa*, Holmberg, 1971) without synaptic ribbons (Holmberg and Öhman, 1975; Holmberg, 1977; Figure 5.3) that appear to make direct contact with the output neurons (retinal ganglion cells), with no obvious retinal bipolar cells or other interneurons (Kobayashi, 1964; Holmberg, 1970, 1971, 1977). Agranular vesicles aggregate near synaptic membrane densities suggesting metabolic activity (Holmberg, 1970), and Müller cells are located at the basal margin of the photoreceptor layer (Holmberg, 1971; Figure 5.3).

The retinal epithelial cells are cuboidal, do not contain melanosomes or myeloid bodies, and make desmosomal intercellular junctions. These cells contain mitochondria, smooth endoplasmic reticulum, and phagosomes, but their basal surface does not display infoldings (Holmberg, 1970, 1971; Figure 5.3). The retinal epithelium and the endothelial cells of

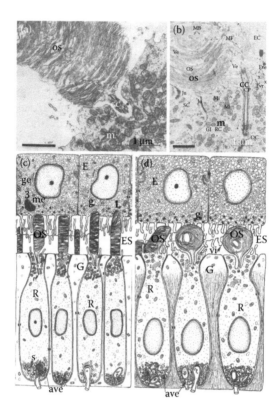

Figure 5.3 Photoreceptors in hagfishes. (a) Electron micrograph of a photoreceptor in *Polistotremata* (*Eptatretus*) *stoutii* showing the central connecting cilium (cc)
and regular stacks of outer segment (os) discs. m, mitochondria. (With kind permission from Springer Science+Business Media: Z Zellfbrsch, The hagfish retina:
electron microscopic study comparing receptor and epithelial cells in the Pacific
hagfish. *Polistotrema stouti* with those in the Atlantic hagfish. *Myxine glutinosa*,
121, 1971, 216–269, K. Holmberg.) (b) Electron micrograph of the whorl-like photoreceptor outer segment (os) of *M. glutinosa* showing the lateral position of the
connecting cilium (cc). (With kind permission from Springer Science+Business
Media: Z Zellforsch Mikrosk Anat, The hagfish retina: fine structure of retinal
cells in *Myxine glutinosa*, L., with special reference to receptor and epithelial
cells, 111, 1970, 519–538, K. Holmberg.) (c and d) Illustrations of the fine structure of the epithelial (E) and photoreceptor (R) layers in *P. stouti* (c) and *M. glutinosa* (d) showing the differences in photoreceptor morphology. ave, agranular
vesicles; ES, extracellular space; g, Golgi membranes; G, glial cells, ge, granular
endoplasmic reticulum; me, membranous inclusions; OS, outer segments; vi, villi
of the epithelial cells. Scale bars, 1 μm (a) and 1 μm (b). (With kind permission
from Springer Science+Business Media: Z Zellfbrsch, The hagfish retina: electron
microscopic study comparing receptor and epithelial cells in the Pacific hagfish.
Polistotrema stouti with those in the Atlantic hagfish. *Myxine glutinosa*, 121, 1971,
216–269, K. Holmberg.)

the choroid together form Bruch's membrane (Holmberg, 1970). The choroid and sclera are not well differentiated (Walls, 1942; Holmberg, 1969).

The optic nerve

The optic nerve (nII) containing the axons of retinal ganglion cells in the Atlantic hagfish, *M. glutinosa*, is thin (~30 μm in diameter), surrounded by a basal lamina, and divided into bundles by glial processes. The axons within the optic nerve are all unmyelinated (0.1–1.4 μm in diameter with a mean of 0.3 μm), as are all nerves in hagfishes (Peters, 1964; Holmberg, 1972; Bullock et al., 1984), and numbered between 810 and 1467 in the two individuals examined at the ultrastructural level by Holmberg (1972). Although low in number compared with gnathostomatous vertebrates, the presence of optic nerve axons projecting to the regions of the brain typically receiving input from the eye (discussed subsequently) indicates a light-sensing function. Interestingly, Vigh-Teichmann et al. (1984) found rhodopsin-containing outer segments of photoreceptor cells located in the tubular lumen of the optic nerve. These extraretinal receptors may also assist in light detection and in setting circadian behaviors in the absence of a pineal organ.

Input from the eyes to the brain

The hagfish brain comprises telencephalic, diencephalic, mesencephalic, and rhombencephalic divisions, but sulci do not always demarcate the boundaries of these brain divisions, and there is no cerebellum. The ventricular system is also almost obliterated by the compression of the brain in the rostro-caudal axis (Wicht and Niewenhuys, 1998). The optic nerves from the left and right eyes of hagfishes cross (decussate) at an optic chiasm but, in contrast to lampreys, this "chiasma opticum" decussates within the hypothalamus (Kusunoki and Amemiya, 1983). Retinal ganglion cell axons of the optic tract in hagfishes project bilaterally to the preoptic, thalamic, and pretectal nuclei and terminate in the mesencephalic optic tectum (Kusunoki and Amemiya, 1983; Wicht and Northcutt, 1990) and may represent the ancestral condition, which has been retained in hagfishes (Fritzsch, 1991). Ipsilateral (nondecussating) input from the eyes to the brain is only approximately 15% in hagfishes (Wicht and Northcutt, 1990). In gnathostomes, the displaced ganglion cells project to the basal optic root and the basal optic nucleus, but this pathway does not exist in hagfishes (Wicht and Northcutt, 1990).

The optic tectum receives the bulk of the afferent input from the eyes but is small in hagfishes, showing a low degree of lamination (Jansen, 1930). The optic tectum is divided into four strata: the stratum ependymale, the stratum periventriculare, the stratum cellulare et fibrosum, and the stratum marginale (Iwahori et al., 1996) as opposed to the seven strata identified in the lamprey optic tectum (Kennedy and Rubinson, 1984). The bulk of the visual input to the tectum terminates within the transitional area between the stratum marginale and the stratum cellulare et fibrosum (Kusunoki and Amemiya, 1983; Wicht and Northcutt, 1990). Visual afferents to the preoptic area may regulate circadian behaviors (Kobayashi et al., 1972; Ooka-Souda et al., 1993).

In hagfishes, two tegmental cell groups (the reticular mesencephalic area and the nucleus M5 of Schober) give rise to centrifugal fibers (fibers that project from the brain to the retina). In gnathostomes, the centrifugal system typically responds to chemical cues, such as sex pheromones, and regulates visually mediated sexual and reproductive behaviors, in addition to altering ganglion cell responses to color contrast, but the role of the efferent system has not yet been examined in hagfishes. The higher proportion of efferent fibers in hagfishes (5%), in contrast to jawed vertebrates (0.5%), is thought to reflect the ancestral condition (Wicht and Northcutt, 1990).

Light-sensing behavior

As hagfishes lack image-forming apparatus (such as lens and iris), they cannot have spatial "vision" as we know it, but they are nevertheless able to sense light to some extent, and they exhibit some responses to light. It is possible that their lateral eyes (or counterparts of the pineal organ) may subserve a function in circadian and/or seasonal time-keeping (discussed subsequently), in shadow detection, or even in control of axial orientation (i.e., roll). Although the relative lack of differentiation of the retina in hagfishes has been interpreted by some to suggest a degenerative state, both *M. glutinosa* and *E. burgeri* respond to changes in illumination by active locomotory movements, possibly ensuring that the animal remains buried during the day or maintains its circadian rhythm of activity (Fernholm, 1974; Kabasawa and Ooka-Souda, 1989). However, the relative contributions of the photoreceptors in the retina and those situated in the cloacal region (*M. glutinosa*, Newth and Ross, 1955; Steven, 1955), the tail (*E. burgeri* and *P. atami*, Patzner, 1978), and the dorsal surface of the trunk (*E. burgeri*, Patzner, 1978) are yet to be assessed.

Patzner (1978) showed that there are behavioral differences in the responses of different species of hagfishes to light. Although *P. atami* stop swimming and settle to the bottom under light intensities of less than 3 lux, shallower water species, such as *E. burgeri*, respond at slightly higher intensities of light (7 lux, Patzner, 1978). *M. glutinosa* respond to light with a withdrawal reflex, culminating in burrowing (Brodal and Fänge, 1963). Steven (1955) found that hagfishes exposed to wavelengths between 500 and 520 nm reacted more quickly than those exposed to light outside of this spectral window. The oceanic water column preferentially transmits this part (blue region) of the visible spectrum (480–490 nm) to the depths occupied by most species of hagfishes.

What we know about the electrical responses of the eyes of hagfishes to light is, at present, confined to a single study by Kobayashi (1964). This study used electroretinography to record from the isolated eye of *Myxine* in response to flashes of light at low light intensities (0.1 lux). The recordings revealed a slow positive wave (reminiscent of the b-wave), together with a hint of a negative wave with bright flashes, but no off response. The measured spectral sensitivity peaked around 500 nm (Kobayashi, 1964), which corresponds to the behavioral sensitivity noted earlier (Steven, 1955). Further electrophysiological study needs to be performed on the photoreceptors of hagfishes to examine the detailed nature of the response to light and, in particular, to determine whether or not cells exhibit the distinguishing feature of rods in gnathostomes, which is why they are able to respond reliably to the absorption of individual photons of light. Although it was recently suggested that the photoreceptors of hagfishes might be expected to respond reliably to single photons (Fain et al., 2010), this has yet to be demonstrated experimentally. Detailed examination of the photoresponses, including kinetics and absolute sensitivity, together with tracing of the neuronal circuitry within the retina in hagfishes, would be the next logical step in characterizing photoreceptive function.

Extraocular photoreception

Given the lack of a lens in the lateral eyes of hagfishes, the retina could be described as a "nonvisual" tissue capable of photoreception and regulating circadian rhythms (Lamb et al., 2007), or simply as a crude organ for detecting light and dark, which would typically be used for phototaxis. However, the eyes and other photoreceptive tissues may also mediate photoentrainment, in which the transition between light and dark (dawn and dusk) is used to adjust circadian phase to local time (Foster and Provencio, 1999). Little is known of these systems in hagfishes, but one may expect a range of nonimage-forming photoreceptive tissues, as has been revealed for jawed vertebrates (Foster and Hankins, 2002; Davies et al., 2011, 2014). As discussed earlier, a high proportion of incident light is transmitted through the skin and neural tissue and, despite some possible selective filtering, would be expected to penetrate deep into the brain (and body) of hagfishes, thereby producing a measure of environmental irradiance and hence time of day.

Given that the eyes of hagfishes are nonimage-forming, but possess well-developed retinofugal and retinopetal systems, it is tempting to hypothesize a prominent role for them in the regulation of circadian rhythms, especially given the absence of the classical "opto-endocrine interface" or pineal gland (Holmgren, 1919; Wicht and Northcutt, 1992;

Lamb et al., 2007). In gnathostomes, the pineal gland typically synthesizes melatonin, but hagfishes lack a pineal (and parapineal) organ, so this function may be mediated via the retina, as is the case for a number of nonmammalian vertebrates (Weichmann, 1986). However, an *in vitro* auto-radiographic study to examine binding of [2-^{125}I]iodomelatonin in the central nervous system of the Atlantic hagfish, *M. glutinosa*, failed to identify any binding site, including in the retina (Vernadakis et al., 1998). This suggests that the gene(s) encoding melatonin receptor protein(s) may have evolved and become widely expressed only after epiphyseal structures were present to supply high levels of circulating melatonin (Gern and Karn, 1983). Alternatively, melatonin receptor proteins may have evolved much earlier (~700 mya), the result of the duplication of the common ancestor of the melatonin and opsin genes in a eumetazoan ancestor (Feuda et al., 2012), and were subsequently lost in hagfishes.

The influence of light on the entrainment of circadian rhythms via the retina has been shown to be substantial in *E. burgeri*. In the absence of a suprachiasmatic nucleus, it appears that the preoptic nucleus (PON) and the hypothalamus have assumed a circadian pacemaker role, because activity patterns under a 12 h light/dark cycle were lost following the ablation of the anterior region of the hypothalamus (Ooka-Souda and Kabasawa, 1988; Ooka-Souda et al., 1993, 1995). However, the regulation of circadian behaviors in hagfishes may also be mediated by other means, such as aggregations of light-sensitive pigment in both the tail and along a line of unpigmented skin running down the back, as revealed in *E. burgeri* (Patzner, 1978), although the latter source of light detection may only be involved in eliciting photoreflex locomotor activity (Ooka-Souda et al., 1988). The hypothalamic and septal nuclei comprising the periventricular cerebrospinal fluid or CSF-contacting neurons (the presumed ancestral protoneurons) (Vigh et al., 2002; David et al., 2003) may also contribute to nonvisual photoreception in hagfishes.

A recent study in hagfishes has confirmed a mechanism for the regulation of circadian behaviors that is well established in gnathostomatous vertebrates. A single type of melanopsin gene (*Opn4m*) has been identified and found to be expressed within the retinal ganglion cells of *E. burgeri* using both *in situ* hybridization and retrograde labeling from the optic nerve (Sun et al., 2014). In cultured cells expressing the hagfish melanopsin, blue light induced an increase in intracellular calcium, revealing that short-wavelength light regulates circadian behaviors in hagfishes as it does for other cyclostomes (the lamprey, *Petromyzon marinus*, Koyanagi and Terakita, 2008) and gnathostomes (Matos-Cruz et al., 2011; Matsuyama et al., 2012). Based on their phylogenetic analysis, Sun et al. (2014) suggest that the last common ancestor of hagfishes and lampreys possessed two melanopsin genes (*Opn4m* and *Opn4x*) and that *Opn4x* was subsequently lost.

Insights into the evolution of the vertebrate eye

It is clear that there is still a great deal the hagfish can teach us about the selective pressures on the evolution of vertebrate eyes. Lacalli (2004) provided evidence of homology between the unpaired frontal eye in *Amphioxus* and the paired lateral eyes of vertebrates (including hagfishes), and recently, Vopalensky et al. (2012) provided powerful evidence for such a homology from the signature expression of transcription factors and the presence of components of a Gi-type phototransduction cascade in the frontal eye of *Amphioxus*. It has been suggested that the ancestral lateral eyes of protovertebrates living between 550 and 500 million years ago may have first served a nonvisual (possibly circadian) role, and only later did the neural processing power and optical and motor components needed for spatial vision evolve in their descendants (Lamb et al., 2007; Lamb, 2013). Embryological evidence supports this interpretation, as the larval phase in lampreys (the ammocoete) possesses paired nonvisual organs buried beneath the skin (as is the case in hagfishes), but during metamorphosis, these organs develop into paired camera-style eyes, with lens, cornea, ocular muscles, and the typical three layers of retinal neurons, so that they then become capable of producing a focussed image (Lamb et al., 2007).

The evolutionary progression from a two-layered (as in hagfishes) to a three-layered retina (as in lampreys) appears to have been realized by the development of bipolar interneurons (that in many ways resemble ciliary photoreceptors) that were integrated into the retinal circuitry between the photoreceptors and the output (or ganglion) cells, providing enhanced visual processing power. In hagfishes, the ciliary photoreceptors synapse directly onto ganglion cell-like projection neurons, in circuitry comparable to that of the pineal in lampreys and gnathostomes (Lamb et al., 2007; Erclik et al., 2009). Although it is often suggested that the eyes of hagfishes have *degenerated* from a camera-style eye in the common ancestor of hagfishes and lampreys, it may instead be the case that in hagfishes the embryological development of the eyes stops at a point corresponding roughly to that reached in the larval stage of lampreys. Hagfishes do not undergo metamorphosis, and so it might be best to view the eye of the hagfish as having resulted from the absence of certain steps in development, rather than from any degeneration of a developed organ. In this case, the eye of the hagfish provides a fascinating and informative window into an early "previsual" phase of vertebrate eye evolution (Lamb et al., 2007; Lamb, 2011, 2013; Figure 5.4).

During early embryonic development, hagfishes possess a lens placode (Allen, 1905; Stockard, 1906), which may assist in specifying the orientation or polarization of the retina and retinal pigment epithelium. However, this placode does not develop into a lens, and so it may be that

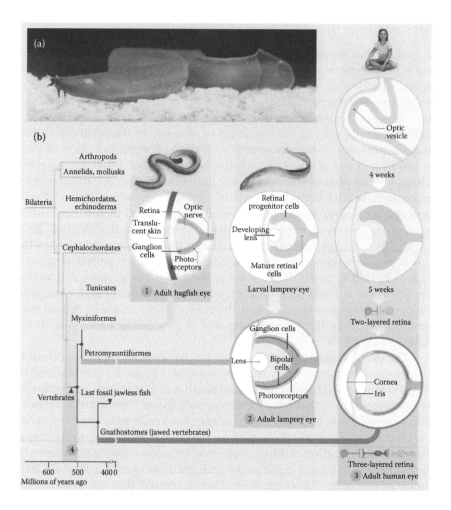

Figure 5.4 (**See color insert.**) (a) Lateral view of *E. stoutii*. (b) Morphological perspective on the evolution of the vertebrate eye drawing on aspects of embryonic development, which may reflect events that occurred during the evolution of that lineage. 1. The hagfish eye which appears degenerate and probably serves to detect light for modulating circadian rhythms. 2. The lamprey eye in early development (ammocoete stage) resembles the hagfish eye before it develops into an image-forming eye complete with an iris and lens. 3. The eye of gnathostomes (including humans) passes through a developmental stage in which the retina comprised two layers before a third layer of cells emerges in the adult phase (see text for further discussion). (Reproduced from Lamb (2011) with permission. Copyright (2011) Scientific American, a Division of Nature America, Inc. All rights reserved. Photograph in (a) by S. E. Temple and S. P. Collin.)

the ancestral function of this "lens" placode was in specifying development of the optic vesicle, rather than anything related to the emergence of
a lens. A situation that may be comparable to that discussed earlier for the
nondevelopment of the eye exists for the lateral line system in eptatretids,
in which there are separate lateral line placodes that give rise to neuromast
primordia but that never develop into true neuromasts and instead form
a single class of flask-shaped receptor cells situated in grooves (Wicht and
Northcutt, 1994; Braun and Northcutt, 1997).

An alternative possibility that we cannot completely reject is that
although the last common ancestor of lampreys and jawed vertebrates
possessed the requisite genes, its eyes might have exhibited a form more
rudimentary than a fully developed camera-style eye, and that the well-
developed eyes of both lampreys and jawed vertebrates occurred as a
result of convergent evolution (Lamb, 2013).

Whatever the sequence of evolutionary events, it is clear that the eyes
of hagfishes hold important clues to the evolution of the vertebrate eye and
that during the Cambrian period of evolutionary history, the selection pressures on light detection mechanisms were very strong (Nilsson and Pelger,
1994; Arendt, 2003). Further studies are critical if we are to fully understand
the photoreceptive capabilities of this key species in vertebrate evolution.

Acknowledgments

We would like to thank the editors Susan Edwards and Greg Goss for the
kind invitation to contribute to this book. We also wish to acknowledge
David M. Hunt, Nathan S. Hart, Wayne I. L. Davies, and Edward N. Pugh
Jr, who have contributed to some parts of the research and theoretical dis-
cussions presented herein. This work was supported by the Australian
Research Council and the Western Australian State Government.

References

Adam, H. and R. Strahan. 1963. Notes on the habitat, aquarium maintenance, and
    experimental use of hagfishes. In: A. Brodal and R. Fänge (eds), *The Biology
    of Myxine*. Scandinavian University Books, Oslo, pp. 33–41.
Allen, B. M. 1905. The eye of *Bdellostoma stouti*. *Anatomischer Anzeiger* 26:208e 211.
Arendt, D. 2003. Evolution of eyes and photoreceptor cell types. *Int J Develop Biol*
    47:563e571.
Auster, P. J. and K. Barber. 2006. Atlantic hagfish exploit prey captured by other
    taxa. *J Fish Biol* 68:618–621.
Braun, C. B. 1996. The sensory biology of the living jawless fishes: A phylogenetic
    assessment. *Brain Behav Evol* 48:262–276.
Braun, C. B. and R. G. Northcutt. 1997. The lateral line system of hagfishes
    (Craniata: Myxinoidea). *Acta Zool (Stockh)* 78:247–268.
Brodal, A. and R. Fänge. 1963. *The Biology of Myxine*. Scandinavian University
    Books, Oslo.

Bullock, T. H., J. K. Moore, and R. D. Fields. 1984. Evolution of myelin sheaths: Both lamprey and hagfish lack myelin. *Neurosci Lett* 48:145–148.

Conway Morris, S. 1989. The persistence of burgess Shale-type faunas: Implications for the evolution of deeper-water faunas. *Trans R Soc Edinburgh: Earth Sci* 80:271–283.

David, C., C. L. Frank, A. Lukats, A. Szél, and B. Vigh. 2003. Cerebrospinal fluid contacting neurons in the reduced brain ventricular system of the Atlantic hagfish, *Myxine glutinosa*. *Acta Biol Hung* 54:35–44.

Davies, W. I. L., R. G. Foster, and M. W. Hankins. 2014. The evolution and function of melanopsin in craniates. In: D. M. Hunt, M. W. Hankins, S. P. Collin, and N. J. Marshall (eds), *The Evolution of Visual and Non-Visual Pigments*. Springer, New York, pp. 23–64.

Davies, W. I., L. Zheng, S. Hughes, T. K. Tamai, M. Turton, S. Halford, R. G. Foster, D. Whitmore, and M. W. Hankins. 2011. Functional diversity of melanopsins and their global expression in the teleost retina. *Cell Mol Life Sci* 68:4115–4132.

Erclik, T., V. Hartenstein, R. R. McInnes, and H. D. Lipshitz. 2009. Eye evolution at high resolution: The neuron as a unit of homology. *Dev Biol* 332:70–79.

Fain, G. L., R. Hardie, and S. B. Laughlin. 2010. Phototransduction and the evolution of photoreceptors. *Curr Biol* 20:R114–R124.

Fernholm, B. 1974. Diurnal variations in the behaviour of the hagfish *Eptatretus burgeri*. *Mar Biol* 27:351–356.

Fernholm, B. and K. Holmberg. 1975. The eyes of three genera of hagfish (*Eptatretus, Paramyxine* and *Myxine*)—A case of degenerative evolution. *Vis Res* 15:253–259.

Feuda, R., S. C. Hamilton, J. O. McInerney, and D. Pisani. 2012. Metazoan opsin evolution reveals a simple route to animal vision. *Proc Natl Acad Sci USA* 109:18868–18872.

Forey, P. and P. Janvier. 1993. Agnathans and the origin of jawed vertebrate. *Nature* 361:129–134.

Foss, G. 1968. Behaviour of *Myxine glutinosa* L. in natural habitat: Investigation of the mud biotype by a suction technique. *Sarsia* 31:1–13.

Foster, R. G. and M. W. Hankins. 2002. Non-rod, non-cone photoreception in the vertebrates. *Prog Ret Eye Res* 21:507–527.

Foster, R. G. and I. Provencio. 1999. The regulation of vertebrate biological clocks by light. In: S. N. Archer, M. B. A. Djamgoz, E. Loew, J. C. Partridge, and S. Vallerga (eds), *Adaptive Mechanisms in the Ecology of Vision*. Kluwer Academic Publishers, Great Britain, pp. 223–243.

Fritzsch, B. 1991. Ontogenetic clues to the phylogeny of the visual system. In: P. Bagnoli and W. Hodos (eds), *The Changing Visual System*. Plenum Press, New York, pp. 33–49.

Gern, W. A. and C. M. Karn. 1983. Evolution of melatonin's function and effects. *Pineal Res Rev* 1:49–90.

Heimberg, A. M., R. Cowper-Sal-lari, M. Sémon, P. C. J. Donoghue, and K. J. Peterson. 2010. MicroRNAs reveal the interrelationships of hagfish, lampreys, and gnathostomes and the nature of the ancestral vertebrate. *Proc Natl Acad Sci USA* 107:19379–19383.

Holmberg, K. 1969. Hagfish eye: Ultrastructure of retinal cells. *Acta Zool* 50:179–183.

Holmberg, K. 1970. The hagfish retina: Fine structure of retinal cells in *Myxine glutinosa*, L., with special reference to receptor and epithelial cells. *Z Zellforsch Mikrosk Anat* 111:519–538.

Holmberg, K. 1971. The hagfish retina: Electron microscopic study comparing receptor and epithelial cells in the Pacific hagfish, *Polistotrema stouti* with those in the Atlantic hagfish, *Myxine glutinosa*. *Z Zellfbrsch* 121:219–269.

Holmberg, K. 1972. Fine structure of the optic tract in the Atlantic hagfish *Myxine glutinosa*. *Acta Zool* 53:165–171.

Holmberg, K. 1977. The cyclostome retina. In: H. Autrum, R. Jung, W. R. Loewenstein, D. M. MacKay, and H. L. Teuber (eds), *Handbook of Sensory Physiology: The Visual System in Vertebrates*. Springer-Verlag, New York, pp. 47–66.

Holmberg, K. and P. Öhman. 1975. Fine structure of retinal synaptic organelles in lamprey and hagfish photoreceptors. *Vis Res* 16:237–239.

Holmgren, N. 1919. Zur Anatomie des Gehirnes von *Myxine*. *Kungl Svenska Vetenskapsakad Handl* 60:1–96.

Iwahori, N., K. Nakamura, and A. Tsuda. 1996. Neuronal organization of the optic tectum in the hagfish, *Eptatretus burgeri*: A Golgi study. *Anat Embryol* 193:271–279.

Jansen, J. 1930. The brain of *Myxine glutinosa*. *J Comp Neurol* 49:359–507.

Janvier, P. 2007. Evolutionary biology: Born again hagfishes. *Nature* 446:622–623.

Jeffrey, W. R. and D. P. Martasian. 1998. Evolution of eye regression in the cavefish *Astyanax*: Apoptosis and the Pax-6 gene. *Am Zool* 38:685–696.

Jensen, D. 1966. The hagfish. *Sci Am* 214:82–90.

Kabasawa, H. and S. Ooka-Souda. 1989. Circadian rhythms in locomotor activity of the hagfish, *Eptatretus burgeri* (IV). The effect of eye ablation. *Zool Sci* 6:135–139.

Kennedy, M. C. and K. Rubinson. 1984. Development and structure of the lamprey optic tectum. In: H. Vanegas (ed.), *Comparative Neurology of the Optic Tectum*. Plenum Press, New York, pp. 1–13.

Kobayashi, H. 1964. On the photo-perceptive function in the eye of the hagfish, *Myxine garmani* Jordan and Snyder. *J Shimonoseki Coll Fish* 13:67–83.

Kobayashi, H., T. Ichikawa, H. Suzuki, and M. Sekimoto. 1972. Seasonal migration of the hagfish, *Eptatretus burgeri*. *Jap J Ichthyol* 19:191–194.

Koyanagi, M. and A. Terakita. 2008. Gq-coupled rhodopsin subfamily composed of invertebrate visual pigment and melanopsin. *Photochem Photobiol* 84:1024–1030.

Kuraku, S. and S. Kuratani. 2006. Timescale for cyclostome evolution inferred with a phylogenetic diagnosis of hagfish and lamprey cDNAs. *Zool Sci* 23:1053–1064.

Kuratani, S., S. Kuraku, and Y. Murakami. 2002. Lamprey and an evo-devo model: Lessons from comparative embryology and molecular phylogenetics. *Genesis* 34:175–195.

Kuratani, S. and K. G. Ota. 2008. Hagfish (Cyclostomata, Vertebrata): Searching for the ancestral developmental plan for vertebrates. *Bioessays* 30:167–172.

Kusunoki, T. and F. Amemiya. 1983. Retinal projections in the hagfish, *Eptatretus burgeri*. *Brain Res* 262:295–298.

Lacalli, T. C. 2004. Sensory systems in amphioxus: A window on the ancestral chordate condition. *Brain Behav Evol* 64:148–162.

Lamb, T. D. 2011. Evolution of the eye. *Sci Am* 305:64–69.

Lamb, T. D. 2013. Evolution of transduction, vertebrate photoreceptors and retina. *Prog Ret Eye Res* 36:52–119.

Lamb, T. D., S. P. Collin, and E. N. Pugh Jr. 2007. Evolution of the vertebrate eye: Opsins, photoreceptors, retina and eyecup. *Nat Rev Neurosci* 8:960–975.

Locket, N. A. and J. M. Jørgensen. 1998. The eyes of hagfishes. In: J. M. Jørgensen, J. P. Lomholt, R. E. Weber, and H. Malte (eds), *The Biology of Hagfishes*. Chapman and Hall, London, pp. 539–556.

Matos-Cruz, V., J. Blasic, B. Nickle, P. R. Robinson, S. Hattar, and E. Halpern. 2011. Unexpected diversity and photoperiod dependence of the zebrafish melanopsin system. *PLoS ONE* 6:e25111.

Matsuyama, T., T. Yamashita, Y. Imamoto, and Y. Shichida. 2012. Photochemical properties of mammalian melanopsin. *Biochemistry* 51:5454–5462.

McInerney, J. E. and D. O. Evans. 1970. Habitat characteristics of the Pacific hagfish *Polisotrema stouti*. *J Fish Res Bd Can* 27:966–968.

Newth, D. R. and D. M. Ross. 1955. On the reaction of light of *Myxine glutinosa* L. *J Exp Biol* 32:4–21.

Nilsson, D. E. and S. Pelger. 1994. A pessimistic estimate of the time required for an eye to evolve. *Proc R Soc Lond B* 256:53–58.

Northcutt, R. G. 1996. The agnathan ark: The origin of craniate brains. *Brain Behav Evol* 48:237–247.

Ooka-Souda, S. and H. Kabasawa. 1988. Circadian rhythms in locomotor activity of the hagfish, *Eptatretus burgeri*. III. Hypothalamus: A locus of the circadian pacemaker? *Zool Sci* 5:437–442.

Ooka-Souda, S., H. Kabasawa, and S. Kinoshita. 1988. Circadian rhythms in locomotor activity of the hagfish, *Eptatretus burgeri* (II). The effect of brain ablation. *Zool Sci* 5:431–435.

Ooka-Souda, S., T. Kadota, and H. Kabasawa. 1993. The pre-optic nucleus: The probable location of the circadian pacemaker of the hagfish, *Eptatretus burgeri*. *Neurosci Lett* 164:33–36.

Ooka-Souda, S., T. Kadota, H. Kabasawa, and H.-A. Takeuchi. 1995. A possible retinal information route to the circadian pacemaker through pre-tectal areas in the hagfish, *Eptatretus burgeri*. *Neurosci Lett* 192:201–204.

Ota, K. G. and S. Kuratani. 2007. Cyclostome embryology and early evolutionary history of vertebrates. *Integr Comp Biol* 47:329–337.

Patzner, R. A. 1978. Experimental studies on the light sense of the hagfish, *Eptatretus burgeri* and *Paramyxine atami* (Cyclostomata). Helgoliinder wiss. Meeresunters 31:180–190.

Peters, A. 1964. An electron microscopic study of the peripheral nerves of the hagfish (*Myxine glutinosa*, L.). *Exp Physiol* 49:35–42.

Price, G. C. 1896. Some points in the development of a myxinoid (*Bdellostoma stouti* Lockington). *Anatomischer Anzeiger* 12 (Suppl):81–86.

Steven, D. M. 1955. Experiments on the light sense of the hag, *Myxine glutinosa* L. *J Exp Biol* 32:22–38.

Stockard, C. R. 1906. The embryonic history of the lens in *Bdellostoma stouti* in relation to recent experiments. *Am J Anat* 6:511e515.

Sun, L., E. Kawano-Yamashita, T. Nagata, H. Tsukamoto, Y. Furutani, M. Koyanagi, and A. Terakita. 2014. Distribution of mammalian-like melanopsin in cyclostome retinas exhibiting a different extent of visual functions. *PLoS ONE* 9:1–8.

Tambs-Lyche, H. 1969. Notes on the distribution and ecology of *Myxine glutinosa* L. *Fisk skrifter serie havundersøkelser* 15:279–284.

Vernadakis, A. J., W. E. Bemis, and E. L. Bittman. 1998. Localisation and partial characterization of melatonin receptors in amphioxus, hagfish, lamprey and skate. *Gen Comp Endocrinol* 110:67–78.

Vigh, B., M. J. Manzano, A. Zadori, C. L. Frank, A. Lukats, P. Röhlich, A. Szel, and C. David. 2002. Non-visual photoreceptors of the deep brain, pineal organs and retina. *Histol Histopathol* 17:555–590.

Vigh-Teichmann, I., B. Vigh, R. Olsson, and T. van Veen. 1984. Opsin-immunoreactive outer segments of photoreceptors in the eye and in the lumen of the optic nerve of the hagfish, *Myxine glutinosa. Cell Tissue Res* 238:515–522.

Vopalensky, P., J. Pergnera, M. Liegertova, E. Benito-Gutierrez, D. Arendt, and Z. Kozmik, 2012. Molecular analysis of the amphioxus frontal eye unravels the evolutionary origin of the retina and pigment cells of the vertebrate eye. *Proc Natl Acad Sci USA* 109:15383–15388. doi: 10.1073/pnas.1207580109.

Walls, G.L. 1942. *The Vertebrate Eye and its Adaptive Radiation.* Cranbrook Press, Bloomfield Hills.

Weichmann, A. F. 1986. Melatonin: Parallels in pineal gland and retina. *Exp Eye Res* 42:507–527.

Wicht, H. and R. Niewenhuys. 1998. Hagfishes, Myxinoidea. In: H. Nieuwenhuys, J. Ten Donkelaar, and C. Nicholson (eds), *The Central Nervous System of Vertebrates,* Vol. 1. Springer-Verlag, Berlin, pp. 497–550.

Wicht, H. and R. G. Northcutt. 1990. Retinofugal and retinopetal projections in the Pacific hagfish, *Eptatretus stouti* (Myxinoidea). *Brain Behav Evol* 36:315–328.

Wicht, H. and R. G. Northcutt. 1992. The forebrain of the Pacific hagfish: A cladistic reconstruction of the ancestral craniate forebrain. *Brain Behav Evol* 40:25–64.

Wicht, H. and R. G. Northcutt. 1994. Observations on the development of the lateral line system in the Pacific hagfish (*Eptatretus stouti,* Myxinoidea). *J Morphol* 32:257–261.

Wicht, H. and R. G. Northcutt. 1995. Ontogeny of the head of the Pacific hagfish (*Eptatretus stouti,* Myxinoidea): Development of the lateral line system. *Phil Trans R Soc Lond B* 349:119–134.

Worthington, J. 1905. Contributions to our knowledge of the myxinoids. *Am Nat* 39:625–663.

Zintzen, V., C. D. Roberts, M. J. Anderson, A. L. Stewart, C. D. Struthers, and E. S. Harvey. 2011. Hagfish predatory behaviour and slime defense mechanism. *Sci Rep* 1:131. DOI: 10.1038/srep00131, pp. 1–6.

chapter six

The hagfish heart

William Davison

Contents

Introduction

Hagfish are regarded as the most primitive of the vertebrates. Current thinking places them at the beginning of the vertebrate phylogenetic tree and indeed places them as prevertebrate chordates rather than true vertebrates (Rasmussen et al., 1998; Cobb et al., 2004; Janvier, 2007), although recent molecular work has reopened this debate (Heimberg et al., 2010). In particular, Heimberg et al. (2010) have indicated that the cyclostomes (hagfish and lampreys) are a monophyletic group and commented that hagfish show a loss of vertebrate characters. This is a significant divergence from the accepted philosophy because much of the research on hagfish hearts has focussed on the premise that hagfish are at the very beginning of the vertebrates and that any changes (especially between hagfish and lampreys) represent forwards rather than backward steps in vertebrate evolutionary history (Forster, 1998; Farrell, 2007).

Structure of the heart

The anatomy of the hagfish circulation has been well described and need not be covered in great detail here. Starting with the descriptions by Cole (1925), subsequent work has refined these, leading to comprehensive descriptions by Johansen (1963), Satchell (1971), Forster (1998), and Farrell (2007), though it should be noted that although there are around 60 species of hagfish worldwide (Lesser et al., 1997; Fernholm, 1998), research on the circulatory system has focussed almost entirely on three species, *Myxine glutinosa* (the Atlantic hagfish), *Eptatretus stoutii* (the Pacific hagfish), and *Eptatretus cirrhatus* (the New Zealand hagfish).

The circulatory system is characterized by low pressures (both arterial and venous). Indeed, Farrell (2007) described the hagfish ventricle as having the poorest pressure-generating ability of all the vertebrates, whereas Forster et al. (1992) called it a low-pressure moderate flow system. Pressures of 1.6 and 1.3 kPa in the ventral and dorsal aortae, respectively (Forster et al., 1992), are certainly modest, compared with pressures seen in teleost fish 3.2 kPa in the dorsal aorta of arctic charr and 4–8 kPa in the ventral aorta of lamprey (DeMont and Wright, 1993; Seth et al., 2013).

A unique feature of the circulatory system is a massive subcutaneous sinus surrounding the whole body which results in hagfish having a very large blood volume (Forster, 1997; Forster et al., 2001). For example, the blood volume of *E. cirrhatus* has been estimated at 177 mL kg^{-1}, with 30% of this contained in the sinus (Forster et al., 1989). This compares with 20–60 mL kg^{-1} for teleosts and 80 mL kg^{-1} for lampreys (Gingerich et al., 1987; Brown et al., 2005). These two features (low pressures and high blood volume) have been described as the reason for another unique feature of hagfish, the presence of multiple hearts. Depending on the definition of a heart, hagfish have up to six hearts; four of them (cranial and caudal hearts) are responsible for moving sinus blood back into the main circulation and are composed of skeletal rather than cardiac muscle, controlled by movements of the animal's body (Satchell, 1984; Forster, 1997). In contrast, the main (systemic) circulation is driven by a true cardiac heart. In addition, there is a portal heart, a true cardiac heart located between the supraintestinal vein draining the gut and the hepatic portal vein proper. It is capable of producing only low systolic pressures (Johnsson et al., 1996) and functions to provide sufficient pressure to force blood through the liver (Forster, 1998). However, Davison (1995) described an example of a single adult specimen of *E. cirrhatus* which lacked a portal heart, but otherwise appeared normal, indicating that although useful it is not absolutely necessary, with the low pressures found in the supraintestinal vein (0.04 kPa, Johnsson et al., 1996) adequate for perfusing the liver. Farrell (2007) suggested that the development of a rigid pericardium in lampreys allowed for the development of improved venous return, thus negating the requirement for accessory hearts.

The systemic circulation of the hagfish is similar to that found in other fishes. A sinus venosus leads to an atrium and a ventricle. There is a structure resembling a conus (Wright, 1984), leading directly into a pair of ventral aortae which supply the lateral gill pouches. However, flow in the ventral aorta is not maintained during diastole, reducing to zero while the ventricle is refilling (Axelsson et al., 1990; Cox et al., 2010). Oxygen is loaded in the gills. Despite the internal location of the gills, there is a very efficient exchange system leading to blood in the efferent branchial arteries being saturated with oxygen (Forster et al., 1992). Blood then passes to the body via a dorsal aorta. Blood returns in cardinal veins. The sinus venosus is contractile (Satchell, 1986) and sets the pace of the whole heart. Wilson and Farrell (2013) showed that the atrium sets the pace for the ventricle. Isolating the two main chambers of the heart using an atrioventricular (AV) ligature had no effect on atrial beat (21.2 bpm), whereas ventricular contraction decreased to 8.4 bpm. Thus the heart is typical in that there are a series of chambers, with the first setting the intrinsic rate and the rate of the other chambers.

Hagfish hearts are unique in having no neural control (Jensen, 1965; Nilsson, 1983; Sandblom and Axelsson, 2011). Ganglia have been found associated with the systemic heart, but appear to be not associated with the function of the heart (Hirsch et al., 1964; Farrell, 2007). In contrast, lampreys have a well-developed autonomic system, with vagal stimulation of the sinus venosus accelerating the heart (Augustinsson et al., 1956; Sandblom and Axelsson, 2011). Certainly, ventilation and cardiac function seem to be controlled very differently in the hagfish. For example, Coxon and Davison (2011) looked at both heart rate and ventilation rate during hypoxia in *E. cirrhatus* and concluded that although changes to heart rate were modest (around 5 bpm) and not "neutrally" driven, the rate of contraction of muscles controlling the velum changed rapidly and markedly, indicating a well-developed neural control. A change in water PO_2 from 20 to 6 kPa produced a change in this muscle from 20 to 90 bpm.

The action potential of the heart of *E. cirrhatus* is long (Davie et al., 1987), with a distinct V wave from depolarization of the sinus venosus and a clear P_T wave showing atrial repolarization occurring well before ventricular depolarization, presumably a feature of all hagfish because it is also seen in *M. glutinosa* (Arlock, 1975). This has been linked to the slow intrinsic rate in these fish (Davie et al., 1987).

Resting and elevated heart rates

Resting heart rate is influenced by many factors including temperature. But what is the resting heart rate of a hagfish? A representative set of data is shown in Table 6.1. The data indicate that heart rate in hagfish at their environmental temperature lies somewhere between 20 and 25 bpm, so an animal such as *M. glutinosa*, living in cold deep water, has this resting heart rate at 10°C or lower, whereas *E. cirrhatus* living in shallower, warmer water has this rate at around 15°C. There does not appear to be a size (allometric) effect with these two species having similar heart rates, despite a 10-fold difference in mass. Interestingly, *E. stoutii* kept at 10°C

Table 6.1 Measurements of heart rate in hagfish at rest and following a number of manipulations

Species		Temperature (°C)	Rate	Source
M. glutinosa	Rest	10	22.3	Axelsson et al. (1990)
	Rest	10	23	Johnsson and Axelsson (1996)
	Following adrenaline	10	25	Axelsson et al. (1990)
	Following increase of preload	10	26	Johnsson and Axelsson (1996)
E. cirrhatus	Rest	17	25	Forster et al. (1988, 1992)
	Rest	13	18	Coxon and Davison (2011)
	Hypoxia	17	27	Forster et al. (1992)
	Active	19	48	Coxon and Davison (2011)
	Swimming	16	29	Forster et al. (1988)
	Hypoxia (6.67 kPa)	13	21	Coxon and Davison (2011)
E. stoutii	Rest	15	26	Jensen (1965)
	Rest	10	10.4	Cox et al. (2010)
	Anoxia	10	8.1	Cox et al. (2010)

had a very low resting heart rate (Cox et al., 2010, 2011). Drazen et al. (2011) found a similar situation in a study on metabolic rate in two species of *Eptatretus*, *E. deani* from cold deep waters and *E. stoutii* from warmer shallow waters. Both heart and ventilation rates are influenced by acute temperature change, as shown by Coxon and Davison (2011). Heart rate in their *E. cirrhatus* (Figure 6.1) changed from 10 bpm at 10°C to 34 bpm at 19°C with a Q_{10} of 2.8, suggesting that the increase was simply a temperature effect, presumably on the cardiac pacemaker cells in the sinus venosus (Harper et al., 1995).

Despite a lack of neural control, hagfish hearts are capable of changing rate. Forster et al. (1992) showed that during hypoxia, *Eptatretus* can increase its cardiac output by 49%. Probably, the most extreme case was reported by Coxon and Davison (2011) who recorded a heart rate of 48 bpm in a single very active hagfish at 19°C, representing an increase of around 15 bpm. Interestingly, this increase was achieved at a very high temperature when it might have been expected that there would be little scope for increasing heart rate. Other reported changes have been much

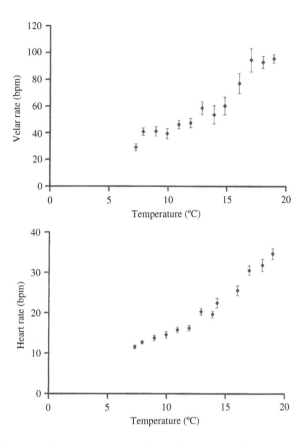

Figure 6.1 Heart and respiratory muscle (velar) rates in *Eptatretus cirrhatus* as a function of temperature. (Adapted from Coxon, S. E. and W. Davison. 2011. *J Fish Biol* 79:280–289.)

more modest. For example, Foster and Forster (2007a) detected a 1 bpm (though significant) change in heart rate, following a change to the osmolality of the external medium. Hypoxia has a stimulatory effect on the heart with Forster et al. (1992) showing a small increase in rate during hypoxia, although there was a larger rise of around 3 bpm during recovery. Hypoxia produced a 6 bpm rise in the hearts of Coxon and Davison's (2011) hagfish. Chapman et al. (1963) recorded a maximum heart rate of 37 bpm, which they regarded as exceptionally high.

Control of the heart

If the hagfish heart is capable of responding to changes in both its internal and external environments, but has no neural control, how is it controlled?

The systemic heart does not have a rigid pericardium as seen in lampreys and higher fish, and as a consequence, filling of the sinus venosus (SV) is dependent on *vis-a-tergo* filling (positive pressure in the veins) rather than *vis-a-fronte* filling (negative pressure in the SV caused by ventricle contraction) (Farrell, 2007). The hagfish heart is a typical heart in that it shows a Frank–Starling mechanism whereby filling pressure, and thus stretch, affects contraction (both rate and stroke volume), showing that this is a basic feature of vertebrate cardiac muscle cells (Jensen, 1958; Axelsson et al., 1990; Johnsson and Axelsson, 1996). Forster (1998) commented that filling of the heart using positive pressure inflation might be a reason for the low intrinsic heart rate in these animals, because the low filling pressure requires adequate time for filling (venous pressure 40 Pa (Johnsson et al., 1996) and atrial diastolic pressure 20 Pa (Satchell, 1986)). Johnsson and Axelsson (1996) showed that increases to preload in *Myxine* increased cardiac output by increasing both heart rate and stroke volume. Although hagfish use positive pressure filling of the sinus venosus, pressures are very low (0.1 kPa in the cardinal veins, Johnsson and Axelsson, 1996) and activity changes preload by contraction of the myotomal muscles, but also by enabling the caudal hearts, thus increasing fluid return from the subcutaneous sinus and increasing venous volume (Davie et al., 1987; Forster et al., 1988).

There is a good deal of information indicating that heart function is regulated by catecholamines, although, as Farrell (2007) points out, information on unanesthetized animals is limited. Catecholamines are contained in chromaffin granules within the heart itself, suggesting a paracrine function rather than an endocrine function. Noradrenaline is dominant in the atrium and portal heart, whereas adrenaline is dominant in the ventricle (Euler and Fänge, 1961). Perfused isolated hearts certainly release catecholamines, although at low levels, emphasizing that catecholamines in the heart are used locally rather than distally (Perry et al., 1993). Depleting cardiac catecholamine stores using reserpine affects contractility and produces a bradycardia (Chapman et al., 1963), adding to the evidence of a paracrine function.

The available data up to 2007 suggested that changes to circulating catecholamines in whole animals are small; for example, Perry et al. (1993) showed that 5 min of exposure to anoxia resulted in elevated plasma noradrenaline, but only from 3 to 10 nM L^{-1} with no changes to adrenaline. This led Farrell (2007) to suggest that humoral control of the hagfish heart is modest, and this was restated in Cox et al. (2010). However, Foster and Forster (2007b) have shown marked changes in circulating levels of catecholamines using a range of stressors. Anesthesia caused large changes to noradrenaline levels (7–818 nM L^{-1}), with no change in adrenaline. Changing the salinity of the medium affected circulating catecholamines. Increasing the salinity produced a small but steady increase in noradrenaline and a large but transient increase in adrenaline. In contrast, decreasing the salinity causes noradrenaline levels to decrease almost to undetectable levels, whereas adrenaline increased to 450 nM L^{-1}. The authors suggest that this is consistent with a vasoconstrictory role for noradrenaline but a vasodilatory role for adrenaline. Interestingly, Forster et al. (1992) showed that an injection of adrenaline caused increases in heart rate and stroke volume leading to an increased cardiac output, though with decreases in both branchial and systemic resistances.

The hagfish heart is a typical vertebrate heart, in that rate is set via pacemaker cells (DiFrancesco, 1993). These cells are characterized by having hyperpolarization-activated cyclic nucleotide-gated (HCN) protein channels in the cell membrane responsible for a slow depolarization of the membrane. The slope of this decrease in potential is responsible for setting the rate of the heart, by determining time taken to reach the threshold for Ca^{2+} channels (DiFrancesco, 1993; Wilson et al., 2013). The sarcoplasmic reticulum plays a minor role in hagfish heart contraction (Wilson and Farrell, 2013), and instead, it is the HCN channels that are important for calcium movements, particularly the finding that zatebradine, a specific HCN inhibitor, stops heartbeat. In a study of mRNA expression in Pacific hagfish, Wilson et al. (2013) discovered six isoforms of HCN, equivalent to HCN 2, 3, and 4 of mammalian muscle. Anoxia caused increases in expression of HCN3a in the ventricle, whereas recovery from anoxia led to a transient increase in expression of HCN3a in the atrium. The authors have suggested that changes in the density of the HCN channels are responsible for changes to resting heart rate during anoxia and recovery. In a companion study, Wilson and Farrell (2013) used a number of membrane channel blockers to investigate the mechanism of depolarization in heart cells. Blocking HCN channels with zatebradine markedly decreased heart rate in the ventricle and stopped heartbeat entirely in the atrium. In contrast, ryanodine had little effect on heart rate, thus showing the minor role for the sarcoplasmic reticulum (Thomas et al., 1996). Blocking Na$^+$ channels with tetrodotoxin had its major effect on the ventricle indicating an action on the AV node, whereas the calcium channel blockers, nickel and nifedipine, had effects mainly on the atrium. Their overall conclusion was that the HCN channels are the important channels regulating depolarization and thus heart rate in the hagfish heart.

Classical physiological work has shown that the hagfish heart is controlled using mechanisms found in all vertebrate hearts, such as a Starling response to increased filling pressure, and circulating, or resident catecholamines. There is also a substantial amount of evidence that the hagfish heart produces natriuretic peptides (Kawakoshi et al., 2006), which may influence the heart. However, the recent use of molecular techniques (Wilson and Farrell, 2013) has demonstrated that there is still much to be discovered about the hagfish heart at the cellular level.

Hypoxia, anoxia, and the hagfish heart

The life style of hagfish is clearly related to the structure and function of their hearts. Many hagfish burrow into mud (Martini, 1998), whereas all hagfish are known to be scavengers, digging into carcasses where oxygen will be limiting or even absent. They produce copious amounts of slime when disturbed which impairs oxygen exchange. In addition, hagfish have been found resident in deep, oxygen-deficient waters (Drazen et al., 2011). Hagfish have low-pressure circulations, and as a consequence, the metabolic rate of the hagfish heart is low. They have a good ability to generate ATP anaerobically, and the heart can generate at least 80% of its energy requirements by converting glucose to lactic acid (Forster, 1991, 1998). There is an assumption that the low arterial pressures in the hagfish circulation are an adaptation for life in hypoxic conditions because the associated low metabolic rate allows the heart to continue to beat anaerobically (Hansen and Sidell, 1983; Farrell and Stecyk, 2007), although Forster (1997) cautioned that this might not be true and that the low pressures could be associated with the presence of the large subcutaneous sinus and the control of blood into this space. Certainly, Hansen and Sidell (1983) showed that the heart is capable of functioning in a true anaerobic capacity because they poisoned the heart with cyanide or azide and showed that it was capable of maintaining output. In addition, Driedzic and Gesser (1994) measured very high levels of myoglobin in hagfish hearts, presumably to scavenge oxygen from the hemoglobin in the blood.

Hagfish are able to withstand very low levels of environmental oxygen (Perry et al., 2009; Cox et al., 2010, 2011; Coxon and Davison, 2011; Drazen et al., 2011) and even anoxia, although at least one species (*E. cirrhatus*) appears to be poorly able to withstand anoxia and will actively try to escape from hypoxic conditions (Forster, 1998; Coxon and Davison, 2011). Exposure to short-term hypoxia does not produce the classic bradycardia seen in teleosts, but instead heart rate is either maintained or even increased (Axelsson et al., 1990; Forster et al., 1992; Coxon and Davison, 2011).

Cox et al. (2010, 2011) examined the ability of hagfish (*E. stoutii*) to withstand not just hypoxia or short-term anoxia, but also prolonged (36 h) periods of anoxia. Routine normoxic cardiac output was 12.3 mL min^{-1} and remained stable for the first 3 h of anoxia. However, by 6 h, it had decreased by 33% and stayed that way for 30 h. Heart rate initially increased as the animals became hypoxic, presumably associated with increased activity. However, very different to earlier studies, it then fell by around 2 bpm by 1 h into the anoxic period and then fell further to around 50% of the normoxic rate. The authors suggest that as oxygen is needed to synthesize catecholamines, exposure to anoxia will lead to a depletion of stored catecholamines and thus impair heart function (Chapman et al., 1963). As a consequence, stroke volume increased during the early stages of anoxia to maintain cardiac output. Interestingly, power output increased early during anoxia associated with activity, but by 3 h had settled back to resting levels and stayed that way for the rest of the 36 h. Cox et al. (2011) showed that despite the animals being anoxic, many of the indicators of anaerobic metabolism did not change for many hours, with some, such as muscle lactate concentrations, not changing until 24 h of anoxia. Based on the levels of oxygen consumption during recovery, the authors suggest that hagfish survive anoxia by showing metabolic depression, although the heart stuff would suggest otherwise.

Conclusion

The hagfish heart is typically described as being primitive, representing the starting point of vertebrate phylogeny. Its lack of neural control would support this, though in most other respects, the hagfish heart is similar to those organs found in higher vertebrates. Its low-pressure-generating ability may be regarded as primitive, but is just as likely to be an adaptation for life in hypoxic or anoxic environments, or even, as Forster (1997) suggests, an adaptation to cope with the presence of a large subcutaneous sinus.

References

Arlock, P. 1975. Electrical activity and mechanical response in the systemic heart and portal vein heart of *Myxine glutinosa*. *Comp Biochem Physiol* 51A:521–522.

Augustinsson, K. B., R. Fänge, A. Johnels, and E. Ostlund. 1956. Histological, physiological and biochemical studies on the heart of two cyclostomes, hagfish (*Myxine*) and lamprey (*Lampetra*). *J Physiol* 131:257–276.

Axelsson, M., A. P. Farrell, and S. Nilsson. 1990. Effect of hypoxia and drugs on the cardiovascular dynamics of the Atlantic hagfish, *Myxine glutinosa*. *J Exp Biol* 151:297–316.

Brown, J. A., C. S. Cobb, S. C. Frankling, and J. C. Rankin. 2005. Activation of the newly discovered cyclostome renin–angiotensin system in the river lamprey *Lampetra fluviatilis*. *J Exp Biol* 208:223–232.

Chapman, C. B., D. Jensen, and K. Wildenthal. 1963. On circulatory control mechanisms in the Pacific hagfish. *Circ Res* 12:427–440.

Cobb, C. S., S. C. Frankling, M. C. Thorndyke, F. B. Jensen, J. C. Rankin, and J. A. Brown. 2004. Angiotensin I-converting enzyme-like activity in tissues from the Atlantic hagfish (*Myxine glutinosa*) and detection of immunoreactive plasma angiotensins. *Comp Biochem Physiol* 138B:357–364.

Cole, F. J. 1925. A monograph on the general morphology of the myxinoid fishes, based on a study of *Myxine*. Part VI. The morphology of the vascular system. *Trans R Soc Edinburgh* 54:309–342.

Cox, G. K., E. Sandblom, and A. P. Farrell. 2010. Cardiac responses to anoxia in the Pacific hagfish, *Eptatretus stoutii*. *J Exp Biol* 213:3692–3698.

Cox, G. K., E. Sandblom, J. G. Richards, and A. P. Farrell. 2011. Anoxic survival of the Pacific hagfish (*Eptatretus stoutii*). *J Comp Physiol* 181B:361–371.

Coxon, S. E. and W. Davison. 2011. Structure and function of the velar muscle in the New Zealand hagfish *Eptatretus cirrhatus*: Response to temperature change and hypoxia. *J Fish Biol* 79:280–289.

Davie, P. S., M. E. Forster, W. Davison, and G. H. Satchell. 1987. Cardiac function in the New Zealand hagfish *Eptatretus cirrhatus*. *Physiol Zool* 60:233–240.

Davison, W. 1995. What is the function of the hagfish portal heart? *N Z Nat Sci* 22:95–98.

DeMont, M. E. and G. Wright. 1993. Elastic arteries in a primitive vertebrate: Mechanics of the lamprey ventral aorta. *Experientia* 1993: 43–46.

DiFrancesco, D. 1993. Pacemaker mechanisms in cardiac tissue. *Ann Rev Physiol* 55:455–472.

Drazen, J. C., J. Yeh, J. Friedman, and N. Condon. 2011. Metabolism and enzyme activities of hagfish from shallow and deep water of the Pacific Ocean. *Comp Biochem Physiol* 159A:182–187.

Driedzic, W. R. and H. Gesser. 1994. Energy metabolism and contractility in ecto-thermic vertebrate hearts: Hypoxia, acidosis and low temperature. *Physiol Rev* 74:221–258.

Euler, U. S. and R. Fänge. 1961. Catecholamines in nerves and organs of *Myxine glutinosa, Squalus acanthias* and *Gadus morhua*. *Gen Comp Endocrinol* 1:191–194.

Farrell, A. P. 2007. Cardiovascular systems in primitive fishes. In D. J. McKenzie, A. P. Farrell, and C. J. Brauner (eds), *Fish Physiology, Vol. 26. Primitive Fishes*. Academic Press, London, pp. 53–120.

Farrell, A. P. and J. A. W. Stecyk. 2007. The heart as a working model to explore themes and strategies for anoxic survival in ectothermic vertebrates. *Comp Biochem Physiol* 147A:300–312.

Fernholm, B. 1998. Hagfish systematics. In J. M. Jørgensen, J. P. Lomholt, R. E. Weber, and H. Malte (eds), *The Biology of Hagfishes*. Chapman & Hall, London, pp. 33–44.

Forster, M. E. 1991. Myocardial oxygen consumption and lactate release by the hypoxic hagfish heart. *J Exp Biol* 156:583–590.

Forster, M. E. 1997. The blood sinus system of hagfish: Its significance in a low pressure circulation. *Comp Biochem Physiol* 116A:239–244.

Forster, M. E. 1998. Cardiovascular function in hagfishes. In J. M. Jørgensen, J. P. Lomholt, R. E. Weber, and H. Malte (eds), *The Biology of Hagfishes*. Chapman & Hall, London, pp. 237–258.

Forster, M. E., P. S. Davie, W. Davison, G. H. Satchell, and R. M. G. Wells. 1988. Blood pressures and heart rates in swimming hagfish. *Comp Biochem Physiol* 89A:247–250.

Forster, M. E., W. Davison, M. Axelsson, and A. P. Farrell. 1992. Cardiovascular responses to hypoxia in the hagfish *Eptatretus cirrhatus*. *Respir Physiol* 88:373–386.

Forster, M. E., W. Davison, G. H. Satchell, and H. H. Taylor. 1989. The subcutane-ous sinus of the hagfish, *Eptatretus cirrhatus*: Its relation to the central blood circulating blood volume. *Comp Biochem Physiol* 93A:607–612.

Forster, M. E., M. J. Russell, D. C. Hambleton, and K. R. Olson. 2001. Blood and extracellular fluid volume in the whole body and tissues of the Pacific hag-fish *Eptatretus stoutii*. *Physiol Biochem Zool* 74:750–758.

Foster, J. M. and M. E. Forster. 2007a. Effects of salinity manipulations on blood pressures in an osmoconforming chordate, the hagfish, *Eptatretus cirrhatus*. *J Comp Physiol* 177B:31–39.

Foster, J. M. and M. E. Forster. 2007b. Changes in plasma catecholamine concentration during salinity manipulation and anaesthesia in the hagfish *Eptatretus cirrhatus*. *J Comp Physiol* 177B:41–47.

Gingerich, W. H., R. A. Pityer, and J. J. Rach. 1987. Estimates of plasma, packed cell and total volume in tissues of the rainbow trout (*Salmo gairdneri*). *Comp Biochem Physiol* 87A:251–256.

Hansen, C. A. and B. D. Sidell. 1983. Atlantic hagfish cardiac muscle: Metabolic basis of tolerance to anoxia. *Am J Physiol* 244:R356–R362.

Harper, A. A., I. P. Newton, and P. W. Watt. 1995. The effects of temperature on spontaneous action potential discharge of the isolated sinus venosus from winter and summer plaice (*Pleuronectes platessa*). *J Exp Biol* 198:137–140.

Heimberg, A. M., R. Cowper-Sallari, M. Semon, P. C. J. Donoghue, and K. J. Peterson. 2010. MicroRNAs reveal the interrelationships of hagfish, lampreys, and gnathostomes and the nature of the ancestral vertebrate. *Proc Natl Acad Sci USA* 107:19379–19383.

Hirsch, E. F., M. Jellinek, and T. Cooper. 1964. Innervation of systemic heart of California hagfish. *Circ Res* 14:212–217.

Janvier, P. 2007. Living primitive fishes and fishes from deep time. In D. J. McKenzie, A. P. Farrell, and C. J. Brauner (eds), *Fish Physiology, Vol. 26. Primitive Fishes*. Academic Press, London, pp. 2–53.

Jensen, D. 1958. Some observations on cardiac automatism in certain animals. *J Gen Physiol* 42:289–302.

Jensen, D. 1965. The aneural heart of the hagfish. *Ann N Y Acad Sci* 127:443–458.

Johansen, K. 1963. The cardiovascular system of *Myxine glutinosa*. In Brodal and R. Fänge (eds), *The Biology of Myxine*. Universitetsforlaget, Oslo, pp. 289–316.

Johnsson, M. and M. Axelsson. 1996. Control of the systemic heart and the portal heart of *Myxine glutinosa*. *J Exp Biol* 199:1429–1434.

Johnsson, M., M. Axelsson, W. Davison, M. E. Forster, and S. Nilsson. 1996. Effects of preload and after load on the performance of the *in situ* perfused portal heart of the New Zealand hagfish, *Eptatretus cirrhatus*. *J Exp Biol* 199:401–405.

Kawakoshi, A., S. Hyodo, M. Nozaki, and Y. Takei. 2006. Identification of a natriuretic peptide (NP) in cyclostomes (lamprey and hagfish): CNP-4 is the ancestral gene of the NP family. *Gen Comp Endocrinol* 148:4–47.

Lesser, M. P., F. H. Martini, and J. B. Heiser. 1997. Ecology of the hagfish, *Myxine glutinosa* L in the Gulf of Maine. I. Metabolic rates and energetics. *J Exp Mar Biol Ecol* 208:215–225.

Martini, F. H. 1998. The ecology of hagfishes. In J. M. Jørgensen, J. P. Lomholt, R. E. Weber, and H. Malte (eds), *The Biology of Hagfishes*. Chapman & Hall, London, pp. 57–77.

Nilsson, G. E. 1983. *Autonomic Nerve Function in the Vertebrates*. Springer Verlag, Berlin.

Perry, S. F., R. Fritsche, and S. Thomas. 1993. Storage and release of catecholamines from the chromaffin tissue of the Atlantic hagfish *Myxine glutinosa*. *J Exp Biol* 183:165–184.

Perry, S. F., B. Vulessevic, M. Braun, and K. M. Gilmour. 2009. Ventilation in Pacific hagfish (*Eptatretus stoutii*) during exposure to acute hypoxia or hypocapnia. *Respir Physiol Neurol* 167:227–234.

Rasmussen, A.-S., A. Janke, and U. Arnason. 1998. The mitochondrial DNA molecule of the hagfish (*Myxine glutinosa*) and vertebrate phylogeny. *J Mol Evol* 46:382–388.

Sandblom, E. and M. Axelsson. 2011. Autonomic control of circulation in fish: A comparative view. *Auton Neurosci Basic Clin* 165:127–139.

Satchell, G. H. 1971. *Circulation in Fishes*. Cambridge University Press, Cambridge.

Satchell, G. H. 1984. On the caudal heart of *Myxine* (Myxinoidea). *Acta Zool* 65:125–133.

Satchell, G. H. 1986. Cardiac function in the hagfish, *Myxine* (Myxinoidea: Cylostomata). *Acta Zool* 67:115–122.

Seth, H., M. Axelsson, H. Sundh, K. Sundell, A. Kiessling, and E. Sandblom. 2013. Physiological responses and welfare implications of rapid hypothermia and immobilisation with high levels of CO_2 at two temperatures in Arctic char (*Salvelinus alpinus*). *Aquaculture* 402–403:146–151.

Thomas, M. J., B. N. Hamman, and G. F. Tibbits. 1996. Dihydropyridine and ryanodine binding in ventricles from rat, trout, dogfish and hagfish. *J Exp Biol* 199:1999–2009.

Wilson, C. M. and A. P. Farrell. 2013. Pharmacological characterization of the heartbeat in an extant vertebrate ancestor, the Pacific hagfish, *Eptatretus stoutii*. *Comp Biochem Physiol* 164A:258–263.

Wilson, C. M., J. A. W. Stecyk, C. S. Couturier, G. E. Nilsson, and A. P. Farrell. 2013. Phylogeny and effects of anoxia on hyperpolarization-activated cyclic nucleotide-gated channel gene expression in the heart of a primitive chordate, the Pacific hagfish (*Eptatretus stoutii*). *J Exp Biol* 216:4462–4472.

Wright, G. M. 1984. Structure of the conus arteriosus and ventral aorta in the sea lamprey *Petromyzon marinus*, and the Atlantic hagfish *Myxine glutinosa*: Microfibrils, a major component. *Can J Zool* 62:2445–2456.

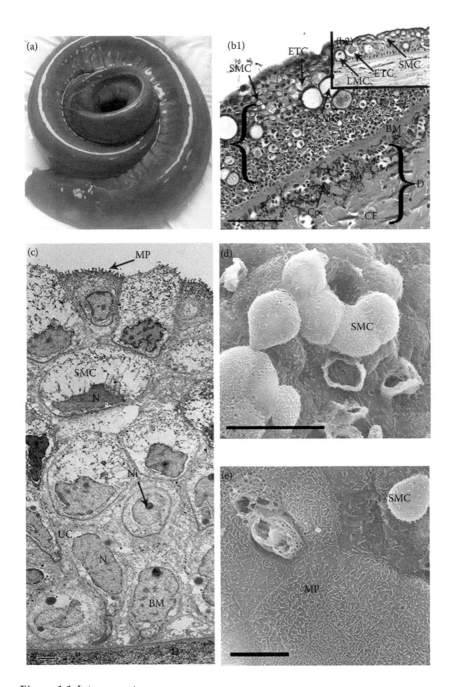

Figure 1.1 Integument.

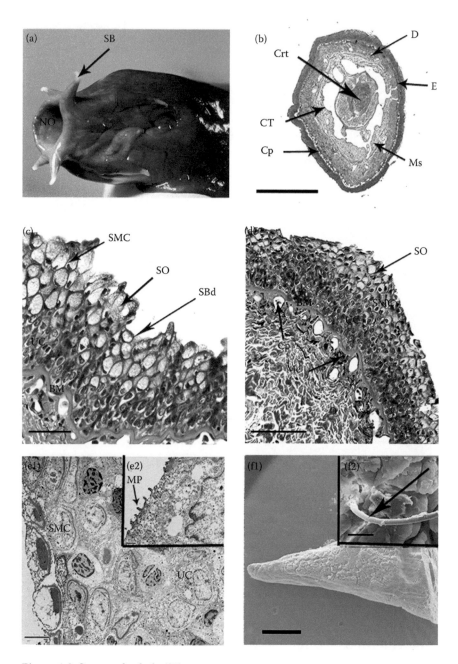

Figure 1.2 Sensory barbels (SB).

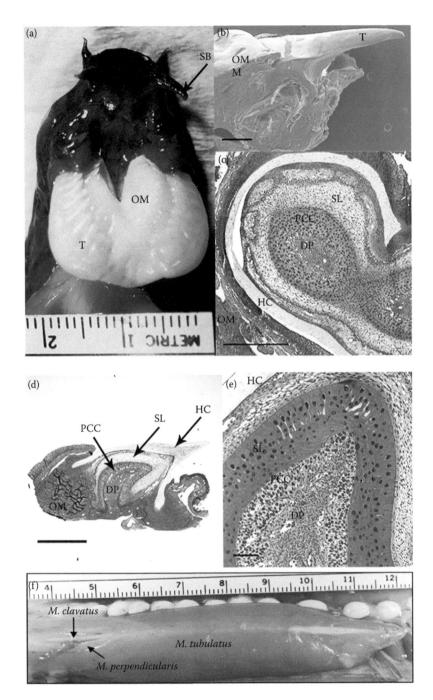

Figure 1.3 Dental plate.

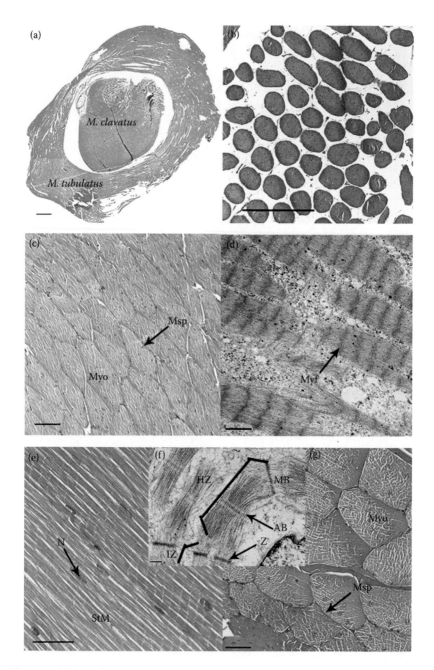

Figure 1.4 Musculature.

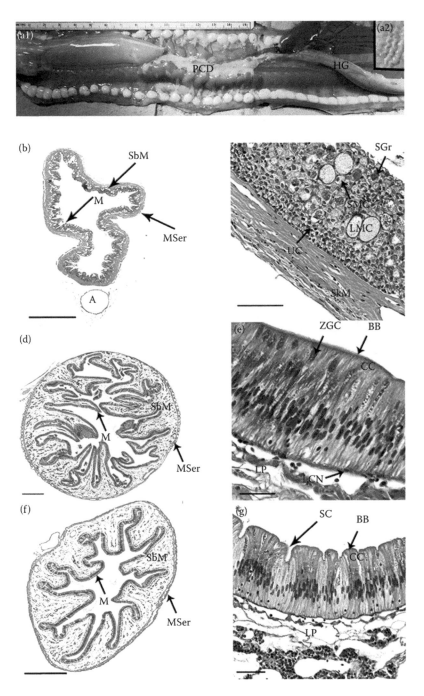

Figure 1.5 Alimentary canal.

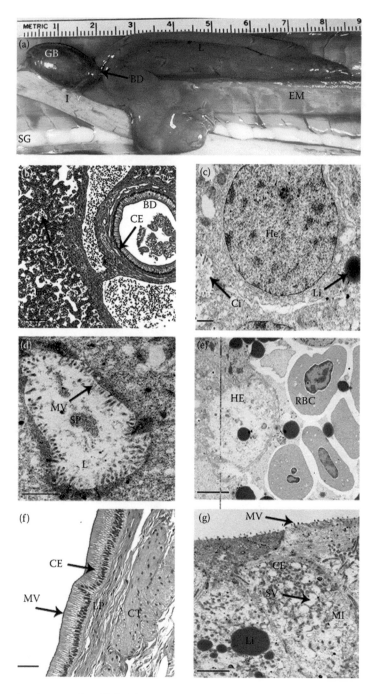

Figure 1.7 Liver and gall bladder.

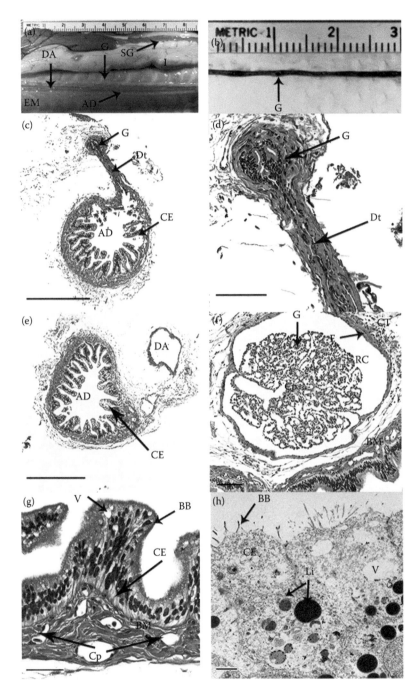

Figure 1.8 Archinephric duct (AD).

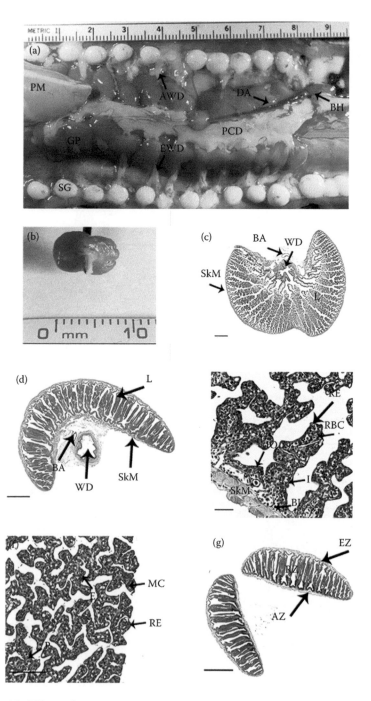

Figure 1.9 Gill pouches.

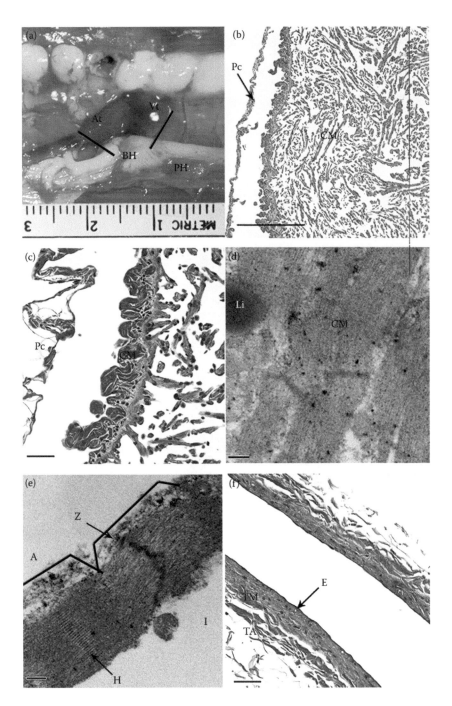

Figure 1.11 Cardiac tissues.

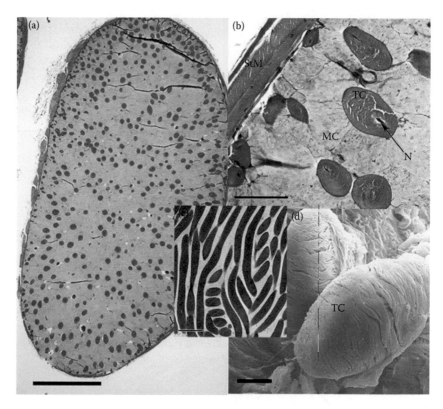

Figure 1.12 Slime glands (SG). (a) Sagittal section of a whole SG. H&E stain. Scale bar = 1 mm. (b) The components of the SG. An outer skeletal muscle (SkM) contracts to expel the inner mucin vesicles (MC) and nucleated (N) thread cells (TC). H&E stain. Scale bar = 100 μm. (c) Transmission electron micrograph detailing the coiling nature of the TC within the SG. Scale bar = 2 μm. (d) Scanning electron micrograph of the tightly coiled TCs. Scale bar = 20 μm.

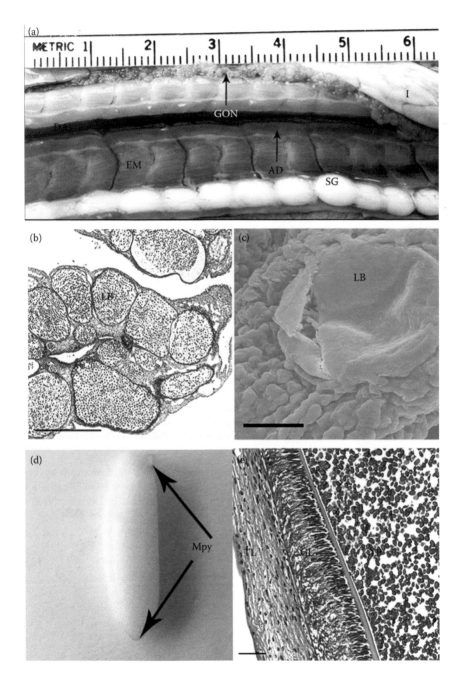

Figure 1.13 Reproductive organs.

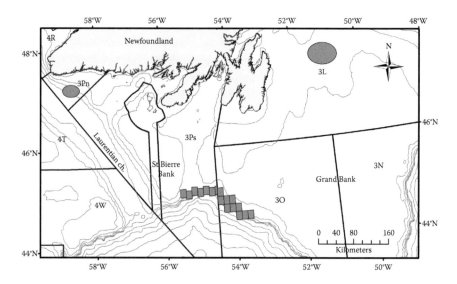

Figure 2.1 Map showing location of Atlantic hagfish fixed grid research survey and commercial fishing areas in NAFO Division 3O and Subdivision 3Ps (shaded squares) and location of exploratory Atlantic hagfish surveys in Division 3L and Subdivision 3Pn (shaded ellipses).

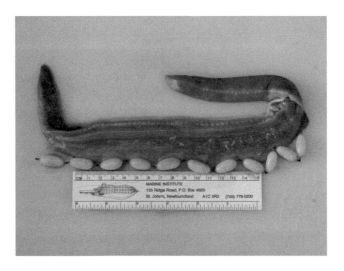

Figure *2.4* Photograph of dissection of female Atlantic hagfish illustrating a chain of mature eggs ovulated from their follicles in preparation for spawning.

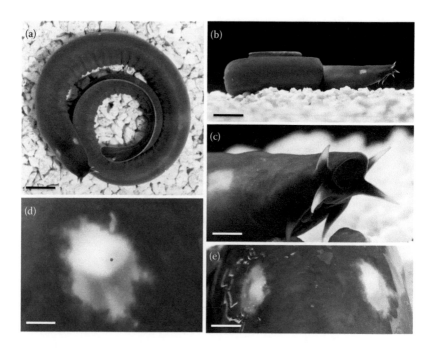

Figure 5.1 Series of photographs of both the Pacific hagfish, *E. stoutii* (a through c), and the New Zealand hagfish, *E. cirrhatus* (d and e). Note the bilateral position of the opaque patches of unpigmented skin overlying the "eyes." Note in *E. cirrhatus* that the skin overlying the eyes is relatively clear, revealing a circular gap in the surrounding adipose tissue (aperture signified by the asterisk in d). Scale bars, 2 cm (a and b), 5 mm (c), 0.5 mm (d), and 1.5 mm (e). Photographs by S. E. Temple and S. P. Collin (a through c) and S. P. Collin and T. D. Lamb (d and e).

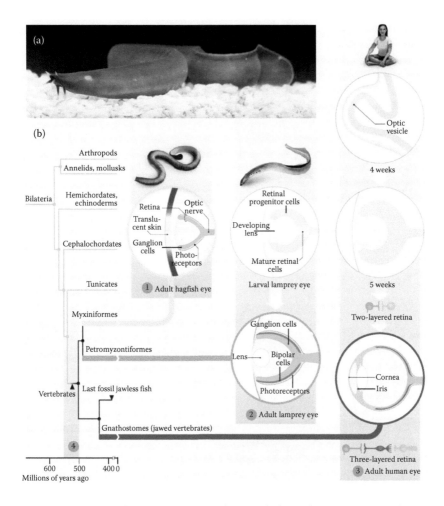

Figure 5.4 (a) Lateral view of *E. stoutii*. (b) Morphological perspective on the evolution of the vertebrate eye drawing on aspects of embryonic development, which may reflect events that occurred during the evolution of that lineage. 1. The hagfish eye which appears degenerate and probably serves to detect light for modulating circadian rhythms. 2. The lamprey eye in early development (ammocoete stage) resembles the hagfish eye before it develops into an image-forming eye complete with an iris and lens. 3. The eye of gnathostomes (including humans) passes through a developmental stage in which the retina comprised two layers before a third layer of cells emerges in the adult phase (see text for further discussion). (Reproduced from Lamb (2011) with permission. Copyright (2011) Scientific American, a Division of Nature America, Inc. All rights reserved. Photograph in (a) by S. E. Temple and S. P. Collin.)

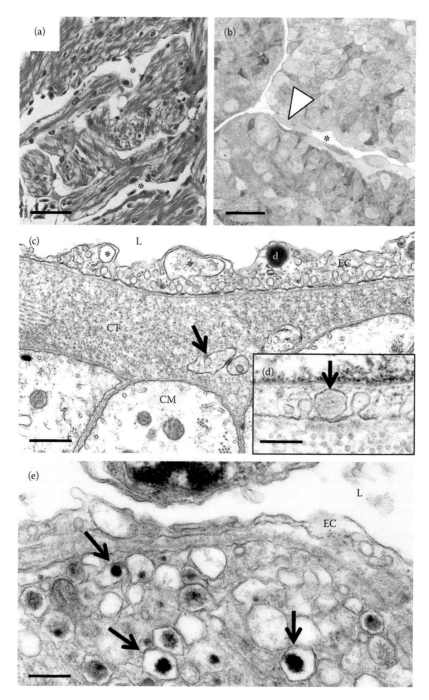

Figure 7.3 Endothelium of the branchial heart in *M. glutinosa*.

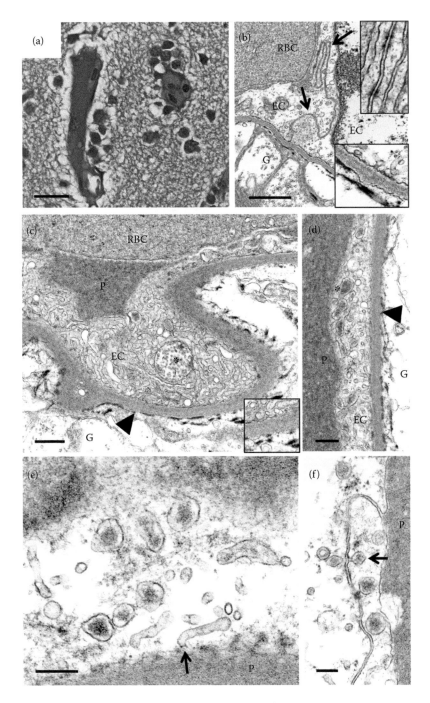

Figure 7.6 Endothelium of brain capillaries in *M. glutinosa*.

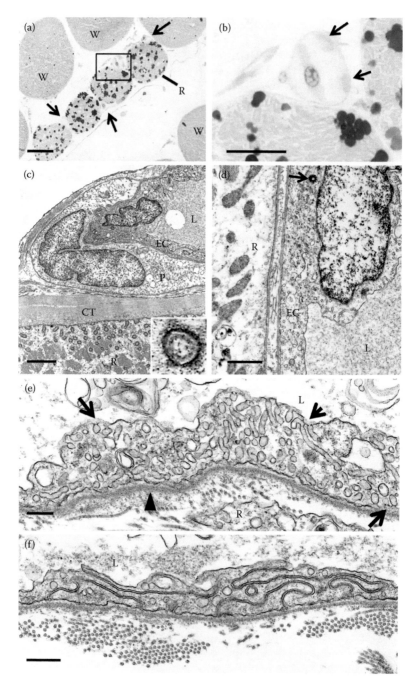

Figure 7.7 Endothelium of skeletal muscle capillaries in *M. glutinosa*.

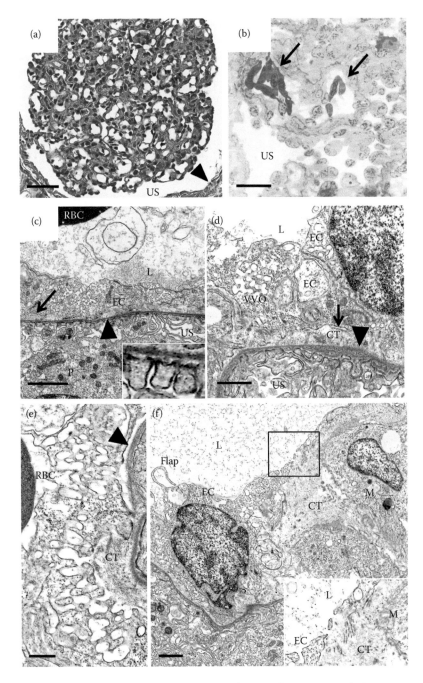

Figure 7.10 Endothelium of kidney glomerular capillaries in *M. glutinosa*.

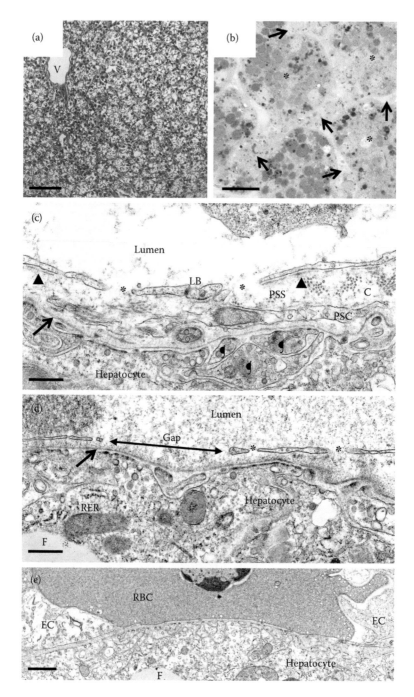

Figure 7.12 Endothelium of liver sinusoids in *M. glutinosa*.

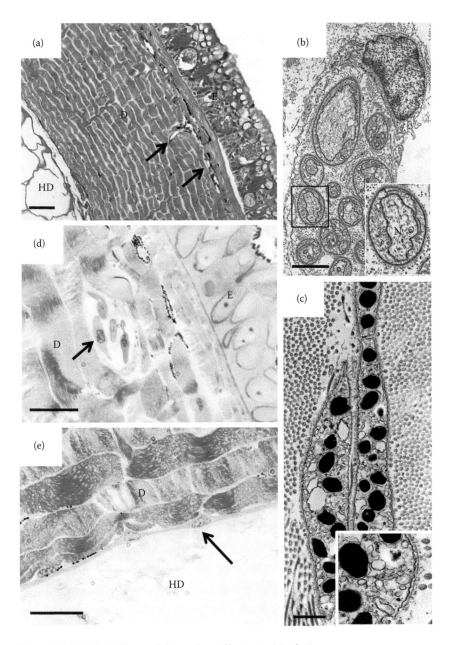

Figure 7.14 Endothelium of dermal capillaries in *M. glutinosa*.

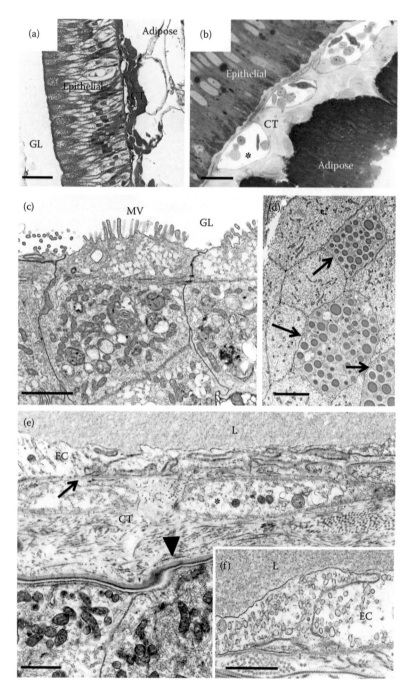

Figure 7.17 Endothelium of intestinal capillaries in *M. glutinosa*.

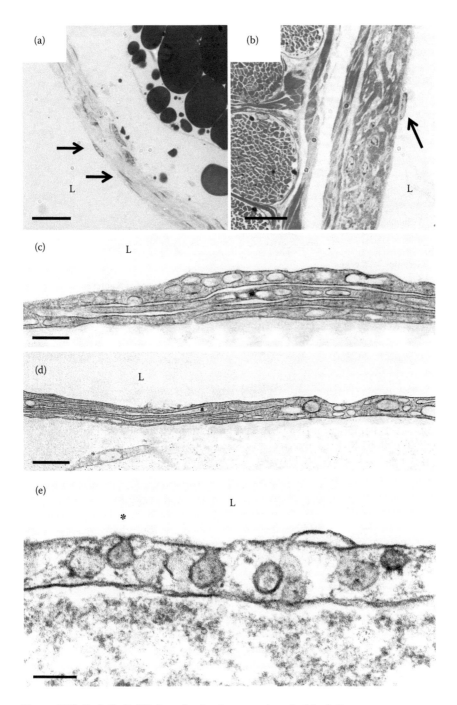

Figure 7.18 Endothelial lining of subcutaneous sinus in *M. glutinosa*.

(a)

Assembled *VLR* gene

(b)

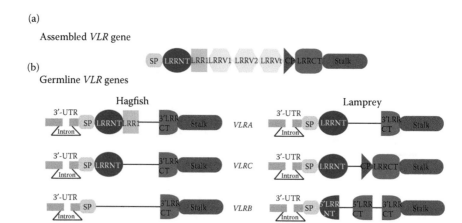

Germline *VLR* genes

Figure 8.2 Lamprey and hagfish *VLR* genes. (a) Assembled "mature" VLR: SP, LRRNT, first LRR1, two variable VLR (LRRV), end LRRV (LRRVt), connecting peptide (CP), LRRCT, stalk region. (b) Hagfish and lamprey germline *VLRA*, *VLRC*, and *VLRB* genes. Lamprey *VLRB* has two large noncoding intervening sequences separating the 5′ LRRNT and 3′ LRRCT, whereas all other *VLR* genes have a single, short intervening sequence (not drawn to scale).

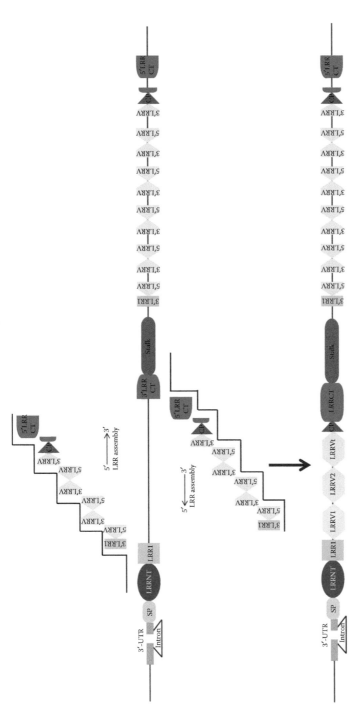

Figure 8.3 VLR gene assembly. The incomplete germline *VLR* genes are flanked by hundreds of donor LRR cassettes. The noncoding intervening sequence between portions of the LRRNT and LRRCT is replaced by LRR fragments that are sequentially copied from randomly selected donor LRR cassettes. The assembly process can be initiated from either the LRRNT or LRRCT end in a stepwise manner that is directed by short sequence homology between the donor and recipient LRR sequences.

VLRA   VLRC   VLRB

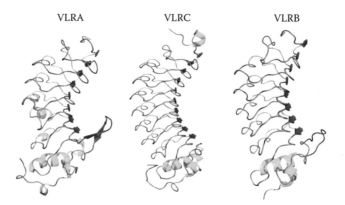

Figure 8.5 Structural comparison of hagfish VLR isotypes. The hagfish VLR proteins are crescent-shaped. Their concave surface is composed of β-strands contributed by the LRRNT, LRR1, variable numbers of LRRV modules, and LRRCP; the LRRCT of VLRA (predicted) and VLRB (PDBID: 2O6S) contain a variable insert loop. VLRC (PDBID: 2O6Q) are structurally similar to VLRA and VLRB, except VLRC does not have an LRRCT insert. β-Sheet and α-helix structures are shown in blue and green colors, respectively. Loops located in the LRRCT portion are indicated in red. (Modified from Li, J. et al. 2013. *Proc Natl Acad Sci USA* 110:15013–15018.)

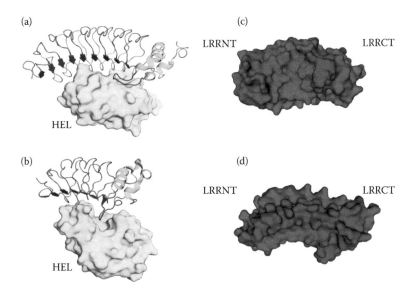

Figure 8.6 VLR antigen-binding sites. (a) Ribbon diagram of lamprey VLRA.R.2.1-HEL complex (PDBID: 3M18). Lamprey VLRA.R.2.1 uses its multiple β-strands on its concave surface and LRRCT loop to bind HEL. (b) Structure of VLRB.2D-HEL complex (PDBID: 3G3A). Note that the LRRCT loop of VLR.2D is inserted into the catalytic cleft of HEL. Upper VLR is shown in ribbon diagram, in which β-sheet and α-helix structures are shown in blue and green colors, respectively. Loops located in the LRRCT portion are indicated in red. HEL is depicted in light blue. (c) Front view of surface diagrams of predicted antigen-binding sites of hagfish VLR, the highly variable residues in the LRR modules labeled in magenta form the binding patch on the concave surface; the protruding loop contributed by the LRRCT insert is labeled in red. (d) Side view of surface diagrams of predicted antigen-binding sites of hagfish VLR, noting the protruding loop.

Figure 9.5 Brown hagfish, *Paramyxine atami*.

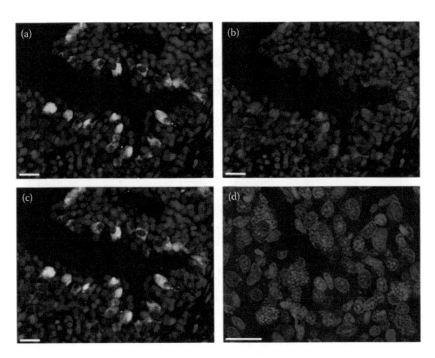

Figure 11.3 VHA- and NKA-rich cells in hagfish gills. Dual immunolocalization of VHA (a, green) and NKA (b, red), with VHA and NKA immunoreactivity found in the same gill cells (c). At higher magnification with optical sectioning, NKA immunoreactivity was visible as punctate staining throughout the edge of cell, indicative of NKA localization along the basolateral membrane (d). Nuclei in blue. Scale bar: 20 μm.

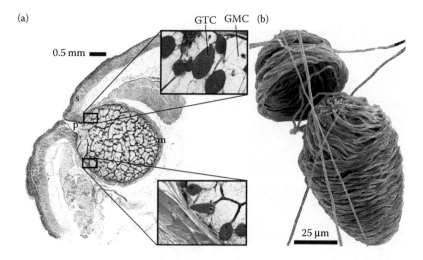

Figure 13.2 Slime gland histology and skein structure. (a) H&E staining of a slime gland section from *E. stoutii* reveals two main cell types in the gland, GTCs and GMCs. The slime gland is located close to the skin (s) and is surrounded by a thin layer of muscle (m), which triggers slime exudate release when it contracts and forces GTCs and GMCs through the narrow gland pore (p). (b) GTCs lose their plasma membrane during holocrine secretion and release a single, elaborately coiled slime thread bundle, called a skein. This skein, viewed with SEM, has cracked open, revealing the staggered arrangement of loops (red) that spiral around the skein and form higher-order structures called conical loop arrangements (purple). (Adapted from Winegard, T., J. Herr, C. Mena, B. Lee, I. Dinov, D. Bird, M. Bernards, Jr. et al. 2014. *Nat Commun* 5:3534.)

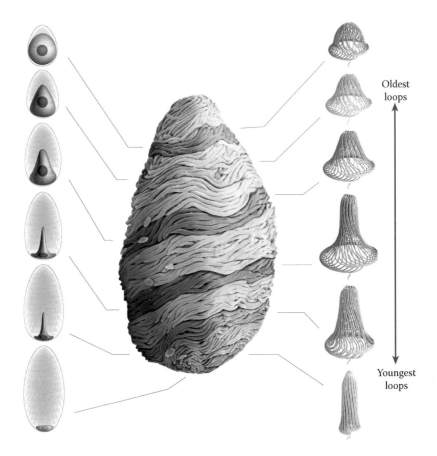

Oldest
loops

Youngest
loops

Figure 13.3 Temporal and spatial model of thread assembly and coiling in GTCs. GTC growth and maturation is characterized by dramatic changes in nuclear size and morphology (left), which correspond with the shape of conical loop arrangements laid down during successive stages of thread production (right). Staggered loops are laid down in the space defined by previous loops and the apical surface of the nucleus, with their morphology changing as the nucleus becomes more spindle-shaped and retreats toward the basal end of the cell. The result is a mature thread skein that can be ejected through the gland pore and unravel to its full extended length of ~150 mm without tangling. (Adapted from Winegard, T., J. Herr, C. Mena, B. Lee, I. Dinov, D. Bird, M. Bernards, Jr. et al. 2014. *Nat Commun* 5:3534.)

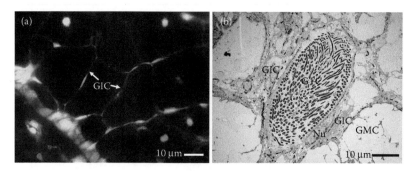

Figure 13.5 Fluorescence and TEM images of GICs within the slime gland of *M. glutinosa*. (a) Nuclei were stained blue with DAPI and IFs were stained green with a pan-cytokeratin antibody. (b) TEM of a slime gland reveals a slender GIC with an elongated nucleus (Nu) wedged between a GTC and a GMC. In the upper left is likely another GIC sandwiched between the GTC and another GMC.

chapter seven

Endothelium in hagfish

Ann M. Dvorak and William C. Aird

Contents

Introduction

The endothelium, which forms the inner lining of vertebrate blood vessels and lymphatics, is a systemically distributed organ. The endothelium was once considered to be little more than an inert layer of nucleated cellophane. However, since the 1950s, it has become increasingly clear that endothelial cells participate in a multitude of physiological functions, including hemostasis, permeability, leukocyte trafficking, innate and acquired immunity, and vasomotor tone.

One feature of the endothelium that is often overlooked is its rich diversity of regional and organ-specific phenotypes. In mammals, endothelial cell heterogeneity has been described at the level of cell morphology, function, gene expression, and antigen composition (Aird, 2007a, b). Endothelial cell phenotypes vary between different organs, between different segments of the vascular loop within the same organ, and between neighboring endothelial cells of the same organ and blood vessel type.

In vivo proteomic approaches have revealed a striking array of vascular bed-specific phenotypes. For example, antibody and subfractionation strategies have been employed to generate monoclonal antibodies that specifically target the caveolae in one vascular bed or another (McIntosh et al., 2002). Others have used phage display peptide libraries to select for peptides that home to specific vascular beds *in vivo* (Pasqualini and Ruoslahti, 1996; Arap et al., 2002). These latter studies have uncovered a vascular address system that allows for site-specific targeting of biologically active compounds, for example, to the endothelial lining of tumor blood vessels (Arap et al., 1998, 2002).

Hagfish and lamprey are often grouped together as the cyclostomes. They are also referred to as agnathans or jawless fish. They lack a hinged jaw, and their skeleton is not mineralized. They are the sole survivors of the agnathan stage in vertebrate evolution and are the closest extant outgroups to all jawed vertebrates (gnathostomes). The immediate nonvertebrate relatives of hagfish and lamprey are the Urochrodata (tunicates) and Cephalochordata (amphioxus). There is uncertainty about the phylogenetic position of hagfish relative to lamprey. Morphological data suggest that hagfish and lamprey are not monophyletic, but rather represent separate divergences, with hagfish being more primitive. On the other hand, certain molecular data favor a sister group relationship between hagfish and lamprey (Heimberg et al., 2010). In either case, the last common ancestor of hagfish and jawed vertebrates [cartilaginous fish (also termed chondrichthyes) and the bony fish (also termed osteichthyes), amphibians, reptiles, birds, and mammals] was also the last common ancestor of all extant vertebrates. Features of endothelial cells that are shared between hagfish and gnathostomes can be inferred to have already been present in this ancestral vertebrate.

Endothelial cells are absent in invertebrates, cephalocordates, and tunicates, but are present in the three major groups of extant vertebrates: hagfish, lampreys, and jawed vertebrates (gnathostomes) (Monahan-Earley et al., 2013). The fact that the endothelium is shared by jawless and jawed vertebrates is evidence that the endothelium was present in the ancestor of these animals. The absence of an endothelium in cephalochordates and tunicates suggests that this structure evolved after the divergence of these groups from the vertebrate lineage, between 540 and 510 million years ago. Previous studies have reported the ultrastructural properties of blood vessels in single organs from hagfish. More recently, we demonstrated that the endothelium of hagfish displays ultrastructural, molecular, and functional heterogeneity across different organs (Feng et al., 2007; Yano et al., 2007). These data suggest that the heterogeneity of the endothelium evolved as an early feature of this cell lineage. In other words, endothelial cell heterogeneity is not simply a descriptor for multiple properties of the endothelium. Rather, endothelial cell heterogeneity is in and of itself a core property of this cell lineage.

We have since carried out a more systematic survey of endothelial ultrastructure in the hagfish, *Myxine glutinosa.** The results, which are presented in this chapter, demonstrate that the extent of phenotypic heterogeneity is even greater than previously appreciated. The data suggest that the endothelium in hagfish and, by extension, the ancestral vertebrate is/was uniquely and specifically adapted to the local needs of the underlying tissue.

Overview of the hagfish cardiovascular system

In this section, we provide a brief overview of the hagfish cardiovascular system. For further details, the reader is referred to Chapter 6 and to several excellent reviews (Johansen, 1960, 1963; Hardisty, 1979; Satchell, 1991; Farrell, 2007). Hagfish, like all vertebrates, have a closed circulation in which blood is confined to spaces lined by an endothelium (Figure 7.1). Similar to other fishes, the branchial (i.e., gill) circulation is in series with the systemic circulation. The arterial system follows the usual pattern observed in other fishes with some notable exceptions that arise from the lack of paired fins, the presence of degenerate eyes, and the unique arrangement of the gill pouches. Moreover, in the branchial region, a single median anterior aorta exists along with the paired anterior aortae (also called common carotid arteries). Caudally, these three vessels fuse to form the common dorsal aorta (also called the posterior or abdominal aorta). The venous system shows the typical craniate pattern of paired anterior and posterior cardinal veins. The posterior cardinal veins are formed by the bifurcation of the caudal vein anterior to the cloaca. The left posterior and left anterior cardinal veins are larger than their right-sided counterparts. The left anterior cardinal vein enters the sinus venosus via a connection with the left posterior cardinal vein (at the point where the common posterior cardinal passes into the sinus venosus) (Figure 7.2). The right anterior cardinal vein, which communicates with the left cardinal vein through the cardinal anastomosis, drains into the portal heart.

The hagfish cardiovascular system has a number of specializations that distinguish it from other vertebrates. First, its pumping mechanism is decentralized, consisting of a number of pulsatory structures. These include two myogenic hearts that contain cardiac muscle (the branchial or systemic heart and the portal heart) and two pairs of nonmyogenic hearts comprised of skeletal muscle (the caudal and cardinal hearts). Satchel regards the systemic, portal, and caudal hearts as true hearts

---

* We employed materials and methods as previously described in Yano et al. (2007). Owing to space restraints, we have omitted a description of certain vascular beds, most notably that of the gill. For details of the branchial circulation and its endothelial lining, see Elger (1987).

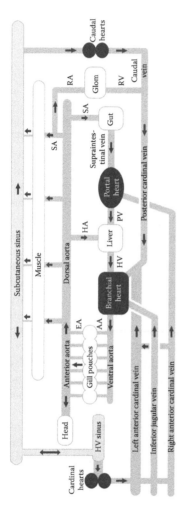

Figure 7.1 Schematic representation of the closed circulation in hagfish. The branchial heart sends deoxygenated blood to the gills via the ventral aorta and afferent branchial arteries (AA). Oxygenated blood from the gills travels from the efferent branchial arteries (EA) to the paired anterior aortae (only one is shown), which anastomose with a median anterior aorta (not shown) to form the dorsal aorta. The dorsal aorta runs posteriorly, sending off branches (segmental arteries, SA) that supply the abdominal gut and the body wall musculature (representative vessels are shown). Renal arteries (RA), which arise from the segmental arteries, supply kidney glomeruli (Glom) (one vessel feeding a single glomerulus is shown). The dorsal aorta ultimately gives rise to the caudal artery (not shown), followed by a capillary bed and the caudal vein. The hepatic artery (HA) arises from the left celiac artery, which in turn is a branch of the dorsal aorta. Venous return from the head region occurs through paired anterior cardinal veins and the inferior jugular vein. The left anterior cardinal vein (which is larger than the right) communicates with the sinus venosus of the branchial heart via the posterior cardinal vein, whereas the right anterior cardinal vein drains into the left cardinal vein and the portal heart. Caudally, blood returns to the sinus venosus of the branchial heart through the caudal vein and paired posterior cardinal veins (only one is shown). Renal veins (represented as a single vessel, RV) drain into the posterior cardinal vein. Venous return from the gut involves passage through the supraintestinal vein, portal heart, common portal vein (PV), liver, anterior and posterior hepatic veins (HV), and finally the sinus venosus of the branchial heart. The hagfish contains many sinuses, the largest of which is the subcutaneous sinus. The caudal hearts pump blood from the caudal end of the sinus into the caudal vein. At the rostral end, the subcutaneous sinus is connected with the hypophysio-velar sinus (HV sinus), which drains into the anterior cardinal veins via the cardinal hearts. Blood flow into the subcutaneous sinus occurs through capillaries at the lateral surface of parietal muscle.

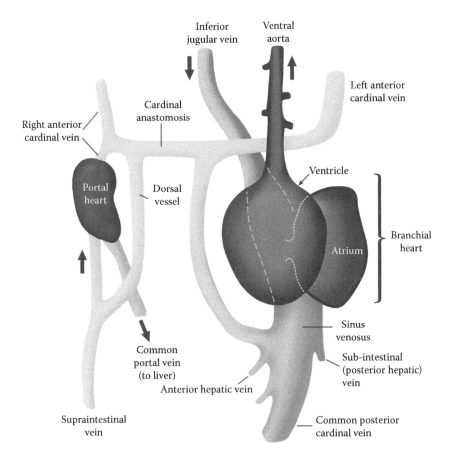

Figure 7.2 Schematic representation of the branchial and portal hearts in hag-fish. The branchial heart is asymmetric, with the atrium and the sinus venosus displaced to the left of the ventricle. Several veins drain into the sinus venosus including the inferior jugular vein, the anterior and posterior hepatic veins, and the common posterior cardinal vein. The left anterior cardinal vein, which is larger than the right, communicates with the branchial heart through the left posterior cardinal vein. The right anterior cardinal vein communicates with the left anterior cardinal vein through the cardinal anastomosis. The portal heart, which contains a single chamber, receives blood from the right anterior cardinal vein and the supraintestinal vein. The dorsal vessel is an anastomosing vessel between the right anterior cardinal vein and supraintestinal vein (in some cases, it connects the right anterior cardinal vein with the anterior section of the right posterior cardinal vein). The portal heart pumps blood to the liver via the common portal vein, which bifurcates to supply the two lobes of the liver. (From Forster, M. E. 1991. *J Exp Biol* 156:583–590.)

because their muscles serve only to power the pump (Satchell, 1991). In contrast, the cardinal heart is considered a propulsor, as it is powered incidentally by skeletal muscles that serve other functions (i.e., production of respiratory movements of the velum). A second unique feature of the hagfish cardiovascular system is the presence of a series of interconnecting blood-containing sinuses, which are in direct communication with systemic vessels (Cole, 1925; Forster, 1991). The most prominent of these is the large subcutaneous vascular sinus located between skeletal muscle and the skin, stretching from the tentacles of the snout to the caudal fin fold. Hagfish can hold up to 30% of their blood volume within the sinus system (Forster, 1997). Consequently, hagfish have higher relative blood volumes compared with other vertebrates.

Hagfish maintain the lowest arterial blood pressure (6–10 mmHg in dorsal aorta) and operate with the lowest afterload of any vertebrate (Foster and Forster, 2007). As a result, myocardial power output of the hagfish heart is also low. Cardiac output in *M. glutinosa* is 8–9 mL/min/kg, with maximal levels reaching 24–26 mL/min/kg (Axelsson et al., 1990; Cox et al., 2010), which approaches the values seen in some teleosts. The estimated basal circulation time is 12 min for *M. glutinosa* and 6 min for *Eptatretus cirrhatus*, approximately fivefold slower compared with teleosts. Based on these features, hagfish have been characterized as having a low-pressure, moderate-flow system (Forster, 1991).

Heart endothelium

The branchial or systemic heart of the hagfish is enclosed in a large, thin-walled sac. The nonrigid pericardium is connected with the perivisceral/abdominal coelom. Unlike other vertebrates, the heart lacks extrinsic neuronal innervation (Augustinsson et al., 1956; Bloom et al., 1961). It consists of a sinus venosus, atrium, and ventricle, with the sinus venosus and atrium displaced well to the left (Figure 7.2). Double-flap valves are present between the sinus venosus and atrium and between the atrium and ventricle. Unlike other fishes, the heart lacks a conus or bulbus arteriosus (i.e., a ventricular outflow track within the pericardium bearing valves). Rather, a paired semilunar valve occurs at the exit from the ventricle. The absence of a well-formed, elastin-containing conus arteriosus in hagfish is consistent with the observation that there is zero flow in the ventral aorta during diastole (Axelsson et al., 1990; Cox et al., 2010) and suggests that there is little need to dampen the small pressures generated by the heart. The heart wall consists of trabeculae of cardiomyocytes surrounded by blood sinuses or lacunae, which in turn are lined by endothelial cells (Figures 7.3a and b). The trabecular pattern promotes the direct nourishment of the myocardium by the blood. The abundance of deep crevices and lacunae between the numerous trabeculae gives the heart a loosely

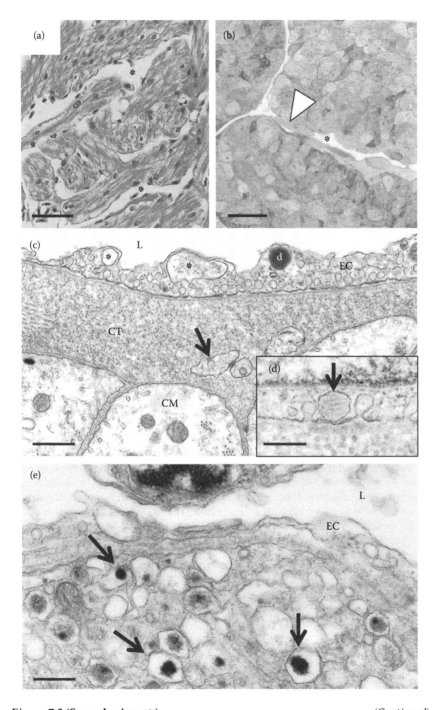

Figure 7.3 **(See color insert.)** *(Continued)*

woven and spongy appearance, as distinct from the compact myocardium observed in many other vertebrates (Bloom et al., 1961; Leak, 1969). There is no arterial blood supply (i.e., coronary circulation). As a result, the heart is supplied solely by mixed venous blood from the sinus venosus that percolates in the intertrabecular spaces. In a previous study of *E. cirrhatus*, the PO_2 of mixed venous blood was only 17.2 mmHg and even lower (3.5 mmHg) in swimming fish (Wells et al., 1986).

Several other factors conspire to create a profoundly hypoxic environment in the hagfish heart. For example, hagfish have a low hemoglobin concentration [e.g., 1.61 ± 0.15 g/dL in one study of *E. cirrhatus* (Wells et al., 1986) and 4.1 ± 0.5 g/dL in *M. glutinosa* (Perry et al., 1993)], which, while reducing viscosity and hence total peripheral resistance and work of the heart, contributes to a low oxygen-carrying capacity of blood (<1 mL/O_2/100 mL blood, compared with 10.7 mL/O_2/100 mL blood for goldfish) (Sidell, 1983). Moreover, hagfish blood has a relatively high affinity for oxygen, with its monomeric hemoglobin exhibiting poor co-operativity and a weak Bohr effect. As a result, P_{50} values are low (4.2–12.9 mmHg PO_2 in *M. glutinosa* and approximately 12 mmHg in *E. cirrhatus*), further limiting oxygen supply to the heart (Bauer et al., 1975; Wells et al., 1986; Perry et al., 1993).

Hagfish maintain an impressive 70% of their normoxic cardiac performance during prolonged anoxia (Cox et al., 2010). Hypoxia tolerance may be explained on two levels. First, power requirements are very low,

---

Figure 7.3 (Continued) Endothelium of the branchial heart in *M. glutinosa*. (a) 5-μm histological section through the ventricle stained with H&E shows trabeculae of cardiomyocytes surrounded by blood sinuses (*), which contain blood cells and are lined by endothelium. (b) 1-μm histological section through the ventricle stained with Giemsa shows a blood sinus (*) containing a red blood cell (arrowhead). (c) Electron micrograph of the ventricle shows a sinus (L, lumen) lined by an endothelial cell (EC). Note the abundant caveolae, most of which open onto the abluminal side of the cell in this image. The opening of many of these caveolae is subtended by a diaphragm. The endothelial cell also contains larger vesicles (*) and a dense body (d), and it rests on a well-formed basal lamina. Underneath the basal lamina is a thick zone of connective tissue (CT) that separates the endothelial cell from the underlying cardiomyocytes (CM). Cellular processes are observed in this space (arrow). These may represent cellular projections of branching chromaffin cells and/or cardiac myocytes. (d) Electron micrograph of an endothelial cell lining a sinus in the ventricle shows caveolae opening on the tissue front of the cell. An immature Weibel–Palade body is also observed (arrow). The basal lamina in this field is not as continuous as it is in (d). (e) Electron micrograph of the ventricle shows a sinus (L, lumen) lined by a portion of an endothelial cell (EC). Lying under the endothelium is a chromaffin-like cell containing dense granules of varying size and shape, consisting of a dense core surrounded by an electron-lucent halo. Granules containing large halos and eccentrically placed dense cores (arrows) are morphologically similar to noradrenalin-containing cells in mammals. Scale bar: 50 μm (a), 20 μm (b), 800 nm (c), 200 nm (d), and 300 nm (e).

which minimizes cardiac adenosine triphosphate demand. Second, the heart has a high anaerobic capacity (Sidell, 1983; Forster, 1991; Forster et al. 1991). Indeed, the heart of *E. cirrhatus* has been shown to fuel up to 50% of its maximum power through anaerobic glycolysis when perfused *in vitro* (Forster, 1991).

The hagfish heart has been shown to store catecholamines in specific cells, which show a resemblance to chromaffin granule-containing cells in the adrenal medulla of mammals (Ostlund et al., 1960; Bloom et al., 1961; Leak, 1969). These highly branched noninnervated cells are located just beneath the trabecular endothelium and are evenly distributed through-out the ventricle and atrium of the branchial heart and throughout the portal heart (Bloom et al., 1961). Previous studies have demonstrated that the portal heart and atrium contain mainly noradrenaline, whereas the ventricle stores primarily adrenaline (Ostlund et al., 1960; Bloom et al., 1961). Chromaffin cells are also found within the walls of the posterior cardinal vein, where they contain predominantly noradrenaline (Perry et al., 1993). The catecholamines in the hagfish heart have a tonic influ-ence on cardiac performance and are released during acute stress, per-haps helping to "kick start" the heart at times of critically low venous return (Bloom et al., 1961; Axelsson et al., 1990; Johnsson and Axelsson, 1996). Other studies have shown that blood flow through the gills of hag-fish is under tonic control by catecholamines (Forster et al., 1992; Sundin et al., 1994). As the gills lack autonomic innervation, it has been suggested that this effect is mediated by heart-derived blood-borne catecholamines (Sundin et al., 1994).

We, and others, have previously described the ultrastructure of endo-thelial cells in the branchial heart (Leak, 1969; Yano et al., 2007). In our studies, the endothelium enveloping the trabeculae formed a thin, atten-uated layer around blood sinuses, with one side facing the lumen and the other side facing the muscle fibers (Figure 7.3c). The cells contained many vesicles and vacuoles (Figures 7.3c and d). On the abluminal side, the endothelium was surrounded by a thin basal lamina followed by a thick connective tissue stroma that contained bundles of collagen fibrils and finally the basal surface of cardiomyocytes, which was also covered by a basal lamina (Figure 7.3c).

Chromaffin-like cells were found in the connective tissue space between endothelial cells and cardiomyocytes, typically lying just below the endothelium. These cells contained many intensely osmiophilic mem-brane-bound granules of varying size and shape (Figure 7.3e). The gran-ules, in turn, contained a central core of high electron density surrounded by a halo of low electron density. In some cases, empty vesicles of similar size were observed. These may represent granules that have been emptied of their contents. The granular cells were surrounded by a basal lamina (Figure 7.3e). Finally, as previously reported (Ostlund et al., 1960; Bloom et al., 1961; Leak, 1969; Helle et al., 1972; Helle and Lonning, 1973), similar dense-cored granules were also observed in the cytoplasm of cardiomyocytes (not shown).

The portal heart, which contains a single chamber consisting of thin trabeculae, collects blood from the right anterior cardinal vein (which curiously allows for passage of blood from the head region through the liver) and supraintestinal vein (also called the portal vein) and pumps it to the liver via the common portal vein, which bifurcates to supply the two lobes of the liver (Figure 7.2). Each of the input and output vessels is protected by a valve. The portal heart beats asynchronously with the systemic heart, but has weaker contractions and lower output (mean pressures are approximately one-fifth of those generated by systemic heart). Like the systemic heart, the portal heart contains high levels of catecholamines and is also under tonic beta-adrenergic control (Johnsson and Axelsson, 1996). The portal heart presumably evolved to compensate for the extremely low venous pressures (a mere 0.5 mmHg in the supraintestinal vein) (Foster and Forster, 2007) and to overcome vascular resistance of the hepatic circulation. The ultrastructure of the endothelial lining of the portal heart was similar to that of the systemic heart (not shown).

Aortic endothelium

As in other vertebrates, the aortic wall of hagfish contains three distinct layers: an innermost endothelium and its underlying basement membrane (intima), a tunica media containing seemingly randomly oriented smooth muscle cells, and an adventitia of dense connective tissue (Wright, 1984). However, in contrast to other vertebrates, hagfish (and lamprey) lack elastin, defined either chemically or histologically (Sage and Gray, 1979, 1980; Wright, 1984). The ultrastructure of the ventral aorta in hagfish was previously reported (Wright, 1984). Here, we describe the fine structure of the dorsal aorta. The endothelium formed a continuous lining, and endothelial cells were generally thicker than those observed in the capillaries of other organs (Figure 7.4a). The cytoplasm contained abundant long tubules and caveolae, many of which opened onto the blood or tissue front (Figure 7.4a). In many cases, the opening of the caveolae and tubules was subtended by a diaphragm (Figure 7.4b). Underlying the endothelium was an incomplete basal lamina followed by a connective tissue layer of variable thickness that contained densely packed slender collagen fibrils. This layer contained patches of hyperdensity that have a similar appearance to elastin in higher vertebrates (Figures 7.4a and c). These foci, which were not described in electron microscopic studies of the ventral aorta (Wright, 1984), presumably contain elastin-like fibers. The media contained smooth muscle cells, which were characterized by the presence of cytoplasmic myofilaments and tubular smooth-surfaced vesicles (similar findings are shown for the

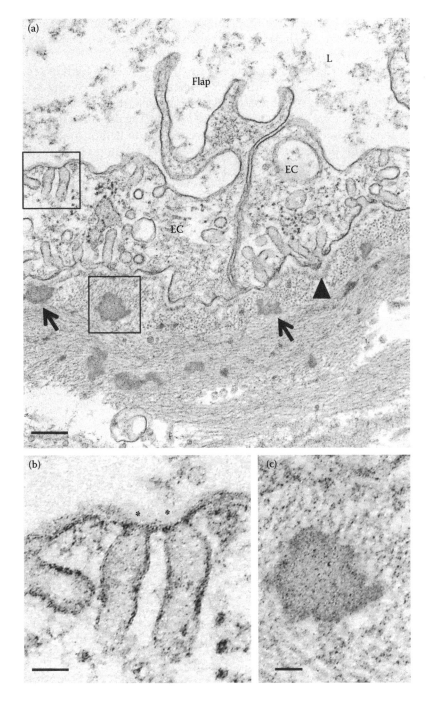

Figure 7.4

(Continued)

arterial wall in Figure 7.5). Smooth muscle cells were surrounded by a well-formed basal lamina. The adventitia consisted primarily of collagen fibers, in addition to fibroblasts and nerve bundles (not shown).

In other vertebrates, elastin provides arteries with a high degree of distensibility so that they can accommodate deformation during systole and provides elastic recoil during diastole. Indeed, the absence of elastin in hagfish may explain why there is zero flow during diastole. However, *in vitro* inflation experiments on hagfish ventral aorta demonstrated elastic behavior with properties of extensibility, hysteresis, and nonlinearity of the stress–strain curves (Davison et al., 1995). It has been suggested that the densely packed collagen fibrils in the tunica media are homologous to the elastic tissue found in the arterial walls of other vertebrates and may provide the blood vessel with compliance (Wright, 1984). The patches of hyperdense fibrils observed in the dorsal aorta, which bear striking similarity to elastin, are likely to have a similar function.

Arteriolar endothelium

Ultrastructural studies of small arteries and arterioles demonstrated the typical arrangement of an endothelium lining the lumen, an incomplete basal lamina, a connective tissue layer, and a layer of smooth muscle cells (Figure 7.5 shows an artery from parietal muscle). The endothelial cells resembled those of the aorta; they were thick and contained many tubules and caveolae. Like the aorta, the connective tissue matrix contained densely packed collagen fibrils. However, no elastin-like foci of hyperdensity were observed.

Brain capillary endothelium

The brain of hagfish is simplified when compared with jawed vertebrates, for example, there is no cerebellum. Previous studies have failed to demonstrate myelin in hagfish (Bullock et al., 1984). Moreover, the adult

---

Figure 7.4 (Continued) Endothelium of the dorsal aorta in *M. glutinosa*. (a) Electron micrograph of a transverse section of the aorta shows two endothelial cells (EC) separated by a lateral border that contains a flap. The cytoplasm possesses abundant tubular profiles, some of which open onto the blood or tissue fronts of the cells. The opening of many tubules is subtended by a diaphragm (see *, inset). The cytoplasm also contains polyribosomes. The endothelial cells rest on a discontinuous basal lamina (arrowhead). Underlying the endothelial cells and their basal lamina is a connective tissue layer that contains patches of hyperdensity (arrows) that have a similar appearance to elastin in higher vertebrates. The smooth muscle layer (not shown) lies just below the connective tissue zone. (b and c) Higher magnification of the boxed regions in (a). Scale bar: 200 nm (a) and 50 nm (b, c).

Figure 7.5 Endothelium of an arteriole in *M. glutinosa*. Electron micrograph of a transverse section of an arteriole from parietal muscle demonstrates a typical arrangement of endothelial cells (EC) lining the lumen (L), an abluminal layer of connective tissue (CT), followed by a layer of smooth muscle cells (SMC). Note that the bulk of the connective tissue layer consists of connective tissue fibrils that are considerably thinner than the cross-striated classical collagen fibrils observed next to the smooth muscle cell. Like the endothelial cell, the smooth muscle cell contains abundant vesicles and tubules. Scale bar: 250 nm.

hagfish has a greatly reduced ventricular system and an absent choroid plexus (Murray et al., 1975). Scanning electron microscopy of microvascular corrosion casts was previously used to characterize the vascularization of the hagfish brain (Cecon et al., 2002). Hagfish have a highly variable arterial supply and venous drainage of the brain. There is no arterial circle resembling the circle of Willis. The luminal diameter of capillaries in the brain was estimated to be 17–22 µm, which is larger than in other vertebrates (Cecon et al., 2002). There are more studies on the ultrastructure (and barrier function) of hagfish brain endothelium compared with any other vascular bed (Mugnaini and Walberg, 1965; Murray et al., 1975; Bundgaard et al., 1979; Bundgaard and Cserr, 1981a). The most consistent finding is the presence of many vesicles and tubules in capillary endothelial cells. This contrasts with the paucity of vesicles in the endothelium of mammalian brain microvasculature (Reese and Karnovsky, 1967).

In histological sections stained with hematoxylin and eosin (H&E), the lumen of brain capillaries was intensely eosinophilic (Figure 7.6a), a feature that was not observed in other organs. In electron microscopy, the plasma within these blood vessels was far more electron-dense compared with the luminal contents of other vascular beds (Figure 7.6c). The capillary walls were irregularly shaped and folded. They were lined by a continuous endothelium. The endothelial cells were remarkable for their abundance of smooth-surfaced circular and tubular profiles (Figure 7.6c and e). Presumably some of the circular profiles represented cross-sections of the tubules, whereas others were caveolae. Many of the vesicles and tubules opened at the blood or tissue fronts of the cell. Other vesicles and tubules appeared isolated in the cytoplasm. In no case was a single vesicle or tubule observed that opened on both fronts of the cell. Many of the vesicles and tubules were partially filled with a substance resembling the contents of the lumen (Figures 7.6d–f). The width of the tubules appeared to be fairly consistent. However, their length, contour, and orientation varied considerably. Tubules occasionally branched or forked. Some tubules were subtended by a diaphragm (Figure 7.6e). The lateral borders between endothelial cells had a very irregular outline, with significant overlapping and interdigitation between cells (Figure 7.6b). Although the plane of sectioning in single sections did not allow a detailed analysis of the entire junctional complex, the intercellular space was consistently narrow and contained several electron-dense junctional specializations (Figure 7.6b). Endothelial cells were surrounded by a thick basal lamina, consisting of moderately dense material (Figures 7.6b–d). Underlying the basal lamina were electron-lucent glial cell processes surrounded by their own basal lamina (Figure 7.6b). In some cases, the basal laminae of the endothelial cell and glial cell were fused (Figure 7.6d). Pericytes were not observed.

In vertebrates, the passage of substances from the blood to the brain is restricted by the blood–brain barrier. In mammals, the blood–brain

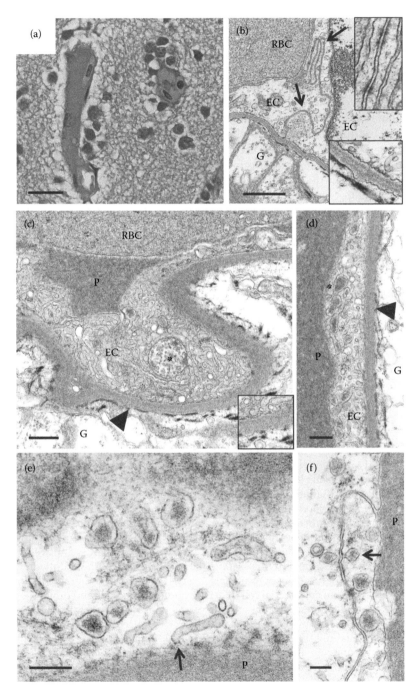

Figure 7.6 **(See color insert.)**   (*Continued*)

barrier is mediated in large part by tight junctions between the endo-thelial cells. Reese and Karnovsky (1967) showed that in capillaries of mouse brains, tight junctions were present in every region of overlap that could be examined in a favorable plane of section and concluded that tight junctions form a continuous belt around every endothelial cell. In higher vertebrates, endothelial cells possess few vesicles and thus vesicular transcellular transport is considered unlikely to play a major role in brain microvessels. In elasmobranchs, the tracer horseradish per-oxidase (40,000 Da) passes through the endothelial layer, but does not permeate the perivascular glial sheath (Bundgaard and Cserr, 1981b), suggesting that the blood–brain barrier of these species is mediated by the glial cells.

Morphological and functional studies of the blood–brain barrier in hagfish have yielded conflicting results. Murray et al. (1975) showed that the intercellular junctions between brain endothelial cells were usually interdigitating and possessed only occasional specializations. Further, they demonstrated a far less restrictive blood–brain barrier in hagfish compared with other vertebrate species as evidenced by increased pen-etration of radiolabeled sucrose, polyethylene glycol, and inulin from plasma into brain (Murray et al., 1975). They proposed that the vesicles and tubules might be involved in transcellular transport of macromolecules.

Figure 7.6 (Continued) Endothelium of brain capillaries in *M. glutinosa*. (a) 5-μm histological section of the brain stained with H&E shows two capillaries (one cut longitudinally, the other cut transversely). The capillary cut in cross-section mea-sures approximately 20 μm in diameter and contains four red blood cells. Note the intense eosinophilic (pink) staining of the plasma in both blood vessels. (b) Electron micrograph of a brain capillary shows two endothelial cells (EC) lying on a basal lamina that is adjacent to but separate from the basal lamina of the underlying glial cells (G) (bottom inset). The two endothelial cells are separated by a highly interdigitating lateral border (arrows), which contains specializations (top inset). A red blood cell (RBC) is observed in the capillary lumen. (c) Electron micrograph of a brain capillary shows a lumen containing a red blood cell (RBC) and highly electron-dense plasma (P). The capillary is lined by an endothelial cell (EC) whose cytoplasm contains many vesicles and tubules. A multivesicular body (*) is also observed. In this section, the basal laminae of the endothelial cell and glial cells are fused (arrowhead and inset). (d) Electron micrograph of a brain capillary shows an endothelial cell (EC) with abundant vesicles and tubules, some of which (*) contain material of similar electron density to that of the plasma (P). Arrowhead, basal lamina; G, glial cell. (e) Electron micrograph of a brain capil-lary endothelial cell shows numerous vesicle and tubules, many of which contain plasma-like contents. A tubule opening onto the luminal side of the cell is sub-tended by a diaphragm (arrow). P, plasma. (f) Electron micrograph of a lateral border between two brain capillary endothelial cells shows a vesicle opening into the intercellular space (arrow). P, plasma. Scale bar: 20 μm (a), 1 m (b), 500 nm (c), 300 nm (d), 200 nm (e), and 150 nm (f).

In contrast, Bundgaard et al. (1979) found that the abluminal part of the intercellular cleft was always obliterated by a punctate pentalaminar junctional complex, where adjacent endothelial membranes were "fused" over a short distance. (We did not carry out a sufficiently comprehensive analysis of the lateral borders to confirm these findings.) Bundgaard et al. (1979) demonstrated that when injected intravascularly, horseradish peroxidase was confined to the lumen and the intercellular cleft, where it always stopped at the contraluminal junction. It never appeared in the basal lamina, arguing against a role for the vesicles and tubules in transendothelial transport of protein (Bundgaard et al., 1979). In a tracer study using microperoxidase (2000 Da), only occasional endothelial vesicles and tubules were labeled (Bundgaard and Cserr, 1981a). Moreover, the penetration of radiolabeled polyethylene glycol (900 Da) from hagfish plasma into the brain was limited to the same extent as in rats. Thus, the authors concluded that there is a blood–brain barrier in hagfish to hydrophilic molecules larger than 900 Da and that the general phenomenon of a blood–brain barrier is a common characteristic of all vertebrates (Bundgaard and Cserr, 1981a).

Bundgaard et al. (1979) failed to observe diaphragms in the tubules and argued that these structures are probably involved in regulating intracellular calcium concentration. In contrast, we observed clear evidence for the existence of diaphragms at the opening of some tubules. This finding, together with the observation that certain tubules contained a substance that had the same appearance as the plasma, suggests that these structures are indeed part of the endocytic pathway.

The thickness of the basal lamina in brain capillaries was striking compared with that of the mammalian brain. Murray et al. (1975) have argued that this feature may contribute to the blood–brain barrier of hagfish. A similar role for the basal lamina has been proposed in mediating glomerular filtration in mammals. It seems that additional studies are needed to resolve the question as to whether the blood–brain barrier of hagfish is different from that of other vertebrates.

The highly electron-dense plasma that we observed in the lumen of the capillaries (and which was reflected by intense eosinophilic staining in H&E) was described as "flocculent material" by Mugnaini and Walberg (1965), but was not noted by other groups (Murray et al., 1975; Bundgaard et al., 1979; Bundgaard and Cserr, 1981a). The reason for this difference in the electron density of plasma in cerebral capillaries compared with capillaries in other organs is not clear.

Skeletal muscle capillary endothelium

In hagfish, the segmental parietal muscle, which is responsible for its undulatory swimming movements, contains three distinct fiber types, which are organized into three–five distinct layers within myomers of each segment: white (fast twitch), red (slow nontwitch), and intermediary. Compared with white fibers, red fibers have a much larger content of glycogen, fat vacuoles, and mitochondria (Korneliussen and Nicolaysen, 1973; Flood, 1979). The blood supply to the skeletal muscle has been previously described (Flood, 1979). In brief, each myomer of the segmental parietal muscle is supplied by its own arterial branch. The arteries give rise to a rich meshwork of relatively straight capillaries (10–15 μm in diameter) between the muscle fibers, which then drain into small venules and veins. Most of the capillaries in skeletal muscle are located in the vicinity of red fibers (Figure 7.7a). Indeed, a previous light microscopic study of segmental parietal muscle in hagfish demonstrated a 150-fold greater capillary density around red fibers compared with white fibers (Flood, 1979).

While a previous report described the ultrastructure of hagfish skeletal muscle fibers (Korneliussen and Nicolaysen, 1973), there are no similar descriptions of the vascular endothelium in this tissue. In our studies, the capillaries from segmental parietal muscle consisted of a continuous layer of relatively flat endothelial cells, surrounded by a well-formed basal lamina and occasional pericytes (Figures 7.7b–f). Endothelial cells were separated from muscle cells by a narrow rim of collagen-containing interstitium (Figures 7.7c and 7.8). The cytoplasm of endothelial cells was packed with tubular structures and vesicles, similar to those seen in brain capillaries (Figures 7.7d and e and 7.8). Many were shown to open on the luminal or abluminal side of the cell and many were subtended by a diaphragm (Figure 7.7e). Although some vesicles were linked, none was seen to span the full distance of the endothelial cell. Occasional Weibel–Palade bodies were observed (Figure 7.8). Lateral borders of endothelial cells were highly interdigitating and, in contrast to brain endothelial cells, contained few discernible specializations (Figures 7.7f and 7.8).

In mammals, capillary endothelium from skeletal muscle is enriched with plasmalemmal vesicles, or caveolae (Bruns and Palade, 1968a). However, tubules have not been reported in mammalian endothelial cells. A previous study in rats demonstrated that ferritin is transported across the endothelium of capillaries in skeletal muscle via plasmalemmal vesicles (Bruns and Palade, 1968b). It seems likely that the vesicles, and possibly the tubules, in the endothelium of hagfish skeletal muscle microvessels (and of several other vascular beds discussed in this chapter) are also involved in transporting "cargoes" of plasma and providing the small amounts of plasma proteins that are required for maintaining tissue health. Caveolae in mammalian endothelium have also been implicated in cell signaling, raising the possibility that these structures play a similar role in hagfish endothelial cells.

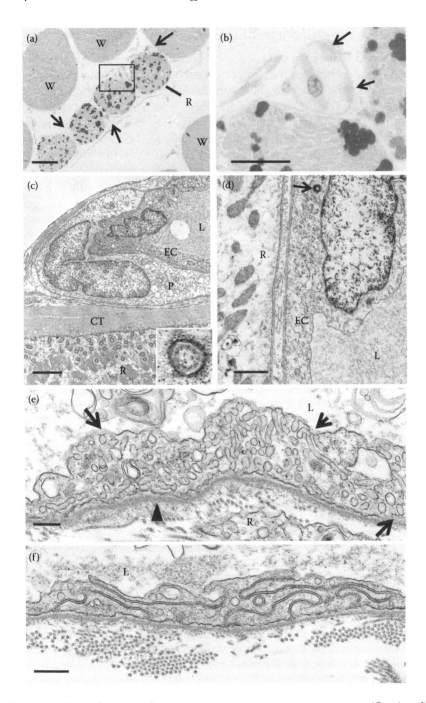

Figure 7.7 **(See color insert.)** *(Continued)*

Kidney glomerular endothelium

The hagfish kidney consists of 30–40 pairs of segmentally arranged large ovoid renal corpuscles that empty on each side into a long ureter (archinephric duct) by way of a short tubule (Heath-Eves and McMillan, 1974; Brown, 1988). The archinephric duct begins as a blind tip at the anterior region at the level of the liver and drains into the paired ureters at the posterior end, which in turn enter the cloaca. The renal vasculature in hagfish has been previously described (Heath-Eves and McMillan, 1974; Brown, 1988). The dorsal aorta gives rise to segmental arteries, which supply the abdominal wall musculature. Renal arteries usually arise from these segmental arteries (Figure 7.9), though they can originate directly from the aorta. The renal arteries do not branch but rather become narrower as they approach and feed the individual glomeruli, which contain a complex system of capillary loops (Figures 7.10a and b). Vascular cast studies have shown no clear distinction between the nonbranching renal artery and afferent arterioles (Brown, 1988). Glomerular capillaries give rise to one or more (typically 2–4) efferent arterioles, which either join the capillary network of the renal tubules and archinephric ducts or the capillaries supplying Bowman's capsule. Blood then drains into irregularly spaced renal veins, which connect to the posterior cardinal vein.

Figure 7.7 (Continued) Endothelium of skeletal muscle capillaries in *M. glutinosa*. (a) 1-µm histological transverse section of parietal muscle stained with Giemsa shows a row of four red muscle fibers (the one at the top is indicated by R) containing multiple, darkly stained fat vacuoles. Capillaries (arrows) are abundant in the vicinity of the red fibers. White fibers (W) are larger than red fibers and lack abundant fat vacuoles. (b) Higher magnification of boxed region in (a) shows muscle capillary adjacent to red muscle fibers and containing a nucleated red blood cell. Endothelial cell nuclei are indicated by arrows. (c) Electron micrograph of parietal muscle shows capillary lined by an endothelial cell (EC). A basal lamina separates the endothelial cell from an underlying pericyte (P). The pericyte, in turn, is separated from a red fiber (R) by a collagen-rich connective tissue space (CT). The red fiber contains numerous bundles of microtubules that demonstrate the "9 + 2" pattern of organization of a cilium (inset). L, capillary lumen. (d) Electron micrograph of parietal muscle shows a capillary lined by an endothelial cell (EC). The endothelial cell contains a centriole (arrow). The endothelial cell rests on a well-formed basal lamina, which is distinct from that of the underlying red fiber (R). L, capillary lumen. (e) Electron micrograph of a parietal muscle capillary shows an endothelial cell with abundant vesicles and tubules, some of which open onto the blood or tissue front and are subtended by a diaphragm (arrows). The endothelial cell has a well-formed basal lamina (arrowhead). The capillary lumen (L) contains membrane fragments. R, red muscle fiber. (f) Electron micrograph of a parietal muscle capillary shows endothelial cells separated by highly interdigitating lateral borders. L, capillary lumen. Scale bar: 50 µm (a), 20 µm (b), 2.5 µm (c), 2 µm (d), 200 nm (e), and 300 nm (f).

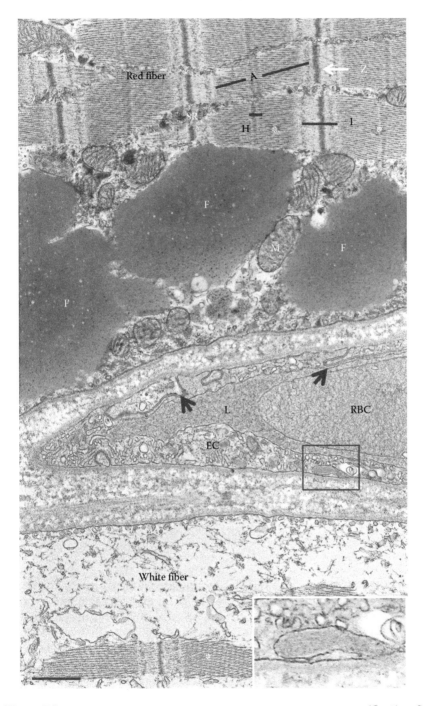

Figure 7.8 (Continued)

Figure 7.8 (Continued) Endothelium of skeletal muscle capillary in *M. glutinosa*. Electron micrograph of a longitudinal section of parietal muscle shows a capillary situated between a red fiber (top) and a white fiber (bottom). The lumen (L) contains a red blood cell (RBC). Lateral borders between endothelial cells are indicated by arrows. A Weibel–Palade body is observed in one of the endothelial cells (shown at higher magnification in inset). The red fiber at the top contains many mitochondria (M) and fat droplets (F). The white fiber at the bottom contains no mitochondria or fat droplets. Contractile fibrils (*) (composed of actin and myosin myofilaments) are more numerous and condensed in the red fiber. Z, Z line; I, I band; A, A band; H, H band; SR, sarcoplasmic reticulum. Scale bar: 600 nm.

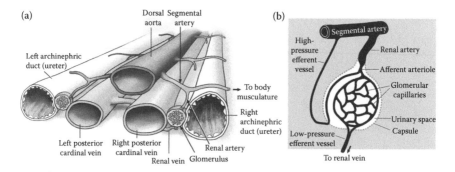

Figure 7.9 Schematic drawing of the kidney in hagfish. (a) The dorsal aorta gives rise to segmental arteries, which then branch to form renal arteries. The renal artery feeds the capillary bed of the glomerulus. Blood drains into the renal vein, followed by the posterior cardinal vein. The urinary space of each renal corpuscle empties into the long archinephric duct (ureter) by way of a short tubule (not shown). (Redrawn from Heath-Eves, M.J. and D.B. McMillan. 1974. *Am J Anat* 139:309–334.) (b) The renal arteries do not branch but rather become narrower as they approach and feed the individual glomeruli. There are no distinct afferent arterioles. Glomerular capillaries give rise to one or more efferent arterioles, which ultimately drain into the renal vein, followed by the posterior cardinal vein. In some cases, arteries bypass the glomerulus, creating high-pressure efferent shunts. (From Riegel, J. A. 1986. *J Exp Biol* 123:359–371.)

In about 43% of the segmental arteries, fine vessels arise outside Bowman's capsule and bypass the glomerulus, creating glomerular bypass shunts (Brown, 1988) (Figure 7.9b). A previous study reported the average hydrostatic pressures in the various vascular segments associated with the hagfish kidney: dorsal aorta 6.8 mmHg, segmental artery 7.5 mmHg, renal artery 5.9 mmHg, glomerular capillary 1.6 mmHg, "low-pressure" efferent vessels 1.1 mmHg, and posterior cardinal vein 0.3 mmHg (Riegel, 1986). Another group of efferent arterioles (presumably the bypass shunts) demonstrated higher pressures, similar to those of the renal

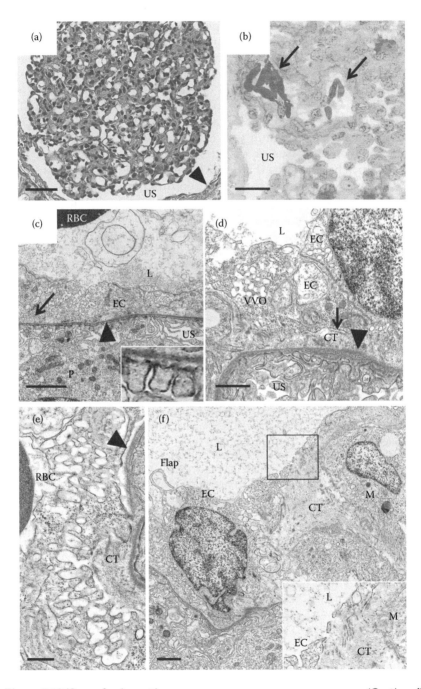

Figure 7.10 (See color insert.)   (*Continued*)

artery. Previous electron microscopic studies of the hagfish kidney have shown that glomerulus contains an unusually large number of mesangial cells (Heath-Eves and McMillan, 1974; Kuhn et al., 1975) and that its endothelium is poorly fenestrated (Heath-Eves and McMillan, 1974).

We found that the cytoplasm of glomerular endothelial cells was generally thicker and more electron-dense compared with capillary endothelial cells from other organs. The basal lamina immediately underlying the endothelium was thin and discontinuous. In contrast, podocyte foot processes or pedicles were observed abutting a well-formed thick basal lamina (Figures 7.10c and d). The foot processes interdigitated with one another to cover the entire surface of the basal lamina. The space between the foot processes formed narrow slits near the basement membrane (inset in Figure 7.10c). Filtration slits were occasionally traversed by a thin diaphragm. Between the endothelium and podocytes, a connective tissue layer was observed, displaying both classical cross-striated collagen fibrils as well as densely packed finer fibrils (Figure 7.10f). Embedded within this connective tissue matrix were mesangial

---

Figure 7.10 (Continued) Endothelium of kidney glomerular capillaries in *M. glutinosa*. (a) 5-μm histological section of a glomerulus stained with H&E shows glomerular tuft (middle) surrounded by a capsule (arrowhead), which contains the parietal epithelial cell layer as well as connective tissue. US, urinary space. (b) 1-μm section through a glomerulus stained with Geimsa shows capillaries containing red blood cells (arrows). US, urinary space. (c) Electron micrograph of a glomerulus shows a capillary lumen (L) containing a portion of a red blood cell (RBC). The capillary wall includes endothelial cells (EC) resting on an incomplete basal lamina (arrow), which is barely visible at this magnification. On the abluminal side of the endothelium are podocytes (P), which form the visceral layer of the urinary space (US) of Bowman's capsule. The podocytes are arranged as numerous extensions or foot processes (inset), which abut onto a prominent basal lamina (arrowhead). (d) Electron micrograph of a glomerulus shows two endothelial cells (EC) separated by a lateral border. The endothelial cytoplasm contains abundant polyribosomes as well as numerous vesicles and vacuoles, some of which are organized as VVOs. The endothelium rests on an incomplete basal lamina (arrow) and is separated from the podocyte foot processes by a connective tissue space (CT). These foot processes are lined by a well-formed basal lamina (arrowhead). L, capillary lumen; US, urinary space. (e) Electron micrograph of a glomerular capillary endothelial cell shows VVOs. RBC, red blood cell. CT, connective tissue matrix. Arrowhead, podocyte basal lamina. (f) Electron micrograph of a glomerulus shows a mesangial cell (M) in the subendothelial connective tissue matrix (CT). Inset (higher magnification of the boxed region) shows endothelial cell (EC) fenestrae, which allow for direct contact of blood plasma with the basal lamina of the podocyte. L, capillary lumen. Scale bar: 50 μm (a), 20 μm (b), 2 μm (c), 800 nm (d), 400 nm (e), and 1 μm (f).

cell processes (Figure 7.10f). The endothelial cells themselves contained many vesicle and vacuoles of varying sizes. Many of these structures had the appearance of vesiculo-vacuolar organelles (VVOs) (Figure 7.10d and e).* No tubules were observed. Several Weibel–Palade bodies were observed. Lateral borders were interdigitating with few areas of specialization. Consistent with a previous report (Heath-Eves and McMillan, 1974), only rare fenestrae were observed (Figure 7.10f). None of these was subtended by a diaphragm. In summary, the glomerular filtration barrier of the hagfish kidney consists of poorly fenestrated, relatively thick endothelial cells with a discontinuous basal lamina, a connective tissue space containing numerous mesangial cells, and podocytes with a well-formed basal lamina.

Although the components of the glomerular filtration barrier are similar in hagfish and higher vertebrates, there are several notable differences. First, glomerular endothelial cells in hagfish are thicker and display a paucity of fenestrae. In other words, the endothelium does not present a porous surface. In contrast, the endothelium of mammalian glomeruli is provided with numerous fenestrae which are seen as repeated interruptions, giving the endothelium a beaded appearance. It has been suggested that fluid loss from glomerular capillaries in hagfish occurs through vesicle-mediated pinocytosis (Heath-Eves and McMillan, 1974). Second, in contrast to mammals in which the basal lamina of endothelium and podocytes is fused, the basal lamina of the glomerular endothelium in hagfish is discontinuous and separated from the basal lamina of podocytes by a subpodocytic space. Third, the abundance of mesangial cells, whose processes intercalate between the endothelial cells, is unique to hagfish. It has been hypothesized that these mesangial cells may "clean" the unusually wide space between endothelial and podocyte basement membranes and/or participate in the reabsorption of proteins (Kuhn et al., 1975). Owing to the thickness of the endothelial cells and the presence of many mesangial cells, the diffusion distance is much greater in hagfish compared with mammals. The average pressure measured in glomerular capillaries is only 1.6 mmHg, which is less than the colloidal pressure (5.4 mmHg in *M. glutinosa*) (Riegel, 1986). As urine formation defies the Starling principle, some mechanism other than pressure filtration must account for the formation of urine. One possibility is that the renal epithelium actively secrets fluid into the urinary space (Riegel, 1999).

---

* VVOs, as described by Dvorak et al. (1996), are grape-like clusters of hundreds of uncoated, trilaminar unit membrane-bound, interconnecting vesicles and vacuoles that extend across the relatively tall cytoplasm of venular endothelium from luminal to abluminal surface.

Liver sinusoidal endothelium

The hagfish liver consists of two distinct lobes: a smaller anterior lobe and a larger posterior lobe. Blood enters the liver through the common portal vein and the hepatic artery. The common portal vein, which lies downstream of the portal heart, bifurcates to supply the two lobes of the liver. The hepatic artery, which arises from the left celiac artery, normally divides into three branches, which supply the gall bladder and the two lobes of the liver (in some cases, it feeds the anterior liver lobe, whereas the right celiac feeds the posterior lobe). Hepatocytes are arranged as tubular structures, with three to eight cells surrounding the central lumen of a bile canaliculus (Mugnaini and Harboe, 1967; Umezu et al., 2012) (Figures 7.11 and 7.12a and b). The tubules drain into bile ductules, which are arranged around the branches of the portal vein. In contrast to the usual portal triad of mammalian livers, the "portal structures" consisting of portal veins and biliary ductules are not accompanied by hepatic arteries (Umezu et al., 2012). Portal venous blood circulates between the tubules in large, irregular hepatic sinusoids or lacunae (Mugnaini and Harboe, 1967)

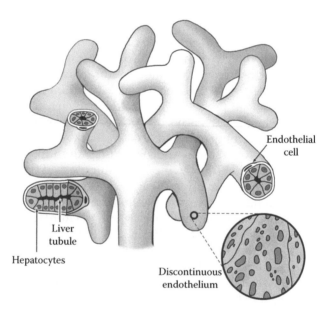

Figure 7.11 Schematic drawing of the liver in hagfish. Hepatocytes are arranged as blind-ended tubular structures with a bile canaliculus in the center. Tubules are surrounded by a layer of discontinuous endothelium (inset). Portal venous blood circulates in the spaces (sinusoids) between the tubules. The blood ultimately drains into anterior and posterior hepatic veins, followed by the sinus venosus. (From Mugnaini, E. and S. B. Harboe. 1967. *Mikrosk Anat* 78:341–369.)

(Figure 7.12b)* and ultimately drains into anterior and posterior hepatic veins, followed by the sinus venosus. Hepatocytes have been shown to contain many fat droplets (Figures 7.12a, b, d, and e). Indeed, a previous study showed that lipid constituted 12.4% of the weight of the hagfish liver (Spencer et al., 1966).

The ultrastructure of the hagfish liver has previously been reported (Mugnaini and Harboe, 1967). In our studies, the endothelium of liver sinusoids was extremely attenuated, sometimes constituting a thin rim within the blood vessel (Figures 7.12c and d). The endothelium contained many gaps. At the site of the gaps, material with the same electron density as plasma was observed in the subendothelial space (Figure 7.12d), suggesting direct access of substances between blood and hepatocytes. In some cases, the gaps were so large that red blood cells abutted directly on the basement membrane of the hepatocytes (Figure 7.12e). The gaps that we observed were not merely artifacts of the preparation, as previously suggested (Mugnaini and Harboe, 1967). Numerous clathrin-coated vesicles were observed in thicker regions of endothelial cells (not shown). Underlying the endothelium was an incomplete basal lamina followed by a narrow peritubular connective tissue space (the perisinusoidal/peritubular space of Disse) containing varying numbers of collagen fibrils (Figure 7.12c). Cell processes were observed in the peritubular spaces (Figure 7.12c). These have been previously referred to as "reticuloendothelial elements" (Mugnaini and Harboe, 1967). The processes did not contain lipid inclusions, as might be expected for a fat-storing Ito cell. Nor did they demonstrate desmosome-like connections with endothelial cells and hepatocytes, as described for a unique perisinusoidal cell found in the liver of teleosts (Nopanitaya et al., 1979). The basal portion of hepatocytes demonstrated irregular infoldings, which created cell processes (previously described as "interdigitating foot processes") (Figure 7.12c) (Mugnaini and Harboe, 1967). These "foot processes" contrast with the presence of microvilli in mammalian hepatocytes. Both the cell processes of hagfish hepatocytes and the microvilli of mammalian hepatocytes likely serve the same function of increasing the surface area for absorption. Hepatocytes were surrounded by a well-developed basal lamina. The latter structure appears to be the only continuous barrier in the wall of the vessels. In contrast to a previous histochemical analysis of hagfish liver (Umezu et al., 2012), we did not observe any cells in the sinusoids that resembled Kupffer cells. However, this may have reflected a sampling error.

* The original description by Cole over 90 years ago is noteworthy for its clarity: "Each tubule is seen to be completely surrounded by a vascular space—in fact so much so that the substance of the liver appears to include more blood tissue than liver tubules ... a course vascular network is disclosed which penetrates everywhere between the liver tubules, and constitutes not a capillary system but a reticular sinus." (Cole 1925, p. 322).

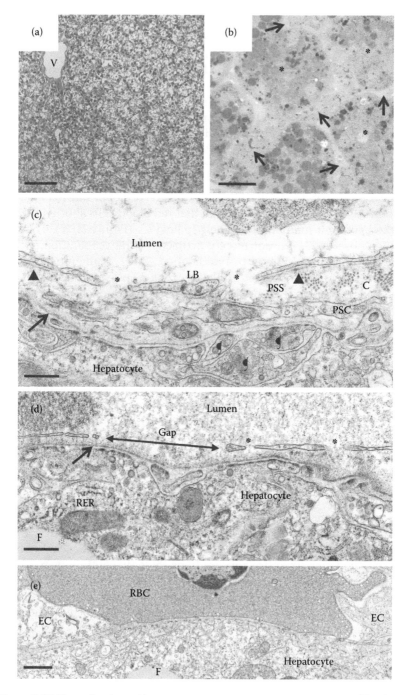

Figure 7.12 **(See color insert.)**   (*Continued*)

The sinusoidal endothelium in mammals functions as a selective sieve. The fenestrae act as a dynamic filter for fluids, solutes, and particles, allowing for passage of small particles (up to medium-sized chylomicrons) from blood to hepatocytes via the space of Disse. The mammalian sinusoidal endothelium contains open nondiaphragmed fenestrations occupying 10% of the surface (Wisse et al., 1996), whereas the endothelium of hagfish sinusoids is permeated by much larger fenestrae/gaps. Indeed, the presence of such wide gaps in the endothelium of hagfish argues against a highly selective sieve function of this vascular bed. On the other hand, the presence of a large number of clathrin-coated pits raises the possibility that these cells, like their mammalian counterparts, are actively involved in endocytosis and perhaps function as scavenger cells to clear the portal venous blood of soluble waste macromolecules.

Dermal capillary endothelium

The skin of the hagfish consists of a cellular epidermis, a dermis with collagen bundles and a hypodermis rich in fat cells (Figures 7.13 and 7.14a). The layers are approximately 70–100, 250, and 240 μm, respectively (Holmberg, 1968; Spitzer and Koch, 1998; Welsch and Potter, 1998).

Figure 7.12 (Continued) Endothelium of liver sinusoids in *M. glutinosa*. (a) 5-μm histological section of the liver stained with H&E. The vein (V) is probably a central or hepatic vein (since it lacks neighboring bile ductules that are characteristic of portal veins). (b) 1-μm section of the liver stained with Giemsa shows tubular structure characterized by groups of hepatocytes surrounding a central lumen (bile canaliculus) (*). The lumen is difficult to see because it is collapsed. Tubules are surrounded by a thin membrane and are separated from one another by blood sinuses (arrows). Note that the hepatocytes contain numerous fat droplets. (c) Electron micrograph of a liver sinusoid shows a thin layer of electron-lucent endothelial cells containing gaps of varying width (*) separating the lumen (L) from the perisinusoidal space (PSS), which contains collagen fibrils (C) and a perisinusoidal cell (PSC). A lateral border is seen between two endothelial cells (LB). The endothelium is associated with an incomplete basal lamina (arrowhead). Liver hepatocyte(s) and liver hepatocyte cell processes (¶) are shown at the bottom, surrounded by a well-formed basal lamina (arrow). The hepatocyte (here and in d, e) contains numerous coated and uncoated vesicles. (d) Electron micrograph of a liver sinusoid shows a large gap within an endothelial cell and several smaller gaps (*). Hepatocyte at the bottom contains rough endoplasmic reticulum (RER) and a fat droplet (F), and is surrounded by a well-formed basal lamina (arrow). (e) Electron micrograph of a liver sinusoid shows a red blood cell (RBC) containing characteristic pinocytotic vesicles along its cell membrane, making direct contact with an underlying hepatocyte through a large gap within an endothelial cell (EC). F, fat droplet. Scale bar: 100 μm (a), 20 μm (b), 500 nm (c, d), and 1 μm (e).

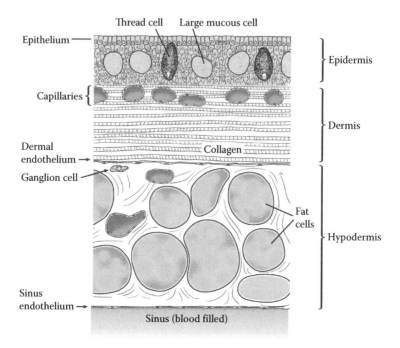

Figure 7.13 Schematic drawing of the skin in hagfish. The epidermis of hagfish is scaleless and contains cells that are involved in mucous production, including thread cells and large mucous cells. The dermis is comprised of layers of densely packed collagen fibers. While the epidermis is avascular, the dermis contains an extensive capillary network, which is best developed in the outer one-third. The dermis also contains melanocytes, fibroblasts, and nonmyelinated nerves (not shown). The hypodermis contains loose connective tissue and fat cells. The dermis is separated from the hypodermis by the dermal endothelium. The subcutaneous sinus lies underneath the skin and is surrounded by the sinus endothelium. (From von During, M. and K. H. Andres. 1998. Skin sensory organs in the Atlantic hagfish *Myxine glutinosa*. In J. M. Jørgensen, J. P. Lomholt, R. E. Weber and H. Malte (eds.), *The Biology of Hagfishes*. Chapman & Hall, Cambridge, pp. 499–511.)

The 6–8-cell layer nonkeratinizing avascular epidermis of hagfish is scaleless and contains cells that are involved in mucous production (Leppi, 1968). The dermis, which lies between the epidermis and hypodermis, is comprised of up to 45 layers of densely packed collagen fibers, which run at 90-degree angles to one another (Welsch and Potter, 1998). The border between the epidermis and the dermis is marked by a thick epidermal basal lamina. The dermis contains melanocytes, fibroblasts, and nonmyelinated nerve fibers enclosed by the cytoplasm of Schwann cells (Holmberg, 1968) (Figures 7.14b and c). In addition, the dermis contains an extensive capillary network, which is best developed in the outer one-third of the dermis, typically in a zone 4–10 μm beneath the basal lamina of the epidermis (Figures 7.14a and d) (Welsch and Potter, 1998).* Most of these capillaries run horizontally, parallel to the surface of the body. However, some are oriented vertically or at an angle through the dermis. The diameter of the capillaries ranges between 7 and 50 μm (Welsch and Potter, 1998). It has been suggested that the high vascular density is required for rapid delivery of precursors used in the biosynthesis of mucous (Potter, 1995; Welsch and Potter, 1998). Additionally, hagfish skin has been shown to take up nutrients (Glover et al., 2011; Schultz et al., 2014), in which case the extensive capillary network might facilitate absorption. It is also possible that the rich network of vessels is involved in cutaneous respiration. Interestingly, the dermal capillary network is far more developed in species of hagfish that burrow (e.g., *M. glutinosa* and *Paramyxine atami*) compared with those that do not (e.g., *Eptatretus stoutii*) (Potter, 1995). It has been argued that these differences reflect the need for burrowing animals to respire through their skin (Potter, 1995). In fact, Steffenssen (1984) reported that as much as 80% of oxygen uptake in Myxine occurs through the skin. However, others have raised questions about the importance of cutaneous respiration, pointing out that the epidermis itself consumes oxygen and that its considerable thickness would limit oxygen diffusion from seawater to blood (Malte and Lomholt, 1998). When burrowed, hagfish skin is exposed to an anoxic environment, while its nasal duct remains close to the surface. Thus, an alternative hypothesis is that the high capillary density in the dermis is an adaptation to supply oxygen to the skin under these conditions. The hypodermis, which underlies the dermis, consists of a loose connective tissue with abundant adipose cells. The dermis is separated from the hypodermis by a continuous flat epithelial cell layer that is termed the *dermal endothelium* (Figure 7.14e). The hypodermis consists mostly of adipocytes and contains many blood vessels.

The ultrastructure of the endothelial cells in dermal capillaries has been previously described by Potter (1995), Welsch and Potter (1998), and Yano et al. (2007). In our studies, the luminal side of dermal capillaries was lined by a single, continuous layer of endothelial cells. As noted previously, endothelial cells varied considerably in their thickness. The endothelium was surrounded by a thin, but well developed and continuous basal lamina (Figure 7.15a). The cytoplasm of the endothelial cells contained many vesicles, vacuoles, and tubules (Figure 7.15a). These were found to open on the apical (luminal) or basal side of the cell. The lateral borders were highly interdigitating and contained specializations consistent with various junctional complexes (Figure 7.15a). Occasional

---

* The vasculature of the dermis is supplied by dorsal branches of the segmental arteries of the body wall musculature, which pass through a strand of tissue (septum) in the dorsal midline into the skin.

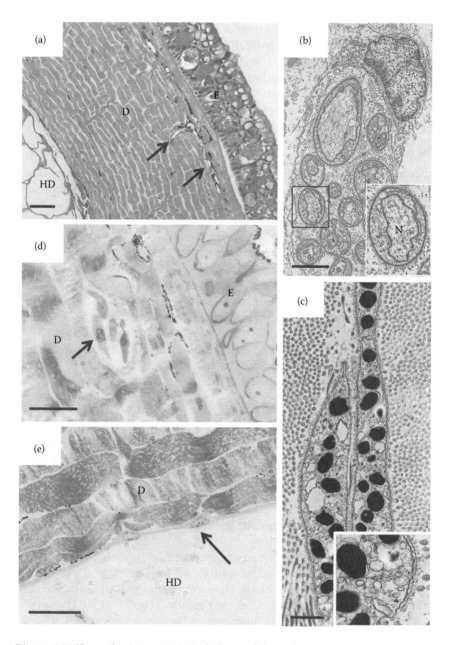

Figure 7.14 **(See color insert.)** Endothelium of dermal capillaries in *M. glutinosa*. (a) 5-μm histological section of integument stained with H&E shows the three layers of the skin: epidermis (E), dermis (D), and hypodermis (HD). Capillaries are observed in the outer layer of the dermis (arrows).   *(Continued)*

pericytes were observed on the abluminal side of the endothelium (Figure 7.15a). Pericytes contained vesicles and tubules that were similar to those observed in the endothelial cells.

The dermal endothelium, which demarcates the border between the dermis and the hypodermis, had a well-developed basal lamina on each side (Figure 7.15b). The cytoplasm contained many vesicles and tubules with a similar appearance to those observed in blood vessel endothelium. Many of these were observed to open on one or the other side of the cell. The cytoplasm also contained abundant intermediate filaments. Thin cell processes were observed on the dermal side of the dermal endothelium. In some locations, bundles of collagen fibrils were seen to penetrate through the dermal endothelium from the dermis to the hypodermis (Figure 7.15c). Although the dermal endothelium presumably forms a barrier between the epidermis and the dermis, its function remains poorly understood.

Intestinal capillary endothelium

The hagfish gut passes straight from the mouth to the anus. There is no obvious regional differentiation or divisions into stomach, duodenum, and intestine. Histologically, the intestinal wall contains four layers: an epithelium, an adipose layer, a smooth muscle cell layer, and a serosa consisting of mesothelial cells (Figure 7.16). The lumen is lined by a single layer of tall columnar epithelial cells with a well-developed brush border (Figure 7.17a). These cells are highly vacuolar in the luminal portion (Spencer et al., 1966). The vacuoles have been shown to contain lipid, as demonstrated by Oil Red O stain (Spencer et al., 1966). The mucosa demonstrates multiple convolutions or folds, which likely represents a

---

Figure 7.14 (Continued) (b) Electron micrograph of a bundle of nonmyelinated nerve fibers in the dermis embedded in the cytoplasm of a Schwann cell, which in turn is ensheathed by the thin elongated processes of another cell, whose nucleus appears at the top (probably a perineural fibroblast). Higher magnification of boxed region (inset) shows a single nerve fiber (N) surrounded by the cytoplasm of a Schwann cell. The Schwann cell, in turn, is surrounded by a basal lamina. (c) Electron micrograph of melanocyte cell processes containing numerous discrete uniformly electron-dense granules (melanosomes). Higher magnification (inset) shows numerous vesicles and tubular profiles, which open at the cell surface. The cells are surrounded by a well-formed basal lamina. Surrounding the melanocyte processes are transversely sectioned collagen fibrils, which are characteristic of the dermis. (d) 1-μm section of the skin stained with Giemsa shows parts of the epidermis (E) and dermis (D). A transversely sectioned capillary (arrow) is present in the dermis. (e) 1-μm section of the skin stained with Giemsa shows the dermis (D) and hypodermis (HD) separated by the dermal endothelium (arrow). Scale bar: 50 μm (a), 4 μm (b), 400 nm (c), and 20 μm (d, e).

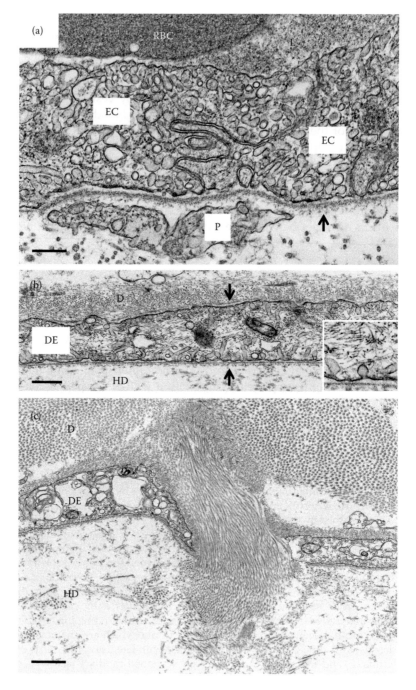

Figure 7.15   (Continued)

Figure 7.15 (Continued) Endothelium of dermal capillaries and dermal endothelium in *M. glutinosa*. (a) Electron micrograph of a dermal capillary shows two endothelial cells (EC) separated by an interdigitating lateral border. The endothelial cells contain many vesicles and vacuoles and are surrounded by a well-formed basal lamina (arrow). A pericyte (P) is observed on the abluminal side of the capillary. L, capillary lumen; RBC, red blood cell. (b) Electron micrograph of the dermal endothelium (DE), which separates the dermis (D) from the hypodermis (HD). The cytoplasm contains many vesicular structures and intermediate filaments (these are shown at higher magnification in the inset). The dermal endothelium is surrounded on both sides by a well-formed basal lamina (arrows). (c) Electron micrograph of the dermal endothelium (DE) shows bundles of collagen from the dermis (D) passing through it to the hypodermis (HD). Scale bar: 300 nm (a), 400 nm (b), and 1 μm (c).

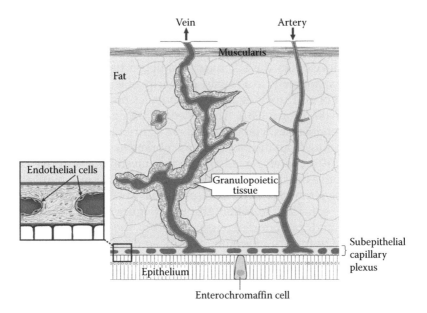

Figure 7.16 Schematic drawing of the intestine in hagfish. The mucosa consists of intestinal absorptive cells and hormone-producing enterocromaffin cells. The submucosa consists of a thin layer of subepithelial connective tissue and a thicker layer of adipose tissue. Segmental arteries arising from the dorsal aorta supply the wall of the abdominal gut, terminating in a subepithelial capillary plexus. The capillaries drain into a submucosal plexus of dilated veins followed by a dense vascular net in the intestinal wall, a pair of segmental veins, and finally the portal vein, which leads to the liver via the portal heart. Nests of hematopoietic cells are found around the plexiform veins, which develop in the submucosal adipose tissue. These hematopoietic foci consist primarily of leukocyte precursors, that is, granulopoietic cells. (From Fange, R. 1998. Hagfish blood cells and their formation. In J. M. Jørgensen, J. P. Lomholt, R. E. Weber and H. Malte (eds.), *The Biology of Hagfishes*. Chapman & Hall, Cambridge, pp. 287–299.)

strategy for increasing the absorptive surface area. The submucosa consists of a thin layer of subepithelial connective tissue (the lamina propria) and a thicker layer of adipose tissue (Figure 7.17a and b). A pair of arteries originate at each myotomal segment from the abdominal aorta, with each artery supplying a corresponding side of the gut wall (Tanaka et al., 1981).[*] These segmental arteries penetrate and encircle the wall, sending off branches toward the lumen at the root of the mucosal folds, where terminal arteries supply a capillary network within the connective tissue layer of the submucosa. Previous ink injection studies have shown that these capillaries anastomose with one another, creating a regular reticular net (Tanaka et al., 1981). Histologically, they appear as thin-walled blood vessels with a diameter of 20–25 µM (Figure 7.17b) (Tanaka et al., 1981). The capillaries drain into a submucosal plexus of dilated veins followed by a dense vascular net in the intestinal wall, a pair of segmental veins and finally the supraintestinal vein, which leads to the liver via the portal heart. Nests of hematopoietic cells are found around the plexiform veins, which develop in the submucosal adipose tissue (Tanaka et al., 1981). These hematopoietic foci consist primarily of leukocyte precursors, with formation of erythrocytes and thrombocytes presumably occurring elsewhere (Tanaka et al., 1981). Thin, slender, elongated hormone-producing (enteroendocrine/zymogen) cells are distributed along the entire intestine, extending from the basement membrane to the luminal surface (Ostberg et al., 1976), interspersed among the intestinal absorptive cells. Like the absorptive cells, the endocrine cells form regular microvilli at the gut lumen. In electron microscopy, these cells have been shown to contain electron-dense, spherical secretion granules (Figure 7.17d). In immunohistochemical and radioimmunoassay studies, the cells demonstrate immunoreactivity toward glucagon, gastrin, somatostatin, cholecystokinin, and substance P, and are thus regarded as a prototype of the endocrine cells observed in the intestines of higher vertebrates (Ostberg et al., 1976; Conlon and Falkmer, 1989). Thus, in addition to its role in digestion/absorption of food and hematopoiesis, the hagfish intestine also functions as an endocrine organ.

The ultrastructure of the endothelium lining the arteries, capillaries, and veins of the gut wall has been previously described (Tanaka et al., 1981). Our study focused on the capillary network in the subepithelial connective tissue. Endothelial cells lining these capillaries were electron-lucent and contained many vesicles and vacuoles (Figures 7.17e and f). These structures are presumably involved in the trafficking of nutrients from the gut lumen to the blood. The cells lacked obvious fenestrations

---

[*] More caudally, other arteries including the celiac and common carotid arteries, supply the gut (Cole, 1925).

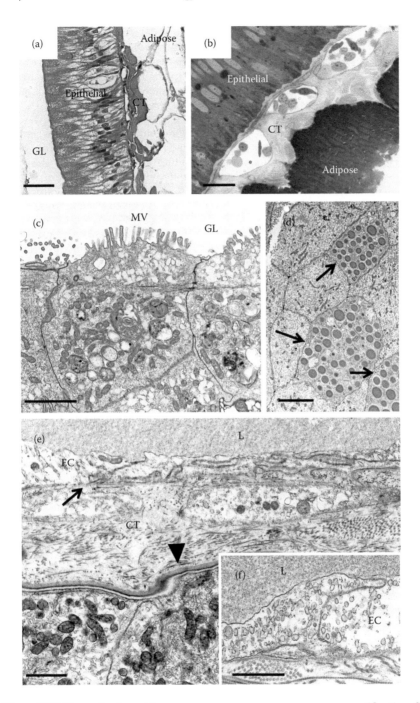

Figure 7.17 **(See color insert.)** (*Continued*)

Figure 7.17 (Continued) Endothelium of intestinal capillaries in *M. glutinosa*. (a) 5-µm histological section of the intestine stained with H&E shows gut lumen (GL) on left, bordered by epithelial cells. The epithelial layer contains both absorptive cells and endocrine cells, which are indistinguishable in this preparation. Underlying the epithelium is a thin connective tissue layer (CT), which contains a capillary plexus, followed by an adipose or fatty layer. (b) 1-µm histological section of the intestine stained with Geimsa shows the epithelial and adipose layers separated by the connective tissue layer (CT). The connective tissue layer contains capillaries (the one on the lower left is indicated by an asterisk). Numerous blood cells are observed in the capillary lumens. (c) Electron micrograph of the apical portion of absorptive epithelial cells shows microvilli (MV) protruding into the lumen of the gut (GL). The cells contain multiple prominent mitochondria and mucous granules. (d) Electron micrograph of the gut mucosa shows three endocrine cells (arrows) interspersed among numerous epithelial cells (enterocytes). (e) Electron micrograph of the submucosa of the gut wall shows electron-lucent endothelial cells (EC) lining a capillary. The endothelium is surrounded by a thin and incomplete basal lamina (arrow). Underlying the endothelium is the connective tissue layer (CT) followed by epithelial cells, which display a prominent basal lamina (arrowhead). The process of a cell (*) in the connective tissue space represents either a pericyte or a fibroblast. L, capillary lumen. (f) Electron micrograph of a capillary endothelial cell (EC) shows numerous vesicles and vacuoles. L, capillary lumen. Scale bar: 50 µm (a), 20 µm (b), 2 µm (c), 5 µm (d), and 1 µm (e, f).

and demonstrated interdigitating lateral borders with few specializations (Figure 7.17e).

Subcutaneous sinus endothelium

Hagfish possess an extensive system of venous sinuses.* The largest of these is the subcutaneous sinus, which spans around the body from the head to the tip of the tail. Blood drains into this sinus from capillary-size vessels at the lateral surface of parietal muscle (Flood, 1979), through anterior branches of the external carotid artery (Cole, 1925) and through the dorsal branch of the segmental arteries that pass through the dorsal midline into the skin. Its return to the systemic circulation is facilitated by auxiliary venous pumps or propulsors, namely the caudal and cardinal

---

* The sinus system is a complex network that includes many sinuses, including the subcutaneous sinus, the rostral sinus, the dorsal oral sinus, the hypophysio-velar sinus, the subhypophysial sinus, the ventral sinus, the dental sinus, the lingual sinus, the subesophageal sinus, the carotid sinus, the cardiac sinus, the peribranchial sinuses, and the subchordal sinus. The sinuses are connected with one another either directly or indirectly. Blood enters the sinus system through arteries and leaves by way of veins. The location of these sinuses and their connections with one another and with the systemic circulation are detailed in Cole (1925).

"hearts," whose function depends on the action of extrinsic striated skeletal muscle (Forster, 1997). The caudal heart, which is an innervated paired
pulsatile vesicle located at the tip of the tail, and shows only intermittent
activity (particularly following the cessation of swimming), returns blood
from the sinus (via a short connecting vessel that penetrates the myotomal muscle layer) to the posterior cardinal vein (Satchell, 1984). The pump
chambers of the caudal heart are considered to be valved swellings of the
left and right halves of the caudal vein (Satchell, 1984). The cardinal heart
is a propulsor that pumps blood from the rostral end of the sinus (via the
hypophysio-velar sinus) into the anterior cardinal vein (Farrell, 2007).[*]
    The hematocrit is lower in blood contained within the subcutaneous
sinus compared with the systemic circulation, suggesting that the red
blood cells are held back ("plasma skimming effect") as blood transits
from systemic blood vessels into the sinus (Johansen et al., 1962; Forster,
1997). Following strenuous activity, the hematocrit in the sinus increases,
suggesting that plasma may redistribute from the subcutaneous sinus to
the central circulation when animals are stressed (Johansen et al., 1962;
Forster, 1997). Satchell (1991) has raised the possibility that the sinuses
function for oxygen transport or as a temporary dump for lactate generated during stressful exercise, but favors the notion that they serve as a
reservoir of fluid, which can increase preload and hence cardiac output.
Lomholt and Franko-Dossar (1998) suggest that the subcutaneous sinus
may serve as a volume buffer that allows the systemic circulation to maintain its volume and pressure in the face of increased total blood volume.
This hypothesis is supported by intravascular pressure measurements in
volume-loaded animals (Foster and Forster, 2007). The endothelium of the
subcutaneous sinus has not been previously described.[†]
    In light microscopic studies, the endothelium formed a continuous
layer of flattened cells (Figure 7.18a and b). On the body-wall side, facing
the parietal muscle, the endothelium was extremely attenuated, with a
poorly formed basal lamina (Figures 7.18c and d). Cells were overlapping
and their lateral borders were highly interdigitating. There were many
regions of overlap between endothelial cells. The interdigitating lateral
borders lacked any evidence of specialization. The cytoplasm of these cells
contained occasional vacuoles but they lacked the abundance of vesicles

---

[*] Paths of entry and exit vary between sinuses. For example, blood enters the peribranchial
sinuses through hollow vascular papillae on the afferent and efferent branchial arteries, which project out from the surface of the vessels into the lumen of the surrounding
sinus and leave the sinus through the anterior cardinal vein. Blood enters the hypophysio-
velar sinus through a branch of the unpaired vertebral artery (see Cole, 1925). Additional
anastomoses between the arterial and sinus system of the gill are described by Elger (1987).
[†] Elger reported that the wall of the sinusoid system of the gill (which connects with the
peribranchial sinus) is composed of a thin endothelial layer that is rarely fenestrated,
displays caveolae, and is lined by a discontinuous basement membrane (Elger, 1987).

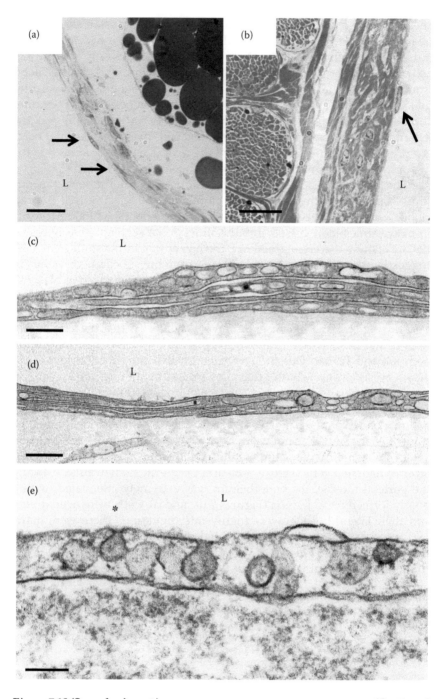

Figure 7.18 **(See color insert.)**   (Continued)

observed on the skin side. Underlying the basal lamina was a thick layer of connective tissue with occasional long slender cytoplasmic processes. On the skin side, the endothelial cells contained many vesicles and vacuoles of many different sizes (Figure 7.18e). Many of the vesicles opened onto one or the other fronts of the endothelial cells and had the classic appearance of clathrin-coated pits and standard caveolae. However, the cells did not contain tubules. The endothelium was surrounded by collagen fibers, which contained occasional cell processes.

Summary

The endothelium of hagfish demonstrates remarkable heterogeneity at the ultrastructural level. These findings suggest that phenotypic diversity appeared as an early feature of the endothelium. The proximate mechanisms of endothelial heterogeneity in hagfish, like that in higher vertebrates, presumably involve a combination of environmentally responsive signaling pathways that depend on the extracellular milieu of the native tissue and epigenetically fixed programs that are mitotically stable. Understanding the relative roles of these processes in mediating vascular bed-specific properties of the endothelium in these animals will require extensive knowledge of hagfish development as well as the functional and molecular characterization of the vascular system, both of which remain largely unexploited areas.

Regardless of the underlying proximate mechanisms, endothelial cell heterogeneity in hagfish almost certainly represents an adaptation of the endothelium to meet the diverse needs of the various organs. For example, the distinction between continuous endothelium in brain capillaries and the discontinuous (almost open) endothelium of liver sinusoids must reflect differences in the need to protect neural tissue from blood and a requirement to provide direct access of blood to liver hepatocytes, respectively.

Figure 7.18 (Continued) Endothelial lining of subcutaneous sinus in *M. glutinosa*. (a) 1-μm histological section of the skin side of the subcutaneous sinus stained with Geimsa shows attenuated endothelium (nuclei indicated by arrows) lining the lumen (L). A lipid-laden fat cell in the hypodermis is observed in the upper right. (b) 1-μm histological section of the body wall (parietal muscle) side of the subcutaneous sinus stained with Geimsa shows attenuated endothelium (nucleus indicated by arrow) lining the lumen (L). Muscle fibers are observed on the left. (c and d) Electron micrographs of the endothelial lining of the body wall (parietal muscle) side of the subcutaneous sinus show attenuated endothelial cells with marked interdigitation. The lateral borders demonstrate no specializations. There is no basal lamina. L, sinus lumen. (e) Electron micrograph of the endothelial lining of the skin side of the subcutaneous sinus shows several vesicles, two of which are open to the lumen (L) of the sinus. Scale bar: 20 μm (a, b), 250 nm (c), 400 nm (d), and 100 nm (e).

Phenotypic heterogeneity of the endothelium also reflects different coping strategies for endothelial cells to survive in different tissue environments. For example, the endothelial cells in the hagfish heart are exposed to profound hypoxia and they must be uniquely adapted in ways that render them phenotypically distinct from the endothelial cells of the dorsal aorta, which are exposed to the highest oxygen levels in the body.

Admittedly, we can only conjecture about structure–function relationships of the endothelium. However, what seems clear from this study is that the endothelium of hagfish is not simply an inert, homogeneous layer of cells that separates blood from underlying tissue, but rather is morphologically (and by extension physiologically) diverse.

Acknowledgments

We thank Tracey Sciuto and Rita Monahan for their technical help with electron microscopy. We thank Steve Moskowitz for his artwork.

References

Aird, W. C. 2007a. Phenotypic heterogeneity of the endothelium: I. Structure, function, and mechanisms. *Circ Res* 100:158–173.

Aird, W. C. 2007b. Phenotypic heterogeneity of the endothelium: II. Representative vascular beds. *Circ Res* 100:174–190.

Arap, W., W. Haedicke, M. Bernasconi, R. Kain, D. Rajotte, S. Krajewski, H. M. Ellerby, D. E. Bredesen, R. Pasqualini and E. Ruoslahti. 2002. Targeting the prostate for destruction through a vascular address. *Proc Natl Acad Sci USA* 99:1527–1531.

Arap, W., M. G. Kolonin, M. Trepel, J. Lahdenranta, M. Cardo-Vila, R. J. Giordano, P. J. Mintz et al. 2002. Steps toward mapping the human vasculature by phage display. *Nat Med* 8:121–127.

Arap, W., R. Pasqualini and E. Ruoslahti. 1998. Cancer treatment by targeted drug delivery to tumor vasculature in a mouse model. *Science* 279:377–380.

Augustinsson, K. B., R. Fange, A. Johnels and E. Ostlund. 1956. Histological, physiological and biochemical studies on the heart of two cyclostomes, hagfish (*Myxine*) and lamprey (Lampetra). *J Physiol* 131:257–276.

Axelsson, M., A. P. Farrell and S. Nilsson. 1990. Effects of hypoxia and drugs on the cardiovascular dynamics of the Atlantic hagfish *Myxine glutinosa*. *J Exp Biol* 151:297–316.

Bauer, C., U. Engels and S. Paleus. 1975. Oxygen binding to haemoglobins of the primitive vertebrate *Myxine glutinosa* L. *Nature* 256:66–68.

Bloom, G., E. Ostlund, E. U. von, F. Lishajko, M. Ritzen and J. Adams-Ray. 1961. Studies on catecholamine-containing granules of specific cells in cyclostome hearts. *Acta Physiol Scand Suppl* 53:1–34.

Brown, J. A. 1988. Glomerular bypass shunts in the kidney of the Atlantic hagfish, *Myxine glutinosa*. *Cell Tissue Res* 253:377–381.

Bruns, R. R. and G. E. Palade. 1968a. Studies on blood capillaries. I. General organization of blood capillaries in muscle. *J Cell Biol* 37:244–276.

Bruns, R. R. and G. E. Palade. 1968b. Studies on blood capillaries. II. Transport of ferritin molecules across the wall of muscle capillaries. *J Cell Biol* 37:277–299.

Bullock, T. H., J. K. Moore and R. D. Fields. 1984. Evolution of myelin sheaths: Both lamprey and hagfish lack myelin. *Neurosci Lett* 48:145–148.

Bundgaard, M. and H. Cserr. 1981a. Impermeability of hagfish cerebral capillaries to radio-labelled polyethylene glycols and to microperoxidase. *Brain Res* 206:71–81.

Bundgaard, M. and H. F. Cserr. 1981b. A glial blood-brain barrier in elasmobranchs. *Brain Res* 226:61–73.

Bundgaard, M., H. Cserr and M. Murray. 1979. Impermeability of hagfish cerebral capillaries to horseradish peroxidase. An ultrastructural study. *Cell Tissue Res* 198:65–77.

Cecon, S., B. Minnich and A. Lametschwandtner. 2002. Vascularization of the brains of the Atlantic and Pacific hagfishes, *Myxine glutinosa* and *Eptatretus stouti*: A scanning electron microscope study of vascular corrosion casts. *J Morphol* 253:51–63.

Cole, F. J. 1925. A monograph of the general morphology of the myxinoid fishes, based on a study of *Myxine*. VI: The morphology of the vascular system. *Trans R Soc Edin* 54:309–342.

Conlon, J. M. and S. Falkmer. 1989. Neurohormonal peptides in the gut of the Atlantic hagfish (*Myxine glutinosa*) detected using antisera raised against mammalian regulatory peptides. *Gen Comp Endocrinol* 76:292–300.

Cox, G. K., E. Sandblom and A. P. Farrell. 2010. Cardiac responses to anoxia in the Pacific hagfish, *Eptatretus stoutii*. *J Exp Biol* 213:3692–3698.

Davison, I. G., G. M. Wright and M. E. DeMont. 1995. The structure and physical properties of invertebrate and primitive vertebrate arteries. *J Exp Biol* 198:2185–2196.

Dvorak, A. M., S. Kohn, E. S. Morgan, P. Fox, J. A. Nagy and H. F. Dvorak. 1996. The vesiculo-vacuolar organelle (VVO): A distinct endothelial cell structure that provides a transcellular pathway for macromolecular extravasation. *J Leukoc Biol* 59:100–115.

Elger, M. 1987. The branchial circulation and the gill epithelia in the Atlantic hagfish, *Myxine glutinosa* L. *Anat Embryol* 175:489–504.

Fange, R. 1998. Hagfish blood cells and their formation. In J. M. Jørgensen, J. P. Lomholt, R. E. Weber and H. Malte (eds.), *The Biology of Hagfishes*. Chapman & Hall, Cambridge, pp. 287–299.

Farrell, A. P. 2007. Cardiovascular systems in primitive fishes. In D. J. McKenzie, A. P. Farrell and C. J. Brauner (eds.), *Primitive Fishes*. Elsevier, London, pp. 53–120.

Feng, J., K. Yano, R. Monahan-Earley, E. S. Morgan, A. M. Dvorak, F. W. Sellke and W. C. Aird. 2007. Vascular bed-specific endothelium-dependent vasomotor relaxation in the hagfish, *Myxine glutinosa*. *Am J Physiol Regul Integr Comp Physiol* 293:R894–R900.

Flood, P. R. 1979. The vascular supply of three fibre types in the parietal trunk muscle of the Atlantic hagfish (*Myxine glutinosa*, L). A light microscopic quantitative analysis and an evaluation of various methods to express capillary density relative to fibre types. *Microvasc Res* 17:55–70.

Forster, M. E. 1991. Myocardial oxygen consumption and lactate release by the hypoxic hagfish heart. *J Exp Biol* 156:583–590.

Forster, M. E. 1997. The blood sinus system of hagfish: Its significance in a low-pressure circulation. *Comp Biochem Physiol* 116:239–244.

Forster, M. E., M. Axelsson, A. P. Farrell and S. Nilsson. 1991. Cardiac function and circulation in hagfishes. *Can J Zool* 69:1985–1992.

Forster, M. E., W. Davison, M. Axelsson and A. P. Farrell. 1992. Cardiovascular responses to hypoxia in the hagfish, *Eptatretus cirrhatus*. *Respir Physiol* 88:373–386.

Foster, J. M. and M. E. Forster. 2007. Effects of salinity manipulations on blood pressures in an osmoconforming chordate, the hagfish, *Eptatretus cirrhatus*. *J Comp Physiol B* 177:31–39.

Glover, C. N., C. Bucking and C. M. Wood. 2011. Adaptations to *in situ* feeding: Novel nutrient acquisition pathways in an ancient vertebrate. *Proc Biol Sci* 278:3096–3101.

Hardisty, M. W. 1979. *Biology of the Cyclostomes*. Chapman & Hall, distributed by Halsted Press, London, New York.

Heath-Eves, M. J. and D. B. McMillan. 1974. The morphology of the kidney of the Atlantic hagfish, *Myxine glutinosa*. *Am J Anat* 139:309–334.

Heimberg, A. M., R. Cowper-Sallari, M. Semon, P. C. Donoghue and K. J. Peterson. 2010. MicroRNAs reveal the interrelationships of hagfish, lampreys, and gnathostomes and the nature of the ancestral vertebrate. *Proc Natl Acad Sci USA* 107:19379–19383.

Helle, K. B. and S. Lonning. 1973. Sarcoplasmic reticulum in the portal vein heart and ventricle of the cyclostome *Myxine glutinosa* (L.). *J Mol Cell Cardiol* 5:433–439.

Helle, K. B., S. Lonning and H. Blaschko. 1972. Observations on the chromaffin granules of the ventricle and the portal vein heart of *Myxine glutinosa*. *Sarsia* 51:97–106.

Holmberg, K. 1968. Ultrastructure and response to background illumination of the melanophores of the Atlantic hagfish, *Myxine glutinosa*, L. *Gen Comp Endocrinol* 1:421–428.

Johansen, K. 1960. Circulation in the hagfish, *Myxine glutinosa* L. *Biol Bull* 118:289–295.

Johansen, K. 1963. The cardiovascular system of *Myxine glutinosa* L. In A. Brodal and R. Fange (eds.), *The Biology of Myxine*. Universitetsforlaget, Oslo, pp. 289–316.

Johansen, K., R. Fange and M. W. Johannessen. 1962. Relations between blood, sinus fluid and lymph in *Myxine glutinosa* L. *Comp Biochem Physiol* 7:23–28.

Johnsson, M. and M. Axelsson. 1996. Control of the systemic heart and the portal heart of *Myxine glutinosa*. *J Exp Biol* 199:1429–1434.

Korneliussen, H. and K. Nicolaysen. 1973. Ultrastructure of four types of striated muscle fibers in the Atlantic hagfish (*Myxine glutinosa* L.). *Z Zellforsch Mikrosk Anat* 143:273–290.

Kuhn, K., H. Stolte and E. Reale. 1975. The fine structure of the kidney of the hagfish (*Myxine glutinosa* L.): A thin section and freeze-fracture study. *Cell Tissue Res* 164:201–213.

Leak, L. V. 1969. Electron microscopy of cardiac tissue in a primitive vertebrate *Myxine glutinosa*. *J Morphol* 128:131–157.

Leppi, T. J. 1968. Morphochemical analysis of mucous cells in the skin and slime glands of hagfishes. *Histochemie* 15:68–78.

Lomholt, J. P. and F. Franko-Dossar. 1998. The sinus system of hagfishes—Lymphatic or secondary circulatory system? In J. M. Jørgensen, J. P. Lomholt, R. E. Weber and H. Malte (eds.), *The Biology of Hagfishes*. Chapman & Hall, Cambridge, pp. 259–272.

Malte, H. and J. P. Lomholt. 1998. Ventilation and gas exchange. In J. M. Jørgensen, J. P. Lomholt, R. E. Weber and H. Malte (eds.), *The Biology of Hagfishes*. Chapman & Hall, Cambridge, pp. 223–234.

McIntosh, D. P., X. Y. Tan, P. Oh and J. E. Schnitzer. 2002. Targeting endothelium and its dynamic caveolae for tissue-specific transcytosis *in vivo*: A pathway to overcome cell barriers to drug and gene delivery. *Proc Natl Acad Sci USA* 99:1996–2001.

Monahan-Earley, R., A. M. Dvorak and W. C. Aird. 2013. Evolutionary origins of the blood vascular system and endothelium. *J Thromb Haemost* 11(Suppl 1):46–66.

Mugnaini, E. and S. B. Harboe. 1967. The liver of *Myxine glutinosa*: A true tubular gland. *Z Zellforsch Mikrosk Anat* 78:341–369.

Mugnaini, E. and F. Walberg. 1965. The fine structure of the capillaries and their surroundings in the cerebral hemispheres of *Myxine glutinosa* (L.). *Z Zellforsch Mikrosk Anat* 66:333–351.

Murray, M., H. Jones, H. F. Cserr and D. P. Rall. 1975. The blood–brain barrier and ventricular system of *Myxine glutinosa*. *Brain Res* 99:17–33.

Nopanitaya, W., J. Aghajanian, J. W. Grisham and J. L. Carson. 1979. An ultrastructural study on a new type of hepatic perisinusoidal cell in fish. *Cell Tissue Res* 198:35–42.

Ostberg, Y., S. Van Noorden, A. G. Everson Pearse and N. W. Thomas. 1976. Cytochemical, immunofluorescence, and ultrastructural investigations on polypeptide hormone containing cells in the intestinal mucosa of a cyclostome, *Myxine glutinosa*. *Gen Comp Endocrinol* 28:213–227.

Ostlund, E., G. Bloom, J. Adams-Ray, M. Ritzen, M. Siegman, H. Nordenstam, F. Lishajko and E. U. von. 1960. Storage and release of catecholamines, and the occurrence of a specific submicroscopic granulation in hearts of cyclostomes. *Nature* 188:324–325.

Pasqualini, R. and E. Ruoslahti. 1996. Organ targeting *in vivo* using phage display peptide libraries. *Nature* 380:364–366.

Perry, S. F., R. Fritsche and S. Thomas. 1993. Storage and release of catecholamines from the chromaffin tissue of the Atlantic hagfish *Myxine glutinosa*. *J Exp Biol* 183:165–184.

Potter, I. C., U. Welsch, G. M. Wright, Y. Honma and A. Chiba. 1995. Light and electron microscope studies of the dermal capillaries in three species of hagfishes and three species of lamprey. *J Zool Lond* 235:677–688.

Reese, T. S. and M. J. Karnovsky. 1967. Fine structural localization of a blood–brain barrier to exogenous peroxidase. *J Cell Biol* 34:207–217.

Riegel, J. A. 1986. Hydrostatic pressures in glomeruli and renal vasculature of the hagfish, *Eptatretus stoutii*. *J Exp Biol* 123:359–371.

Riegel, J. A. 1999. Secretion of primary urine by glomeruli of the hagfish kidney. *J Exp Biol* 202 (Pt 8):947–955.

Sage, H. and W. R. Gray. 1979. Studies on the evolution of elastin—I. Phylogenetic distribution. *Comp Biochem Physiol B* 64:313–327.

Sage, H. and W. R. Gray. 1980. Studies on the evolution of elastin. II. Histology. *Comp Biochem Physiol B* 66:13–22.

Satchell, G. H. 1984. On the caudal heart of *Myxine* (Myxinoidea: Cyclostomata). *Acta Zool (Stockholm)* 65:125–133.

Satchell, G. H. 1991. *Physiology and Form of Fish Circulation*. Cambridge University Press, Cambridge, New York.

Schultz, A. G., S. C. Guffey, A. M. Clifford and G. G. Goss. 2014. Phosphate absorption across multiple epithelia in the Pacific hagfish (*Eptatretus stoutii*). *Am J Physiol Regul Integr Comp Physiol* 307: R643–R652.

Sidell, B. D. 1983. Cardiac metabolism in the Myxinidae: Physiological and phylogenetic considerations. *Comp Biochem Physiol A Comp Physiol* 76:495–505.

Spencer, R. P., R. L. Scheig and H. J. Binder. 1966. Observations on lipids of the alimentary canal of the hagfish *Eptatretus stoutii*. *Comp Biochem Physiol* 19:139–144.

Spitzer, R. H. and E. A. Koch. 1998. Hagfish skin and slime glands. In J. M. Jørgensen, J. P. Lomholt, R. E. Weber and H. Malte (eds.), *The Biology of Hagfishes*. Chapman & Hall, Cambridge, pp. 109–132.

Steffenssen, J. F. 1984. Ventilation and oxygen consumption in the hagfish, *Myxine glutinosa* L. *J Exp Mar Biol Ecol* 84:173–178.

Sundin, L., M. Axelsson, S. Nilsson, W. Davison and M. Forster. 1994. Evidence of regulatory mechanisms for the distribution of blood between the arterial and the venous compartments in the hagfish gill pouch. *J Exp Biol* 190:281–286.

Tanaka, Y., Y. Saito and H. Gotoh. 1981. Vascular architecture and intestinal hematopoietic nests of two cyclostomes, *Eptatretus burgeri* and ammocoetes of *Entosphenus reissneri*: A comparative morphological study. *J Morphol* 170:71–93.

Umezu, A., H. Kametani, Y. Akai, T. Koike and N. Shiojiri. 2012. Histochemical analyses of hepatic architecture of the hagfish with special attention to periportal biliary structures. *Zool Sci* 29:450–457.

von During, M. and K. H. Andres. 1998. Skin sensory organs in the Atlantic hagfish *Myxine glutinosa*. In J. M. Jørgensen, J. P. Lomholt, R. E. Weber and H. Malte (eds.), *The Biology of Hagfishes*. Chapman & Hall, Cambridge, pp. 499–511.

Wells, R. M., M. E. Forster, W. Davison, H. H. Taylor, P. S. Davie and G. H. Satchell. 1986. Blood oxygen transport in the free-swimming hagfish, *Eptatretus cirrhatus*. *J Exp Biol* 123:43–53.

Welsch, U. and I. C. Potter. 1998. Dermal capillaries. In J. M. Jørgensen, J. P. Lomholt, R. E. Weber and H. Malte (eds.), *The Biology of Hagfishes*. Chapman & Hall, Cambridge, pp. 273–283.

Wisse, E., F. Braet, D. Luo, R. De Zanger, D. Jans, E. Crabbe and A. Vermoesen. 1996. Structure and function of sinusoidal lining cells in the liver. *Toxicol Pathol* 24:100–111.

Wright, G. M. 1984. Structure of the conus arteriosus and ventral aorta in the sea lamprey, *Petromyzon marinus*, and the Atlantic hagfish, *Myxine glutinosa*: Microfibrils, a major component. *Can J Zool* 62:2445–2456.

Yano, K., D. Gale, S. Massberg, P. K. Cheruvu, R. Monahan-Earley, E. S. Morgan, D. Haig, U. H. von Andrian, A. M. Dvorak and W. C. Aird. 2007. Phenotypic heterogeneity is an evolutionarily conserved feature of the endothelium. *Blood* 109:613–615.

chapter eight

The adaptive immune system of hagfish

**Jianxu Li, Sabyasachi Das, Brantley R. Herrin,
Masayuki Hirano, and Max D. Cooper**

Contents

Introduction

Protective immunity in all vertebrates is mediated by both innate and adaptive immune systems (AIS). Innate immune systems are also found in invertebrates, whereas a lymphocyte-based AIS only exists in the vertebrates (Figure 8.1). As the first line of host defense, the innate immune system is mediated by phagocytic cells, such as dendritic cells and macrophages, that typically employ a limited number of germline-encoded pattern-recognition receptors, including Toll-like receptors, retinoic acid-inducible gene I-like receptors, and nucleotide-binding oligomerization domain-like receptors, to sense conserved pathogen-associated molecular patterns expressed by microbial pathogens (Kawai and Akira, 2009). Innate

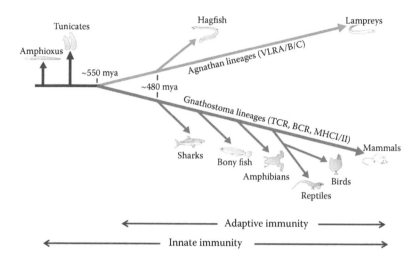

Figure 8.1 Evolution of innate and adaptive immunity. Innate immunity is found across the animal kingdom, whereas lymphocyte-based adaptive immunity is only found in the vertebrates. The LRR-based *VLR* genes have only been found in jawless vertebrates, whereas Ig-based *BCR*, *TCR*, and *MHCI/II* genes are limited to the extant jawed vertebrates. Neither type of antigen-recognition receptors have been found in amphioxus and tunicates. These observations suggest that the LRR-based VLRs and the Ig-based BCRs and TCRs have undergone convergent evolution, which attests to the survival advantage and strong selective pressure for the development of AISs.

immunity is an ancient defense strategy that both invertebrate and vertebrate species use to protect against infection by potential pathogens in a relatively nonspecific manner. In contrast, both jawed and jawless vertebrates have an AIS that serves as a second line of defense by employing a vast repertoire of anticipatory receptors on lymphocytes to recognize specific antigens. A distinguishing feature of the AIS is the capacity for memory of previous antigen encounters, which allows for a more rapid response to a previously encountered pathogen (Dunn-Walters and Ademokun, 2010).

The jawed vertebrates have clonally diverse T and B lymphocyte lineages which employ immunoglobulin (Ig) domain-based T cell receptors (TCRs) and B cell receptors (BCRs), to recognize and eliminate infectious microorganisms. The highly diverse *TCR* and *BCR* genes are generated primarily through the recombinatorial assembly of Ig V-(D)-J gene segments, and the BCRs are further diversified by somatic hypermutation. T cells coordinate the cellular immune response via cytokines and chemokines, and B cells mediate humoral immunity by secreting their antigen receptors as antibodies. Although the innate and adaptive immune systems have distinct features, they act in concert to protect the host (Litman et al., 1999; Dudley et al., 2005; Hirano et al., 2011).

Hagfish and lampreys, which are the only two surviving jawless vertebrates (agnathans or cyclostomes), occupy a critical juncture in evolution at the interface of invertebrates and vertebrates. Due to their unique phylogenetic position, hagfish and lampreys have been a focal point in the search for the evolutionary origins of adaptive immunity (Pancer and Cooper, 2006; Litman et al., 2007). Anatomically and physiologically more primitive than lampreys, hagfish are thought to have diverged earlier in vertebrate evolution (Janvier, 1996). In light of this distinction, knowledge of the immunological capacity of hagfish has the potential to provide fundamental insights into the origin and evolution of adaptive immunity.

Early studies of the hagfish adaptive immune response

The defining features of an AIS in vertebrates are the clonally diverse receptors on lymphocytes. They have molecular specificity for antigen recognition and allow the induction of clonal amplification during an initial immune response to provide memory for that specific antigen (Dunn-Walters and Ademokun, 2010). Many investigators have conducted studies on the immune responsiveness of hagfish since the 1960s, and the earlier studies suggested the existence of a rudimentary AIS in this basal vertebrate (Linthicum and Hildemann, 1970; Thoenes and Hildemann, 1970; Raison et al., 1987).

Humoral response

Many investigators have been able to demonstrate specific agglutinin responses to microorganisms and mammalian cells in lampreys, but it has been difficult to induce humoral immune responses in hagfish. Low-titer, naturally occurring agglutinins that are broadly reactive with microbial antigens have been reported in hagfish. Under optimized housing conditions and with frequent immunization over an extended time period, hagfish were found to generate agglutinins against both soluble and cellular antigens. Studies by Thoenes and Hildemann indicated that when maintained in unusually warm water, hagfish can produce high-titer, circulating agglutinins to keyhole limpet hemocyanin (Linthicum and Hildemann, 1970). Repetitive immunization with sheep red blood cells over 3 months could induce agglutinins at titers of up to 1:1024. Gel filtration and immunoelectrophoresis studies of immune hagfish serum suggested an agglutinin molecular weight similar to that of IgM antibodies in jawed vertebrates (Linthicum and Hildemann, 1970). However, biochemical evidence for Ig-domain-based antibodies could not be obtained in these studies, and the molecular basis for the immune response in this agnathan remained elusive.

Cellular response

Allogeneic recognition has been demonstrated in representatives of almost every phylum of life, although information about histocompatibility genes is available only for three animal taxa: colonial cnidarians (*Hydractinia symbiolongicarpus*) (Nicotra et al., 2009; Rosa et al., 2010), colonial tunicates (*Botryllus schlosseri*) (De Tomaso et al., 2005), and jawed vertebrates. Each of these allorecognition systems employs a different histocompatibility determinant. A notable distinction between invertebrate and vertebrate allorecognition systems is that invertebrate histocompatibility reactions are not regarded as adaptive immune responses due to the absence of evidence for immunological memory. The basis for allorecognition has been well defined only in jawed vertebrates, where somatically diversified $\alpha\beta$ T cells recognize peptides in the context of self-MHC molecules, and strenuous clonal selection for self-peptides occurs during T cell development in the thymus (Starr et al., 2003). Although *MHC* genes and a thymus-equivalent tissue have not been found in Hagfish, they have been shown to mount cellular responses to allogeneic stimuli and to recognize and reject skin allografts (Hildemann and Thoenes, 1969). The histopathological features of these allograft reactions, which include lymphocyte infiltration, hemorrhage, and pigment cell destruction, are much like those found in jawed vertebrates. Although accelerated secondary responses were observed, insufficient "third-party" grafts did not allow for clear interpretation of the results. Hagfish leukocytes were also shown to proliferate in response to *in vitro* stimulation with allogeneic cells (Raison et al., 1987). The responding cells were small leukocytes, whereas the stimulating cells were found in the granulocyte/monocyte population; however, the limited allogeneic pairings assessed in these studies precluded resolution of the extent and cellular mechanisms involved.

Lymphoid tissues

In jawed vertebrates, clonally diverse T and B cells are generated in primary lymphoid tissues, where they also undergo selection to eliminate self-reactive clones from the repertoire. T cells develop in the thymus in all jawed vertebrate species, and B cells develop in hematopoietic tissues, such as the bone marrow in mammals and the bursa of Fabricius in birds. In mammals, the spleen, lymph nodes, and other secondary lymphoid tissues are well-defined sites where antigen-driven lymphocyte clonal expansion and antibody affinity maturation occur. Thymus and bone marrow equivalents have not been found in hagfish, nor has a spleen equivalent been identified. Yet, despite the lack of such defined tissues, hagfish are able to mount immunological responses, including allograft rejection and production of agglutinins to soluble and cellular antigens as noted above. The pharyngeal velar muscles of hagfish contain collections of small mononuclear cells and these have been suggested to represent a primordial thymus (Riviere et al., 1975). An island-like lymphoid structure containing an abundance of granulocytes has been described in the intestinal submucosa of hagfish (Tanaka et al., 1981). The pronephros is the major hematopoietic organ in teleost fish, and the pronephros of hagfish notably contains a large number of basophilic lymphocyte-like cells and plasma cells (Zapata et al., 1984).

Discovery of VLRs

In the jawed vertebrates, T and B lymphocytes are primarily responsible for cell-mediated and humoral immunity, respectively, and they work together with phagocytic cells and other cell types to mediate an effective adaptive immune response. Cells with similar morphological features and much of the molecular machinery of lymphocytes in jawed vertebrates are also found in lampreys and hagfish (Najakshin et al., 1999; Mayer et al., 2002; Nagata et al., 2002; Uinuk-Ool et al., 2002). These findings coupled with earlier observations that lampreys and hagfish produce specific agglutinins following immunization with bacteria and foreign red blood cells (Finstad and Good, 1964; Linthicum and Hildemann, 1970; Pollara et al., 1970) suggested that agnathans have an AIS. However, the molecular basis for the agnathan AIS was not discovered until 2004.

As the nearest living phylogenetic relatives of jawed vertebrates, it was anticipated that lampreys and hagfish would have orthologous genes responsible for adaptive immunity. Transcriptome analysis of lamprey and hagfish lymphocytes indeed revealed several genes orthologous to those that jawed vertebrate lymphocytes use for cellular migration, proliferation, differentiation, and intracellular signaling (Mayer et al., 2002; Nagata et al., 2002; Uinuk-Ool et al., 2002; Pancer et al., 2004b; Rothenberg and Pant, 2004; Suzuki et al., 2004). A lamprey *TCR-like* gene with both Ig V (variable) and J (joining) sequences was found, but unlike the jawed vertebrate *TCR* genes, the lamprey *TCR*-like gene is a single copy gene composed of a single exon encoding both V- and J-like sequences and two functional immunoreceptor tyrosine-based inhibitory motifs in the cytoplasmic domain (Pancer et al., 2004b; Yu et al., 2009). A VpreB-like gene expressed by lymphocyte-like cells in lampreys (Cannon et al., 2005) and a family of paired-Ig-like receptor genes encoding transmembrane proteins with activating and inhibitory potential, named agnathan paired receptor resembling Ag receptors (*APAR*), were identified in hagfish (Suzuki et al., 2005). However, no *MHC*, *TCR*, *BCR*, and *RAG* orthologs could be found in the agnathans.

In a revised strategy to search for antigen receptor genes in jawless vertebrates, lamprey larvae were stimulated by an antigen and mitogen mixture to survey the transcriptome of activated lamprey lymphocytes. These experiments were based on the assumption that expression of immune-related genes would be increased during an immune response. The population of large lymphoblastoid cells induced by the stimulating mixture were then sorted by the light scatter characteristics of the activated cells and used to construct a cDNA library, which was then subtracted by myeloid and erythrocyte cDNAs (Pancer et al., 2004a). *TCR*, *BCR*, and *MHC* genes were not detected still, but the lymphoblastoid cDNA-enriched library contained a large number of transcripts that encoded for highly diverse LRR proteins. These were named variable lymphocyte receptors (VLRs) because of their lymphocyte-restricted expression and sequence diversity (Pancer et al., 2004a). Each *VLR* transcript encoded a conserved signal peptide (SP) followed by highly variable LRR modules: a 27–38 residue N-terminal LRR (LRRNT), the first 18-residue LRR (LRR1), up to eight 24-residue variable LRRs (LRRV), one 24-residue terminal LRRV (LRRVt), one 13-residue connecting peptide (CP), and a 48–65 residue C-terminal LRR (LRRCT) (Figure 8.2). The invariant threonine/proline-rich stalk region contained a glycosyl-phosphatidyl-inositol (GPI) cleavage site, and the phospholipase cleavage of a recombinant VLR from the surface of transduced mouse cell line was indicative of its GPI membrane anchorage.

After the discovery of the first lamprey *VLR* (now called *VLRB*), two hagfish *VLR* homologs, *VLRA* (now renamed as *VLRC*) and *VLRB*,

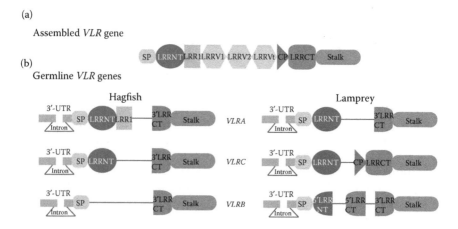

Figure 8.2 (**See color insert.**) Lamprey and hagfish *VLR* genes. (a) Assembled "mature" VLR: SP, LRRNT, first LRR1, two variable VLR (LRRV), end LRRV (LRRVt), connecting peptide (CP), LRRCT, stalk region. (b) Hagfish and lamprey germline *VLRA*, *VLRC*, and *VLRB* genes. Lamprey *VLRB* has two large noncoding intervening sequences separating the 5′ LRRNT and 3′ LRRCT, whereas all other *VLR* genes have a single, short intervening sequence (not drawn to scale).

were identified in an expressed sequence tag database of hagfish leuko-cyte transcripts (Pancer et al., 2005). A lamprey *VLRA* was identified in a subsequent search of the draft sequence database of the sea lamprey genome (Rogozin et al., 2007). Later, a third *VLR* gene known as *VLRC* was discovered in lamprey (Kasamatsu et al., 2010) and, more recently, a third *VLR* gene was also identified in hagfish (Li et al., 2013). Based on comparative analysis of the three types of jawless vertebrate *VLR* genes, we have proposed a new nomenclature for the hagfish *VLRs* because the novel third *VLR* in hagfish is the true lamprey *VLRA* counterpart and the previously identified hagfish "*VLRA*" is orthologous to the lam-prey *VLRC*.

Hagfish VLR genes: Structure and comparison with lamprey VLR genes

All germline *VLR* genes in hagfish and lampreys are incomplete, and they typically have coding sequences only for the leader sequence, incom-plete amino- and carboxy-terminal LRR subunits and for the stalk region (Figure 8.2) (Li et al., 2013). There are two exons, the first of which encodes only a portion of the 5′ untranslated region. The second exon typically contains the rest of the 5′ untranslated region, an SP, a 5′ portion of the LRRNT, a 3′ portion of the LRRCT, and the stalk region. For hagfish *VLRA*, *VLRB* and *VLRC* and for lamprey *VLRA* and *VLRC*, the 5′ LRRNT sequence is separated from the 3′ LRRCT sequence by a short noncod-ing intervening sequence that lacks canonical splice donor and acceptor sites. The first coding region encodes the SP, LRRNT, and LRR1 modules, whereas the second coding region encodes the 3′-end of LRRCT and the C-terminus domain (Figure 8.2). The lamprey *VLRB* gene is more complex in that it has a 5′ LRRCT coding sequence located between two interven-ing sequences. Each germline *VLR* gene is flanked by hundreds of *LRR*-encoding sequences, which are used as randomly selected templates to add the missing *LRR* cassettes needed for completion of a mature *VLR* gene. The demonstration of three orthologous *VLR* genes in both lam-preys and hagfish suggests that this anticipatory receptor system evolved in a common ancestor of the two cyclostome lineages around 480 million years ago (Figure 8.1).

Hagfish VLR assembly mechanism and sequence diversity

In both lamprey and hagfish, a gene conversion-like mechanism has been postulated for the complex *VLR* assembly process (Cooper and Alder, 2006; Nagawa et al., 2007) in which the intervening sequence is replaced in a stepwise, piecewise manner by the random usage of flanking *LRR* cassettes as templates for adding the necessary sequences to form a complete *VLR* gene (Figure 8.3). The assembly process can be initiated at either the 5′ LRRNT or the 3′ LRRCT ends (Alder et al., 2005; Nagawa et al., 2007) and appears to depend on short segments of nucleotide homology (10–30 bp) between the donor and acceptor sequences to direct the copying of flanking LRR-encoding genomic donor cassettes into the germline gene (Nagawa et al., 2007) (Figure 8.3). Analysis of the lamprey genome indicated that the germline *VLRC* gene is flanked by five types of donor LRR-encoding cassettes which contribute to the stepwise assembly of mature *VLRC* genes: (i) 3′ LRRNT–5′ LRR1 cassettes; (ii) 3′ LRR1–5′ LRRV cassettes; (iii) 3′ LRRV–5′ LRRV cassettes; (iv) 3′ LRRV-CP–5′ LRRCT cassettes; (v) and LRRCT cassettes (Das et al., 2013). Through the recombinatorial usage of these donor cassettes, a potential repertoire of >10^{14} distinct receptors could be achieved, a diversity that is comparable to that of the Ig-based antigen receptors (Alder et al., 2005). Notably, in keeping with their lack of recombination signal sequences and the apparent absence of lamprey *RAG1* and *RAG2* genes, the genomic LRR donor cassettes are not rearranged during the *VLR* assembly process. Instead, LRR cassette sequences are copied into the germline *VLR* gene. The agnathan *VLR* genes are assembled on one allele at a time, and allelic exclusion is the general rule for *VLR* gene assembly (Pancer et al., 2004a; Nagawa et al., 2007).

Although the hagfish *VLRB* and *VLRC* loci are located on the same chromosome, they are not close to each other, thereby facilitating their function as separate genetic units (Kasamatsu et al., 2007). In contrast, the *VLRA* and *VLRC* genes in European lamprey are located in close proximity in the genome (Das et al., 2014). Notably, the four types (3′ LRRNT–5′ LRR1, 3′ LRR1–5′ LRRV, 3′ LRR1–5′ LRRV, and 3′ LRRV-CP–5′ LRRCT) of the genomic donor cassettes may be shared in the assembly of the *VLRA* and *VLRC* genes in lampreys (Figure 8.4a). The hagfish genome sequence is currently unavailable, but analysis of a large number of mature *VLRA* and *VLRC* sequences in hagfish indicates that, as in lampreys, some genomic donor cassettes are shared for the assembly of *VLRA* and *VLRC* genes (Das et al., 2014). In contrast, donor cassette sharing is not evident between mature *VLRA* and *VLRB* genes or between the mature *VLRC* and *VLRB* genes in either lampreys or hagfish. Among the different types of genomic *LRR* cassettes, the 3′LRRV–5′ LRRV donor sequences are most frequently shared by *VLRA*s and *VLRC*s in both of these agnathans. The comparative analysis of their mature *VLRA* and *VLRC* sequences suggests that cassette sharing for *VLRA* and *VLRC* assemblies is more frequent in hagfish than in lamprey (Das et al., 2014). A similar phenomenon occurs in the *TCRα/δ* locus in the jawed vertebrates, whereby *TCRα* and *TCRδ* gene assemblies may

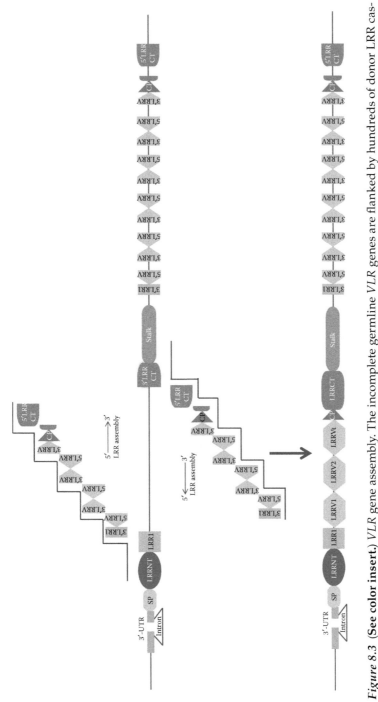

Figure 8.3 (**See color insert.**) *VLR* gene assembly. The incomplete germline *VLR* genes are flanked by hundreds of donor LRR cassettes. The noncoding intervening sequence between portions of the LRRNT and LRRCT is replaced by LRR fragments that are sequentially copied from randomly selected donor LRR cassettes. The assembly process can be initiated from either the LRRNT or LRRCT end in a stepwise manner that is directed by short sequence homology between the donor and recipient LRR sequences.

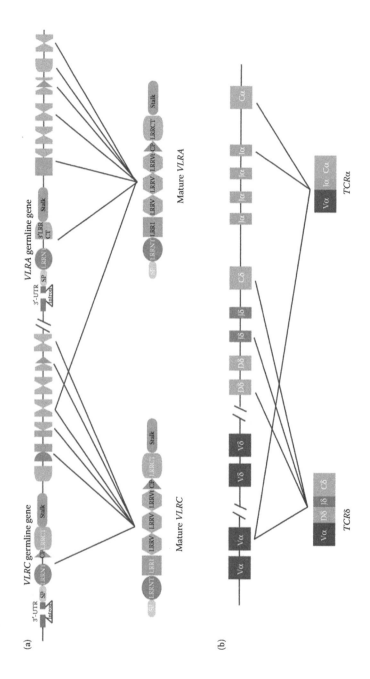

Figure 8.4 Cassette sharing in jawless vertebrates and V region sharing in jawed vertebrates. (a) Genomic organizations of germline genes and LRR donor cassettes in the *VLRA/C* loci of lampreys. Note the shared LRR cassettes during VLRA and VLRC assembly. (b) Genomic organization of constant and variable regions in the *TCRα/δ* loci of jawed vertebrates. Variable regions can be shared during the recombination process. Note the similarities of cassette and V region sharing in these two vertebrate lineages. (Modified from Das, S. et al. 2014. *Proc Natl Acad Sci USA* 111:14828–14833.)

share members of the overlapping pool of *V* genes (Kubota et al., 1999; Krangel et al., 2000). The close proximity of the *VLRA* and *VLRC* loci, which allows some of the donor *LRR* cassettes used for *VLRA* and *VLRC* assembly, is reminiscent of the organization of *TCRα/δ* locus in jawed vertebrates (Figure 8.4b), wherein the *TCRα* and *TCRδ* gene segments are interspersed.

The molecules and the mechanistic details involved in *VLR* gene assembly have not yet been elucidated. However, two activation-induced cytidine deaminase (AID)/apolipoprotein B mRNA editing catalytic component family orthologs, cytidine deaminase 1 (CDA1) and 2 (CDA2) (Rogozin et al., 2007), have been identified in lamprey and these enzymes are thought to be involved in the *VLR* assembly. Furthermore, *CDA1* is expressed during development of the VLRA and VLRC lymphocytes, whereas *CDA2* expression is restricted to the VLRB lymphocyte lineage (Guo et al., 2009). These findings are consistent with the hypothesis that CDA1 may catalyze *VLRA* and *VLRC* gene assembly, whereas CDA2 may play a similar role in *VLRB* gene assembly. As AID is essential for somatic hypermutation and affinity maturation of antibodies in jawed vertebrates, a similar role could also be envisioned for CDA2 in the diversification of *VLRBs* in jawless vertebrates.

Structure of the VLR antigen-binding domain

The crystal structures for the LRR portions of hagfish VLRB and VLRC have been solved (Kim et al., 2007). The crystal structure for hagfish VLRA is not yet available, although its 3D structure has been modeled by using lamprey VLRA as a template (Li et al., 2013). All three hagfish VLR isotypes have crescent-shaped solenoid structures that are characteristic of other LRR family proteins (Figure 8.5). The inner concave surface is formed by β-strands in the LRRNT, LRR1, LRRV, and LRRCP components. The convex surface consists of more diverse secondary structure elements, such as loops, α-helices, and 3_{10} helices. The LRRV components of all VLR proteins contain 24 amino acid repeats of consensus sequence $XL^2XXL^5XXL^8XL^{10}XXN^{13}XL^{15}XXL^{18}P^{19}XXXF^{23}X$ (where X can be any amino acid, and L, N, P, and F represent the conserved leucine, asparagine, proline, and phenylalanine, respectively). The leucines and phenylalanine residues form the hydrophobic core of the solenoid structure. The hydrophobic core is also protected by capping N-terminal and C-terminal LRR modules that contain two intramolecular disulfide bonds to support the integrity of the solenoid structure. In the LRRCT portion of VLRA and VLRB, a protruding loop projects toward the concave surface, which makes the full structure of VLRA and VLRB resemble a cupped hand with parallel β-strand lining the palm; the protruding loop is contributed by LRRCT inserts that are highly variable in

VLRA   VLRC   VLRB

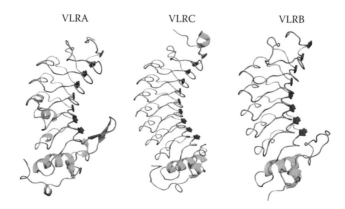

Figure 8.5 (**See color insert.**) Structural comparison of hagfish VLR isotypes. The hagfish VLR proteins are crescent-shaped. Their concave surface is composed of β-strands contributed by the LRRNT, LRR1, variable numbers of LRRV modules, and LRRCP; the LRRCT of VLRA (predicted) and VLRB (PDBID: 2O6S) contain a variable insert loop. VLRC (PDBID: 2O6Q) are structurally similar to VLRA and VLRB, except VLRC does not have an LRRCT insert. β-Sheet and α-helix structures are shown in blue and green colors, respectively. Loops located in the LRRCT portion are indicated in red. (Modified from Li, J. et al. 2013. *Proc Natl Acad Sci USA* 110:15013–15018.)

amino acid composition and in length. The length of the insert ranges from 9 to 13 amino acids in VLRAs, whereas the insert length in VLRBs is more variable, ranging from 0 to 13 amino acids. Due to their very short LRRCT inserts (only 2–3 amino acids), the VLRCs do not form such loops (Li et al., 2013).

The crystallographic structures for lamprey VLRA-HEL and several VLRB-antigen complexes have been solved (Velikovsky et al., 2009; Deng et al., 2010). These studies indicate that hypervariable patches on the concave surface contribute to antigen recognition. In addition, the protruding LRRCT loop makes important antigen contacts in all of the structures that have been characterized so far. In the VLRA.R.2.1-HEL complex, the 10 residue LRRCT insert is critical for complex formation by packing against one side of the antigen (Figure 8.6a). In the VLRB.2D-HEL complex, the 6-residue LRRCT insert projects deep into the catalytic cleft of the enzyme (Figure 8.6b). Hagfish VLR-antigen complexes have not yet been reported, but in accord with lamprey VLRB antibodies, sequence analysis indicates that variable amino acids are mainly distributed on the concave surface and clustered together to form a large hypervariable patch (Figures 8.6c and d). PPI-Pred model prediction and structural comparison imply that this hypervariable patch is likely the antigen-binding site (Kim et al., 2007).

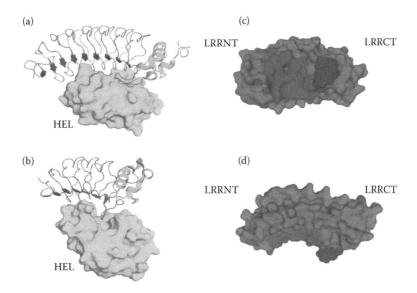

Figure 8.6 (**See color insert.**) VLR antigen-binding sites. (a) Ribbon diagram of lamprey VLRA.R.2.1-HEL complex (PDBID: 3M18). Lamprey VLRA.R.2.1 uses its multiple β-strands on its concave surface and LRRCT loop to bind HEL. (b) Structure of VLRB.2D-HEL complex (PDBID: 3G3A). Note that the LRRCT loop of VLR.2D is inserted into the catalytic cleft of HEL. Upper VLR is shown in ribbon diagram, in which β-sheet and α-helix structures are shown in blue and green colors, respectively. Loops located in the LRRCT portion are indicated in red. HEL is depicted in light blue. (c) Front view of surface diagrams of predicted antigen-binding sites of hagfish VLR, the highly variable residues in the LRR modules labeled in magenta form the binding patch on the concave surface; the protruding loop contributed by the LRRCT insert is labeled in red. (d) Side view of surface diagrams of predicted antigen-binding sites of hagfish VLR, noting the protruding loop.

Hagfish VLR expression in B- and T-like lymphocyte lineages

Lampreys have a B cell-like lineage of lymphocytes that express VLRB and two T cell-like populations that express VLRA and VLRC, respectively (Pancer et al., 2004a; Alder et al., 2008; Guo et al., 2009; Hirano et al., 2013). The VLRA[+] and VLRC[+] cells develop in a thymus-equivalent region located at the tips of the gill folds (Bajoghli et al., 2011). Both types of cells resemble T cells in the jawed vertebrates in that they express their receptor only on the cell surface, express similar cytokine, chemokine and transcription factor gene profiles, and preferentially respond to the classic T cell mitogen, phytohemaglutinin (Guo et al., 2009; Hirano et al., 2013). Conversely, VLRB[+] cells develop in the typhlosole, a hematopoietic tissue, and differentiate into plasma cells that secrete multimeric VLRB antibodies upon antigenic stimulation (Alder et al., 2005, 2008; Herrin et al., 2008; Guo et al., 2009; Hirano et al., 2013). Thus, they closely resemble the B cells in jawed vertebrates.

The hagfish cell lineages that express VLRA, VLRB, and VLRC have not yet been well characterized. Single-cell polymerase chain reaction analysis indicates that hagfish VLRB and VLRC are expressed by two distinct lymphocyte populations, and analysis of hagfish serum indicates that only VLRB is secreted (Takaba et al., 2013; Li et al., unpublished data). These findings indicate that hagfish VLRB+ lymphocytes are B-like cells and VLRC+ lymphocytes are T-like cells, thereby resembling their lamprey counterparts (Figure 8.7). However, the tissue distribution and gene expression profiles of hagfish VLRA+, VLRB+, and VLRC+ cells remain to be elucidated, and the site(s) in which they are generated are presently unknown.

Functional potential of the agnathan AIS

The selective pressure for the evolution of an alternative VLR-based AIS in hagfish and lampreys was most likely the advantage offered by

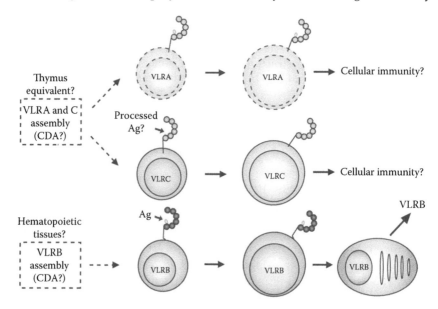

Figure 8.7 Model of B- and T-like lymphocytes in hagfish. Antigens stimulate lymphocytes including VLRC and VLRB cells, but whether or not receptors of VLRC type, see native or processed and presented antigens is unknown. Antigen-stimulated VLRB cells differentiate into VLRB-secreting plasma cells, whereas activated VLRC cells do not secrete their antigen receptors. Dashed lines highlight some of the presently unresolved features of this hypothetical model.

acquisition of the capacity for specific recognition, response, and memory of pathogen threats. However, we lack information about the infectious agents that agnathans encountered more than 450 million years ago. Even now, the data on the types of infections that agnathans have is limited, especially for hagfish, although several reports suggest that bacterial infections are not uncommon in lampreys (Eissa et al., 2006; Merivirta et al., 2006; Faisal et al., 2007; Sobecka et al., 2009). The protective value of VLR-based immunity has also not been explored extensively. An effector function has been demonstrated for lamprey VLRB antibodies; together with lamprey C1q and C3, VLRB antibodies can kill bacteria and tumor cells (Wu et al., 2013). Additionally, in response to lamprey challenge with *Escherichia coli*, VLRB antibodies have been observed on circulating granulocytes (Nathanael McCurley, unpublished data), but the basis for the VLRB binding to the granulocytes and its functional consequences has not been elucidated. The future analysis of the functional interplay between adaptive and innate immune system components should be most informative. An extensive definition of the humoral and cellular components of the innate immune system in hagfish and lampreys is also needed to address the larger issue of how a lymphocyte-based, antigen-specific AIS evolved to enhance vertebrate survival.

Conclusion

Hagfish and lampreys are the only extant cyclostome representatives and they last had a common ancestor ~480 million years ago (Kuraku et al., 2009). Hagfish are considered the most basal vertebrate representative and thus represent a pivotal point for the study of AIS evolution. Studies in jawed and jawless vertebrates have revealed the convergent evolution of surprisingly different recombinatorial strategies for generating clonally diverse receptors in order to achieve specific adaptive immunity. This new phylogenetic insight strongly attests the fitness value of an AIS and favors the view that lymphocyte lineage bifurcation preceded the evolutionary split of the two vertebrate lineages, rather than the formerly held view that a primordial Ig V gene was pivotal to the evolution of T and B cells. The fundamental AIS strategy of functionally interactive T and B lymphocytes may reflect the essential need to counter the inevitable threat of autoimmunity posed by the development of diverse and randomly generated receptor repertoires expressed by lymphocytes with proinflammatory potential. Recent studies in lampreys indicate that VLRA and VLRC lymphocytes are generated in the thymoid gill region and that their gene expression profiles, ordered receptor assembly, functional characteristics, and homing patterns resemble those for mammalian γδ and αβ T cells. Due to the basal position of hagfish in vertebrate phylogeny, future studies of the hagfish AIS will bring us closer to an understanding of how and why the lymphocyte bifurcations along T and B cell lineages occurred during evolution. Identification of the allorecognition system components that elicit hagfish T cell responses is of special importance, because this information is the key to resolving the issues of what the cyclostome TCRs recognize and how they discriminate between self and non-self.

Acknowledgments

This work was supported by NIH grants AI072435, GM100151, and GM108838 and the Georgia Research Alliance.

References

Alder, M. N., B. R. Herrin, A. Sadlonova, C. R. Stockard, W. E. Grizzle, L. A. Gartland, G. L. Gartland, et al. 2008. Antibody responses of variable lymphocyte receptors in the lamprey. *Nat Immunol* 9:319–327.

Alder, M. N., I. B. Rogozin, L. M. Iyer, G. V. Glazko, M. D. Cooper and Z. Pancer. 2005. Diversity and function of adaptive immune receptors in a jawless vertebrate. *Science* 310:1970–1973.

Bajoghli, B., P. Guo, N. Aghaallaei, M. Hirano, C. Strohmeier, N. McCurley, D. E. Bockman, M. Schorpp, M. D. Cooper and T. Boehm. 2011. A thymus candidate in lampreys. *Nature* 470:90–94.

Cannon, J. P., R. N. Haire, Z. Pancer, M. G. Mueller, D. Skapura, M. D. Cooper and G. W. Litman. 2005. Variable domains and a VpreB-like molecule are present in a jawless vertebrate. *Immunogenetics* 56:924–929.

Cooper, M. D. and M. N. Alder. 2006. The evolution of adaptive immune systems. *Cell* 124:815–822.

Das, S., M. Hirano, N. Aghaallaei, B. Bajoghli, T. Boehm and M. D. Cooper. 2013. Organization of lamprey variable lymphocyte receptor C locus and repertoire development. *Proc Natl Acad Sci USA* 110:6043–6048.

Das, S., J. Li, S. J. Holland, L. M. Iyer, M. Hirano, M. Schorpp, L. Aravind, M. D. Cooper and T. Boehm. 2014. Genomic donor cassette sharing during VLRA and VLRC assembly in jawless vertebrates. *Proc Natl Acad Sci USA* 111:14828–14833.

Deng, L., C. A. Velikovsky, G. Xu, L. M. Iyer, S. Tasumi, M. C. Kerzic, M. F. Flajnik, L. Aravind, Z. Pancer and R. A. Mariuzza. 2010. A structural basis for antigen recognition by the T cell-like lymphocytes of sea lamprey. *Proc Natl Acad Sci USA* 107:13408–13413.

De Tomaso, A. W., S. V. Nyholm, K. J. Palmeri, K. J. Ishizuka, W. B. Ludington, K. Mitchel and I. L. Weissman. 2005. Isolation and characterization of a protochordate histocompatibility locus. *Nature* 438:454–459.

Dudley, D. D., J. Chaudhuri, C. H. Bassing and F. W. Alt. 2005. Mechanism and control of V(D)J recombination versus class switch recombination: Similarities and differences. *Adv Immunol* 86:43–112.

Dunn-Walters, D. K. and A. A. Ademokun. 2010. B cell repertoire and ageing. *Curr Opin Immunol* 22:514–520.

Eissa, A. E., E. E. Elsayed, R. McDonald and M. Faisal. 2006. First record of *Renibacterium salmoninarum* in the sea lamprey (*Petromyzon marinus*). *J Wildl Dis* 42:556–560.

Faisal, M., A. E. Eissa and E. E. Elsayed. 2007. Isolation of *Aeromonas salmonicida* from sea lamprey (*Petromyzon marinus*) with furuncle-like lesions in Lake Ontario. *J Wildl Dis* 43:618–622.

Finstad, J. and R. A. Good. 1964. The evolution of the immune response. 3. Immunologic responses in the Lamprey. *J Exp Med* 120:1151–1168.

Guo, P., M. Hirano, B. R. Herrin, J. Li, C. Yu, A. Sadlonova and M. D. Cooper. 2009. Dual nature of the adaptive immune system in lampreys. *Nature* 459:796–801.

Herrin, B. R., M. N. Alder, K. H. Roux, C. Sina, G. R. Ehrhardt, J. A. Boydston, C. L. Turnbough, Jr. and M. D. Cooper. 2008. Structure and specificity of lamprey monoclonal antibodies. *Proc Natl Acad Sci USA* 105:2040–2045.

Hildemann, W. H. and G. H. Thoenes. 1969. Immunological responses of Pacific hagfish. I. Skin transplantation immunity. *Transplantation* 7:506–521.

Hirano, M., S. Das, P. Guo and M. D. Cooper. 2011. The evolution of adaptive immunity in vertebrates. *Adv Immunol* 109:125–157.

Hirano, M., P. Guo, N. McCurley, M. Schorpp, S. Das, T. Boehm and M. D. Cooper. 2013. Evolutionary implications of a third lymphocyte lineage in lampreys. *Nature* 501:435–438.

Janvier, P. 1996. *Early Vertebrates*. Oxford University Press, Oxford.

Kasamatsu, J., Y. Sutoh, K. Fugo, N. Otsuka, K. Iwabuchi and M. Kasahara. 2010. Identification of a third variable lymphocyte receptor in the lamprey. *Proc Natl Acad Sci USA* 107:14304–14308.

Kasamatsu, J., T. Suzuki, J. Ishijima, Y. Matsuda and M. Kasahara. 2007. Two variable lymphocyte receptor genes of the inshore hagfish are located far apart on the same chromosome. *Immunogenetics* 59:329–331.

Kawai, T. and S. Akira. 2009. The roles of TLRs, RLRs and NLRs in pathogen recognition. *Int Immunol* 21:317–337.

Kim, H. M., S. C. Oh, K. J. Lim, J. Kasamatsu, J. Y. Heo, B. S. Park, H. Lee, O. J. Yoo, M. Kasahara and J. O. Lee. 2007. Structural diversity of the hagfish variable lymphocyte receptors. *J Biol Chem* 282:6726–6732.

Krangel, M. S., M. T. McMurry, C. Hernandez-Munain, X. P. Zhong and J. Carabana. 2000. Accessibility control of T cell receptor gene rearrangement in developing thymocytes. The TCR alpha/delta locus. *Immunol Res* 22:127–135.

Kubota, T., J. Wang, T. W. Gobel, R. D. Hockett, M. D. Cooper and C. H. Chen. 1999. Characterization of an avian (*Gallus gallus domesticus*) TCR alpha delta gene locus. *J Immunol* 163:3858–3866.

Kuraku, S., A. Meyer and S. Kuratani. 2009. Timing of genome duplications relative to the origin of the vertebrates: Did cyclostomes diverge before or after? *Mol Biol Evol* 26:47–59.

Li, J., S. Das, B. R. Herrin, M. Hirano and M. D. Cooper. 2013. Definition of a third VLR gene in hagfish. *Proc Natl Acad Sci USA* 110:15013–15018.

Linthicum, D. S. and W. H. Hildemann. 1970. Immunologic responses of Pacific hagfish. 3. Serum antibodies to cellular antigens. *J Immunol* 105:912–918.

Litman, G. W., M. K. Anderson and J. P. Rast 1999. Evolution of antigen binding receptors. *Annu Rev Immunol* 17:109–147.

Litman, G. W., L. J. Dishaw, J. P. Cannon, R. N. Haire and J. P. Rast. 2007. Alternative mechanisms of immune receptor diversity. *Curr Opin Immunol* 19:526–534.

Mayer, W. E., T. Uinuk-Ool, H. Tichy, L. A. Gartland, J. Klein and M. D. Cooper. 2002. Isolation and characterization of lymphocyte-like cells from a lamprey. *Proc Natl Acad Sci USA* 99:14350–14355.

Merivirta, L. O., M. Lindström, K. J. Björkroth and H. J. Korkeala. 2006. The prevalence of *Clostridium botulinum* in European river lamprey (*Lampetra fluviatilis*) in Finland. *Int J Food Microbiol* 109:234–237.

Nagata, T., T. Suzuki, Y. Ohta, M. F. Flajnik and M. Kasahara. 2002. The leukocyte common antigen (CD45) of the Pacific hagfish, *Eptatretus stoutii*: Implications for the primordial function of CD45. *Immunogenetics* 54:286–291.

Nagawa, F., N. Kishishita, K. Shimizu, S. Hirose, M. Miyoshi, J. Nezu, T. Nishimura et al. 2007. Antigen-receptor genes of the agnathan lamprey are assembled by a process involving copy choice. *Nat Immunol* 8:206–213.

Najakshin, A. M., L. V. Mechetina, B. Y. Alabyev and Taranin, A. V. 1999. Identification of an IL-8 homolog in lamprey (*Lampetra fluviatilis*): Early evolutionary divergence of chemokines. *Eur J Immunol* 29:375–382.

Nicotra, M. L., A. E. Powell, R. D. Rosengarten, M. Moreno, J. Grimwood, F. G. Lakkis, S. L. Dellaporta and L. W. Buss. 2009. A hypervariable invertebrate allodeterminant. *Curr Biol* 19:583–589.

Pancer, Z., C. T. Amemiya, G. R. Ehrhardt, J. Ceitlin, G. L. Gartland and M. D. Cooper. 2004a. Somatic diversification of variable lymphocyte receptors in the agnathan sea lamprey. *Nature* 430:174–180.

Pancer, Z. and M. D. Cooper. 2006. The evolution of adaptive immunity. *Annu Rev Immunol* 24:497–518.

Pancer, Z., W. E. Mayer, J. Klein and M. D. Cooper. 2004b. Prototypic T cell receptor and CD4-like coreceptor are expressed by lymphocytes in the agnathan sea lamprey. *Proc Natl Acad Sci USA* 101:13273–13278.

Pancer, Z., N. R. Saha, J. Kasamatsu, T. Suzuki, C. T. Amemiya, M. Kasahara and M. D. Cooper. 2005. Variable lymphocyte receptors in hagfish. *Proc Natl Acad Sci USA* 102:9224–9229.

Pollara, B., G. W. Litman, J. Finstad, J. Howell and R. A. Good. 1970. The evolution of the immune response. VII. Antibody to human "O" cells and properties of the immunoglobulin in lamprey. *J Immunol* 105:738–745.

Raison, R. L., P. Gilbertson and J. Wotherspoon. 1987. Cellular requirements for mixed leucocyte reactivity in the cyclostome, *Eptatretus stoutii*. *Immunol Cell Biol* 65 (Pt 2), 183–188.

Riviere, H. B., E. L. Cooper, A. L. Reddy and W. H. Hildemann. 1975. In search of the hagfish thymus. *Am Zool* 15:39–49.

Rogozin, I. B., L. M. Iyer, L. Liang, G. V. Glazko, V. G. Liston, Y. I. Pavlov, L. Aravind and Z. Pancer. 2007. Evolution and diversification of lamprey antigen receptors: Evidence for involvement of an AID-APOBEC family cytosine deaminase. *Nat Immunol* 8:647–656.

Rosa, S. F., A. E. Powell, R. D. Rosengarten, M. L. Nicotra, M. A. Moreno, J. Grimwood, F. G. Lakkis, S. L. Dellaporta and L. W. Buss. 2010. Hydractinia allodeterminant alr1 resides in an immunoglobulin superfamily-like gene complex. *Curr Biol* 20:1122–1127.

Rothenberg, E. V. and R. Pant. 2004. Origins of lymphocyte developmental programs: Transcription factor evidence. *Semin Immunol* 16:227–238.

Sobecka, E., J. Moskal and B. Więcaszek. 2009. Checklist of the pathogens of lamprey species of Poland. *Oceanol Hydrobiol Stud* 38:129–137.

Starr, T. K., S. C. Jameson and K. A. Hogquist. 2003. Positive and negative selection of T cells. *Annu Rev Immunol* 21:139–176.

Suzuki, T., I. T. Shin, A. Fujiyama, Y. Kohara and M. Kasahara 2005. Hagfish leukocytes express a paired receptor family with a variable domain resembling those of antigen receptors. *J Immunol* 174:2885–2891.

Suzuki, T., I. T. Shin, Y. Kohara and M. Kasahara. 2004. Transcriptome analysis of hagfish leukocytes: A framework for understanding the immune system of jawless fishes. *Dev Comp Immunol* 28:993–1003.

Takaba, H., T. Imai, S. Miki, Y. Morishita, A. Miyashita, N. Ishikawa, H. Nishizumi and H. Sakano. 2013. A major allogenic leukocyte antigen in the agnathan hagfish. *Sci Rep* 3:1716.

Tanaka, Y., Y. Saito and H. Gotoh. 1981. Vascular architecture and intestinal hematopoietic nests of two cyclostomes, *Eptatretus burgeri* and ammocoetes of *Entosphenus reissneri*: A comparative morphological study. *J Morphol* 170:71–93.

Thoenes, G. H. and W. H. Hildemann. 1970. Immunological responses of Pacific hagfish. II. Serum antibody production to soluble antigens. In: J. Sterzl and I. Riha (eds.), *Developmental Aspects of Antibody Formation and Structure*, Vol. II. Academia Publishing House of the Czechoslovak Academy of Science, Prague, pp. 711–726.

Uinuk-Ool, T., W. E. Mayer, A. Sato, R. Dongak, M. D. Cooper and J. Klein. 2002. Lamprey lymphocyte-like cells express homologs of genes involved in immunologically relevant activities of mammalian lymphocytes. *Proc Natl Acad Sci USA* 99:14356–14361.

Velikovsky, C. A., L. Deng, S. Tasumi, L. M. Iyer, M. C. Kerzic, L. Aravind, Z. Pancer and R. A. Mariuzza. 2009. Structure of a lamprey variable lymphocyte receptor in complex with a protein antigen. *Nat Struct Mol Biol* 16:725–730.

Wu, F., L. Chen, X. Liu, H. Wang, P. Su, Y. Han, B. Feng, X. Qiao, J. Zhao, N. Ma, H. Liu, Z. Zheng and Q. Li. 2013. Lamprey variable lymphocyte receptors mediate complement-dependent cytotoxicity. *J Immunol* 190:922–930.

Yu, C., G. R. Ehrhardt, M. N. Alder, M. D. Cooper and A. Xu. 2009. Inhibitory signaling potential of a TCR-like molecule in lamprey. *Eur J Immunol* 39:571–579.

Zapata, A., R. Fange, A. Mattisson and Villena, A. 1984. Plasma cells in adult Atlantic hagfish, *Myxine glutinosa*. *Cell Tissue Res* 235:691–693.

chapter nine

Hypothalamic–pituitary–gonadal endocrine system in the hagfish

Masumi Nozaki and Stacia A. Sower

Contents

The hypothalamic–pituitary system, which is specific to vertebrates, is considered to be an evolutionary innovation and seminal event that emerged prior to or during the differentiation of the ancestral jawless vertebrates (agnathans) (Sower et al., 2009). Such an evolutionary innovation is one of the key elements, leading to physiological divergence, including reproduction, growth, metabolism, stress, and osmoregulation in subsequent evolution of jawed vertebrates (gnathostomes). The control and the integration of reproductive processes are generally regarded as probably the oldest and original functions of the vertebrate hypothalamic–pituitary system. Of all vertebrate species, both extant and extinct, hagfishes are considered to be the earliest divergent (Forey and Janvier, 1993, 1994), and possess the most primitive hypothalamic–pituitary system. Accordingly, hagfishes are of particular importance in understanding the evolution of the hypothalamic–pituitary system related to vertebrate reproduction. However, until the recent identification of a functional gonadotropin (GTH) in the hagfish and its effects on the gonads (Uchida et al., 2010), it had not been established on whether the hagfish pituitary gland contained tropic hormones of any kind and whether the pituitary gland regulated gonadal functions in the hagfish. Previous studies showed a correlation of the hypothalamic gonadotropin-releasing hormone (GnRH) associated with reproductive cycle in hagfish (Kavanaugh et al., 2005). Additional recent studies on the hagfish steroidogenic hormones and enzymes have revealed that the gonadal functions are under the control of the pituitary gonadotropic activities. Even with these recent data, the hypothalamic–pituitary–gonadal system of the hagfish is still the least understood of all the vertebrates. This chapter summarizes the recent findings on the hypothalamic–pituitary–gonadal endocrine system involved in the reproduction in the hagfish.

Hagfish hypothalamic–pituitary system

Background

The pituitary gland is present in all vertebrates from agnathans (jawless fishes) to mammals and consists of the same two principal elements, the neurohypophysis and adenohypophysis. The neurohypophysis develops from the floor of the diencephalon as an infundibular extension, whereas the adenohypophysis develops from the epithelium that comes in contact with this infundibulum. The enigma of the pituitary gland is that evolution of a composite organ with such a complex double developmental origin must have been associated with some functionally adaptive value and probably resulted due to whole genome duplications that occur early in vertebrate evolution (Smith et al., 2013). Yet, the demonstration of this adaptive value in one of the agnathans, hagfish, has remained elusive. Extensive molecular, biochemical, and physiological studies have shown that lamprey, the other member of extant agnathans, has the conserved hypothalamic–pituitary–gonadal axis with some distinct differences (Sower et al., 2009). The pituitary gland along with the three major pituitary hormone families, glycoprotein hormone family that includes the GTHs and thyroid-stimulating hormone (TSH), growth hormone (GH) family that includes GH and prolactin and the hormones derived from pro-opiomelanocortin such as adrenocorticotropin (ACTH) and melano-phore-stimulating hormone (MSH), and their receptors emerged during the early evolution of the ancestral vertebrates (Kawauchi and Sower, 2006; Decatur et al., 2013). In protochordates, neither *Ciona* nor amphioxus has these pituitary hormone genes in their genomes (Holland et al., 2008).

The hagfish possesses the most primitive hypothalamic–pituitary system. The neurohypophysis of the hagfish is a flattened sac-like structure, whereas the adenohypophysis consists of a mass of clusters of cells embedded in connective tissue below the neurohypophysis (Holmes and Ball, 1974; Hardisty, 1979) (Figures 9.1a and b). The adenohypophysis and the neurohypophysis are completely separated by a layer of connective tissue, and there is no or little anatomical relationship between them (Gorbman et al., 1963; Kobayashi and Uemura, 1972) (Figures 9.1a,b). In addition, there is no clear cytological differentiation between the pars distalis and the pars intermedia (Holmes and Ball, 1974; Hardisty, 1979)

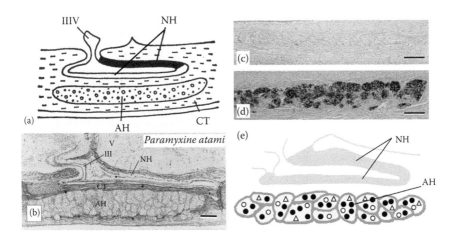

Figure 9.1 (a) Diagrammatically sagittal section of the hagfish pituitary gland. Dark area of the neurohypophysis (NH) shows posterior part of the dorsal wall, where ir-GnRH nerve fibers and AVT nerve fibers are densely accumulated. (b) Nearly mid-sagittal section of the pituitary gland of the brown hagfish, stained with hematoxylin and eosin. (c and d) GTHβ-like immunoreaction in the ade-nohypophysis of the juvenile (c) and sexually mature (d) brown hagfish stained with anti-hagfish GTHβ. Note that GTH-positive cells are almost absent in c, whereas they are abundant in d. (e) Diagrammatically sagittal section of the hag-fish pituitary gland showing the topographic distribution of adenohypophysial cells. Closed circle, GTH cell; open circle, ACTH cell; open triangle, undifferenti-ated cell and possible GH cell. AH, adenohypophysis; CT, connective tissue; IIIV, third ventricle. Scale bars: 100 μm. (From Nozaki M. 2013. *Front. Endocrinol.* 4:200. doi:10.3389/fendo. 2013.00200.)

(Figure 9.1b). The question arises whether the simplicity of the hagfish pituitary gland is a primitive or a degenerate feature. For example, some authors have claimed that the pars intermedia seem to have been lost via a secondary degenerative process (Hardisty, 1979; Gorbman et al., 1983). Moreover, until recent identification of a functional GTH in the hagfish pituitary (Uchida et al., 2010), it had not been established whether the hagfish pituitary gland contained adenohypophysial hormones of any kind (Matty et al., 1976; Gorbman, 1983). Because of the simplicity and primitiveness of the pituitary morphology and equivocal data on the adenohypophysial hormones in the hagfish, many researchers had questioned whether there were any functional adenohypophysial hormones (Matty et al., 1976; Gorbman, 1983). On the other hand, arginine vasotocin (AVT), as a single neurohypophysial hormone, was identified in the hagfish (Suzuki et al., 1995). In addition, the presence of GnRH has been suggested in the hagfish hypothalamus by both radioimmunoassay and immunohistochemistry (Braun et al., 1995; Sower et al., 1995; Oshima et al., 2001; Kavanaugh et al., 2005) (Figure 9.2). Thus, the hagfish appears to have neurohypophysial and hypothalamic hormones similar to those of other vertebrates.

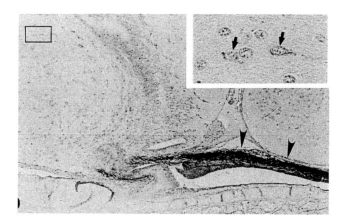

Figure 9.2 A nearly mid-sagittal section through the neurohypophysis of the Atlantic hagfish, *Myxine glutinosa*, showing the accumulation of ir-GnRH in the dorsal wall of the neurohypophysis (arrowheads). This section was stained with anti-salmon GnRH. Inset, An enlargement of the rectangular area showing GnRH-positive neuronal cells. Arrows show GnRH-positive cell bodies. Scale bars: 100 μm; inset, 20 μm. (From Oshima, Y., K. Ominato and M. Nozaki. 2001. Distribution of GnRH-like immunoreactivity in the brain of lampreys and hagfish. Annual Activity Reports of the Sado Marine Biological Station, Niigata University, 31:4–5.)

At present, the adenohypophysis of the hagfish is the least under-stood of all the vertebrates. However, recent immunohistochemical stud-ies provided the first clear-cut evidence for the presence of GTH and ACTH in the hagfish (Nozaki et al., 2005, 2007; Miki et al., 2006) (Figure 9.3). Although not conclusive, Nozaki et al. (2005) also suggested the pres-ence of GH in the hagfish. In addition, these three adenohypophysial hor-mones were suggested to be the ancestral adenohypophysial hormones that have maintained their original functions throughout vertebrate evo-lution (Nozaki, 2008). On the other hand, the later derived hormones, such as prolactin and TSH, may have contributed to the expansion of verte-brates into new environments, as suggested by Kawauchi et al. (2002) and Kawauchi and Sower (2006) in the lamprey. Moreover, it has been fur-ther revealed that GTH cells, ACTH cells and unidentified cells that were assumed to include both undifferentiated cells and GH cells were packed together in the same cell cluster of the hagfish adenohypophysis, and thus each cluster appeared to serve as a separate functional unit (Nozaki et al., 2007; Nozaki, 2008) (Figures 9.1e and 9.3). If the absence of the pars inter-media is the most ancestral vertebrate pituitary characteristic as found in the hagfish, MSH activity seems to be gained secondarily together with the differentiation of the pars intermedia. Further studies are needed to clarify this possibility.

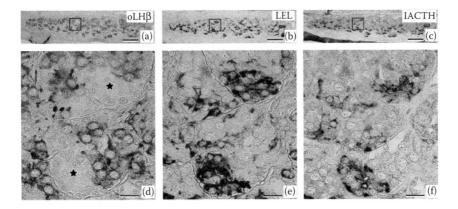

Figure 9.3 Three successive sagittal sections (a–c) through the adenohypophysis of an adult brown hagfish with developing gonads, stained with anti-ovine LHβ (oLHβ), biotin-conjugated *Lycopersicon esculentum* lectin (LEL), and anti-lamprey ACTH (lACTH), respectively. Boxes in a–c are shown enlarged in d–f, respec-tively. Asterisks in d show mass of cells, which are negative to anti-oLHβ, but are stained with LEL and anti-lamprey ACTH. These cells are assumed to be ACTH cells. Scale bars: (a–c) 200 μm; (d–f) 20 μm. (From Nozaki, M., T. Shimotani and K. Uchida. 2007. *Cell Tissue Res* 328:563–572. doi:10.1007/s00441-006-0349-3.)

Glycoprotein hormones (GPHs)

GTHs, in response to hypothalamic GnRH, are released from the pituitary and act on the gonads to regulate steroidogenesis and gametogenesis. Two GTHs, follicle-stimulating hormone (FSH), and luteinizing hormone (LH), together with TSH form a family of pituitary hormones (Figure 9.4). They are heterodimeric glycoproteins consisting of two subunits, an α-subunit and a unique β-subunit. These glycoprotein hormones (GPHs) are believed to have evolved from a common ancestral molecule through duplication of β-subunit genes and subsequent divergence (Dayhoff, 1976;

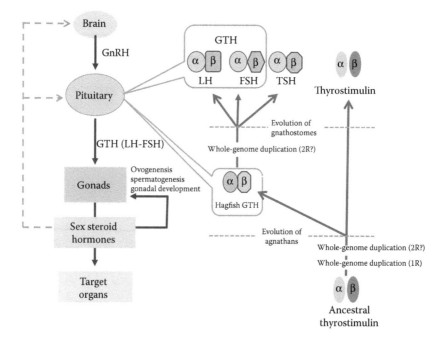

Figure 9.4 Schematic diagram of the evolution of glycoprotein hormones in the hypothalamic–pituitary–gonadal axis. Ancestral thyrostimulin (α and β) existed before divergence of vertebrates. An ancestral thyrostimulin (α and β) diverged into GTH (α and β) and thyrostimulin (α and β) during the early phase of agnathan divergence. The GTH (α and β) formed a heterodimer in the pituitary and acted as the first adenohypophysial gonadotropic hormone during the evolution of agnathan species. This GTH dimer further diverged into three functional units of adenohypophysis, LH and FSH as two gonadotropins, and TSH as a thyrotropin, in the lineage to gnathostomes. (From Nozaki M. 2013. *Front. Endocrinol.* 4:200. doi:10.3389/fendo. 2013.00200 with minor modification.)

Kawauchi and Sower, 2006; Dos Santos et al., 2011) (Figure 9.4). Two GTHs have been identified in all taxonomic groups of gnathostomes, including actinopterygians (Kawauchi et al., 1989; Quérat et al., 2000), sarcopterygians (Quérat et al., 2004), and chondrichthyans (Quérat et al., 2001), but not in agnathans.

A single β-subunit of GTH was identified from the sea lamprey pituitary gland after extensive and exhaustive research that took over 20 years (Sower et al., 2006; Kawauchi and Sower, 2006). Using heterologous probes, there were many physiological and morphological evidence supporting the presence of a GTH in the lamprey (Larsen and Rothwell, 1972; Hardisty and Baker, 1982; Sower, 1998; Sower et al, 2006; Nozaki et al., 2008). However, despite extensive molecular and biochemical studies, the α-subunit of lamprey GTH was not identified which is now supported by synteny analysis of the genes from the lamprey genome (Sower et al., 2015). Instead, there is a thyrostimulin type alpha called A2 (see subsequent paragraph). It is suggested by these authors that the typical alpha subunit was lost in the lamprey lineage. It is now reported that lampreys have an ancestral non-classical, pituitary heterodimer GPH consisting of the thyrostimulin A2 subunit (GPA2) with the classical β subunit in the sea lamprey (Sower et al., 2015). The current hypothesis is that this one glycoprotein hormone may be acting as both a GTH and TSH hormone in some unknown differential manner (Sower et al., 2015).

Recently, a fourth heterodimeric GPH has been discovered in the human genome and termed "thyrostimulin" due to its thyroid-stimulating activity (Nakabayashi et al., 2002). The thyrostimulin α-subunit, called glycoprotein α-subunit 2 (GPA2), is homologous but not identical to the common α-subunit (GPHα or GPA1). With the discovery of GPA2 and glycoprotein β-subunit 5 (GPB5, thyrostimulin beta), homologs are present not only in other vertebrates, but also in invertebrates including fly, nematode and sea urchin (Sudo et al., 2005; Park et al., 2005), it is proposed that ancestral glycoprotein subunits existed before the divergence of vertebrates/invertebrates, and later gene duplication events in vertebrates produced the thyrostimulin (GPA2 and GPB5) and GTH/TSH [GPHα and GPHβ(LHβ/FSHβ/TSHβ)] (Sudo et al., 2005) (Figure 9.4). The basal lineage of chordates, such as tunicates and amphioxus, contains GPA2 and GPA5 in their genomes but not GPHα or GPHβ (Holland et al., 2008; Dos Santos et al., 2009, 2011; Tando and Kubokawa, 2009a, b). Lamprey also has GPA2 and GPB5 genes in addition to the canonical GTHβ (Sower et al., 2006, 2009; Dos Santos et al., 2009; Decatur et al., 2013, Sower et al., 2015). At present, no information is available as to the presence or absence of thyrostimulin GPA2/GPB5 in the hagfish.

Identification of hagfish GTH

A single GTH, which comprises α- and β-subunits, was recently identified in the pituitary of the brown hagfish, *Paramyxine* (= *Eptatretus*) *atami,* one of the Pacific hagfish (Uchida et al., 2010) (Figures 9.4 and 9.5). Both subunits of GTH are produced in the same cells of the adenohypophysis, providing definitive evidence for the presence of a heterodimeric GTH in the hagfish. GTH increases at both the gene and protein levels corresponding to the reproductive stages of the hagfish (Figures 9.1c and d). Moreover, purified native GTH induces the concentrations of sex steroids (estradiol-17β and testosterone) from cultured testis in a dose-dependent manner. From the phylogenetic analysis, the hagfish GTHα forms a clade with the gnathostome GPHαs. The hagfish GTHβ forms a clade with the TSHβs, however the bootstrap values are low and hagfishes evolved prior to the gnathostomes. The sea lamprey GTHβ also groups with the GPHβs but appears to be one of the outgroups of the LHβs. These results clearly show that the GTH identified in the hagfish acts as a functional GTH. From these data, the pituitary and its hormone early in vertebrate evolution are showing an intermediate stage as evident by only one GTH-like hormone in the agnathans versus the classical two GTHs found in gnathostomes (jawed vertebrates).

Now, it is quite clear that hagfish has a functional GTH in the pituitary. Then, a question arises as to the failure to detect any clear evidence for pituitary gonadotropic activity in the earlier study in hypophysectomized *Eptatretus stoutii* (Matty et al., 1976). In the study, gametogenesis appeared to be unaffected by hypophysectomy. On the other hand, in *Eptatretus burgeri,* the only hagfish known to have a definite breeding season (Ichikawa et al., 2000; Nozaki et al., 2000) (Figures 9.6 and 9.7), there

Figure 9.5 (**See color insert.**) Brown hagfish, Paramyxine atami.

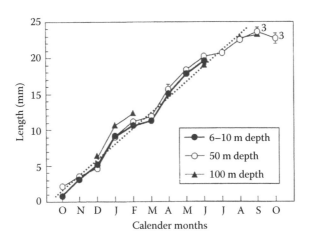

Figure 9.6 Annual growth curve of developing eggs of the female hagfish (*Eptatretus burgeri*) caught in three different locations: water of 6–10 m depths (closed circles), water of 50 m depth (open circles) and water of 100 m depth (triangles). Numbers of animals studied were 20–40 otherwise indicated in the figure. In most cases, ranges of the standard errors are within the size of the mean data points. Broken line indicates a regression line, which is made by the mean data points of all three locations. (From Nozaki, M., T. Ichikawa, K. Tsuneki and H. Kobayashi. 2000. *Zool Sci* 17:225–232.)

have been some indications that gonadal development and spermatogenesis may be retarded after hypophysectomy (Patzner and Ichikawa, 1977). There is at least one possible explanation for the difference of results on hypophysectomy between two species. In the hagfish, the process of gametogenesis may be relatively autonomous, and may not be completely arrested by hypophysectomy. In both species, gametogenesis may still proceed, although at a slower pace, after hypophysectomy. If so, since *E. burgeri* exhibits clear seasonal changes in gonadal development, it is possible to detect the difference of the gonadal conditions between hypophysectomized and sham-control animals. On the other hand, in *E. stoutii*, developmental conditions of gonads vary among individuals, and thus it is difficult to detect clear difference between two groups. In support of this possibility, hypophysectomy experiments have been conducted extensively in lampreys, in which limited gonadal growth still continued after hypophysectomy (Evennett and Dodd, 1963; Larsen, 1973; Gorbman, 1983).

Feedback regulation of hagfish GTH functions by sex steroid hormones

Gonadal steroid hormones and hypothalamic hormones play major roles in controlling the synthesis and release of LH and FSH in gnathostomes

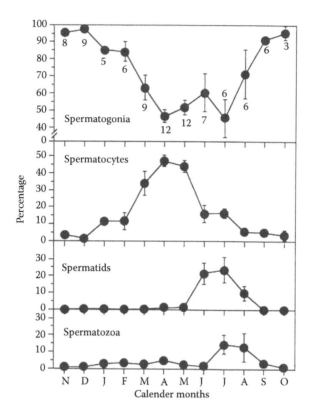

Figure 9.7 Annual changes in percentage of follicles containing each developmental stage of spermatogenic cells in the testis of hagfish (*Eptatretus burgeri*). Solid circles and bars indicate the mean and standard errors, respectively. Numbers indicate number of individuals studied. (From Nozaki, M., T. Ichikawa, K. Tsuneki and H. Kobayashi. 2000. *Zool Sci* 17:225–232.)

(Figure 9.4). Both positive and negative feedback effects of gonadal steroids have been demonstrated in teleosts, depending on modes of administration and reproductive stages of animals. In general, in sexually mature fish, sex steroids are considered to regulate gonadal maturation and recrudesce, whereas in juvenile fish, sex steroids are considered to regulate puberty. Thus, negative feedback effects of estradiol and testosterone are evident during the latter stages of gonadal development; specifically, it has been shown that gonadal removal increases LH secretion in salmon (Larsen and Swanson, 1997), goldfish (Kobayashi and Stacey, 1990), and African catfish (Habibi et al., 1989). The observed increases in LH levels can be suppressed by the treatment with estradiol, testosterone or both. FSH is also controlled by steroid-dependent negative feedback loops in rainbow trout (Saligaut et al., 1998), salmon (Dickey and Swanson, 1998), and goldfish (Kobayashi et al., 2000). The negative feedback effects of steroids may be mediated primarily at the levels of the hypothalamic GnRH neurons (Vacher et al., 2002; Levavi-Sivan et al., 2006; Banerjee and Khan, 2008), because both *in vivo* and *in vitro* studies have shown that the expression of LHβ mRNA or FSHβ mRNA is often unchanged or increases following the exposure to estradiol, testosterone or both (Saligaut et al., 1998; Huggard-Nelson et al., 2002; Levavi-Sivan et al., 2006). However, in sexually immature teleosts, sex steroids appear to exert primarily a positive feedback effect that acts directly at the level of the pituitary and via effects on the GnRH system (Huggard-Nelson et al., 2002; Aroura et al., 2007). LH content and LH mRNA levels of the pituitary in juvenile fish increase in response to estrogens and aromatizable androgens (Huggard et al., 1996; Saligaut et al., 1998).

Estradiol treatment in the juvenile brown hagfish resulted in the marked accumulation of both immunoreactive (ir)-GTHα and ir-GTHβ in the pituitary (Nozaki et al., 2013). However, mRNA levels of GTHα and GTHβ in the pituitary were not, or only transiently, increased by the estradiol treatment (Nozaki et al., 2013). The latter results suggest that syntheses of both α- and β-subunits of GTH were not, or only transiently, affected by the estradiol treatment. Accordingly, the marked accumulation of both ir-GTH subunits could be attributed to the suppression of GTH secretion from the pituitary. From the study, it shows that the feedback effects of estradiol appeared to be inhibitory rather than stimulatory, and also mediated by the possible suppression of the secretion of GTH from the pituitary in these juvenile hagfish. These conditions in juvenile hagfish resembled to those in adults, but not in juveniles, of teleosts (Saligaut et al., 1998; Huggard-Nelson et al., 2002; Levavi-Sivan et al., 2006). Such suppression of GTH secretion in the hagfish is probably regulated by the hypothalamic factors including GnRH, as mentioned below.

On the other hand, testosterone treatment in the juvenile brown hagfish had no effect on the staining intensities of the ir-GTHα and ir-GTHβ in the pituitary (Nozaki et al., 2013). Nevertheless, testosterone treatment resulted in the suppression of mRNA expressions of both GTHα and GTHβ in the pituitary (Nozaki et al., 2013). Therefore, testosterone probably acts to suppress both the synthesis and the secretion of GTH. This conclusion follows from the reasoning that if the secretion of GTH was not suppressed, the intensities of immunoreactions of both GTHα and GTHβ would have decreased due to decreased levels of mRNA expressions in both GTH subunits. Thus, it seems likely that estradiol and testosterone differ with regard to their roles in the regulation of synthesis and secretion of GTH in the pituitary of the hagfish.

Gonads and gonadal sex steroids

Gonads and germ cells

Reproductive organs and hormones of hagfishes have been mainly studied in four species of hagfish: Atlantic hagfish (*Myxine glutinosa*), Eastern Pacific Hagfish (*E. stoutii*), and Western Pacific hagfish (=Japanese hagfish) (*E. burgeri* and *P. atami*) (for reviews, see Hardisty, 1979; Gorbman, 1983; Patzner, 1998). Among these hagfishes, only *E. burgeri* live in shallower water less than 50 m in depth, and show a seasonal migration and a seasonal development of gonads (Kobayashi et al., 1972; Patzner and Ichikawa, 1977; Tsuneki et al., 1983; Ichikawa et al., 2000; Nozaki et al., 2000) (Figures 9.6 and 9.7). However, a subpopulation of *E. burgeri* is also found in water of greater than 50 m depth throughout the year, which exhibits seasonal development of gonads similar to that of migrating population (Ichikawa et al., 2000; Nozaki et al., 2000) (Figure 9.6).

The gonads of hagfishes are situated in the peritoneal cavity. The anlage for the ovary is found in the anterior part of the gonads, the one for the testis is in the posterior part close to the anus. The problem of sex differentiation in hagfishes was discussed in detail by Gorbman (1990), who concluded that gonadal differentiation in *E. stoutii* is juvenile progynous. The characteristic features of the hagfish gonads are the scanty number of mature eggs (less than about 40) and the small amount of sperm (Patzner, 1998). In *M. glutinosa*, most individuals are females or hermaphrodites, and males with no female tissue are rarely present. For example, 58% of *M. glutinosa* examined ($n = 1080$) contained only female gonadal tissue, 41% were hermaphrodites with both male and female tissue and 0.05% were males with no female tissue (Powell et al., 2004). On the other hand, the sex ratio of the Pacific hagfish is nearly 1:1 (*E. stoutii*, Conel, 1931; *E. burgeri*, Ichikawa et al., 2000; *P. atami*, Miki et al., 2006).

Plasma levels of sex steroid hormones in the hagfish gonads

Sex steroids in vertebrate gonads have crucial roles in reproductive phenomena including sex differentiation, gametogenesis and gamete maturation. The physiological active gonadal steroids of higher vertebrates are progesterone from the corpus luteum, androgen from the testis and estrogen from the ovary.

In the hagfish, sex steroid hormones such as estradiol and testosterone have been detected in the circulating plasma (Matty et al., 1976; Weisbart et al., 1980; Schützinger et al., 1987; Nishiyama et al., 2013) or in the gonads (Hirose et al., 1975; Gorbman and Dickhoff, 1978; Powell et al., 2004, 2006), but their concentrations were very low near the lower limits of assay sensitivities. Among these studies, Schützinger et al. (1987) reported that plasma estrogen content increased in relation to the stages of ovarian development in female Atlantic hagfish, *M. glutinosa*. Powell et al. (2004, 2005) also reported using *in vitro* organ-cultured ovaries that the number of females with large eggs increased following estradiol peaks in January in *M. glutinosa*. Nishiyama et al. (2013) further observed in *P. atami* that among estradiol-17β, testosterone, and progesterone only plasma levels of estradiol-17β showed the significant correlation to ovarian development (Figure 9.8). In another study, Yu et al. (1981) demonstrated that the synthesis of hepatic vitellogenin was inducible by estrogens, estradiol and

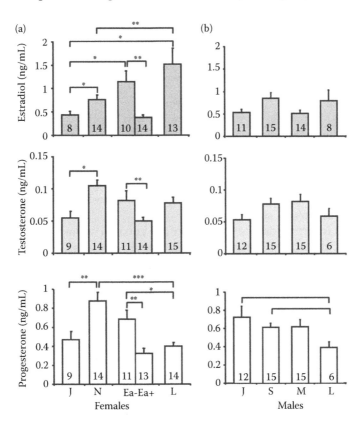

Figure 9.8 Plasma concentrations of estradiol, testosterone, and progesterone in females (left panels) and males (right panels) of the brown hagfish. Numbers in each column indicate number of animals examined, and bars indicate standard errors, respectively. In the left panels, J, juveniles; N, non-vitellogenic adults; Ea-, early vitellogenic adults possessing only normal follicles; Ea+, early vitellogenic adults possessing atretic follicles as well as normal follicles; L, late vitellogenic adults possessing only normal follicles. In the right panels, J, juveniles; S, adults with small GSI; M, adults with medium GSI; L, adults with large GSI. *$P < 0.05$; **$P < 0.01$; ***$P < 0.001$. (From Nishiyama, M., H. Chiba, K. Uchida, T. Shimotani and M. Nozaki. 2013. *Zool Sci* 30:967–974.)

estrone, in *E. stoutii*. Based on these results, estrogenic control of ovarian development and hepatic vitellogenesis seems to have arisen early in vertebrate evolution.

In males, however, no clear relationships were observed between plasma estradiol or testosterone concentrations and testicular development, while plasma progesterone concentrations showed a significant inverse relationship with testicular development (Nishiyama et al., 2013) (Figure 9.8). The failure to correlate with the circulating levels of sex steroid hormones and gonadal developments in the male hagfish is discussed in relation to *CYP11A* mRNA expressions in the testis of the hagfish (see below).

Expression of steroidogenic enzymes in the hagfish gonads

The biosynthetic enzymes of sex steroids have been well studied in gnathostomes. Typically, there are three cytochrome P450 enzymes (CYP) such as P450 side chain cleavage (CYP11A), P450 17α-hydroxylase (CYP17), and P450 aromatase (CYP19) and two types of hydroxylated dehydrogenases (HSDs), such as 3β-HSD and 17β-HSD. Among these enzymes, CYP11A is an enzyme that regulates the conversion from cholesterol to pregnenolone by its side chain cleavage activity, and it is the first and the essential enzyme of steroidogenesis.

Some information is available on the gonadal steroidogenesis in the hagfish (Patzner, 1998). Recently, based on EST analysis of the testis of the brown hagfish (*P. atami*), *CYP11A* was cloned (Nishiyama et al., 2015). Following the real-time PCR analysis, *CYP11A* mRNA expression levels were clearly correlated with the developmental stages of gonads in both sexes of the brown hagfish (Nishiyama et al., 2015) (Figure 9.8). These results are consistent with those in more advanced gnathostomes. For example, transcript for *CYP11A* in the gonads increased in correlation with gonadal development in both sexes of rainbow trout (Nakamura et al., 2005; Kusakabe et al., 2006) and female Japanese eel (Kazeto et al., 2006). Previously, it has been shown that plasma concentrations of estradiol-17β increased in correlation with the gonadal development in female brown hagfish (Nishiyama et al., 2013). Moreover, Yu et al. (1981) demonstrated that the synthesis of hepatic vitellogenin in *E. stoutii* was induced by estrogens, estradiol, and estrone. Thus, CYP11A is suggested to play a crucial role in the synthesis of estradiol-17β, which in turn acts on ovarian development and hepatic vitellogenesis in the female hagfish.

In male hagfish, transcriptional levels of *CYP11A* increased in accordance with the developmental stages of testis, as well as in females (Nishiyama et al., 2015) (Figure 9.8). Moreover, in the testis incubated with hagfish GTH (5 µg/mL), *CYP11A* mRNA expression levels were significantly higher than those incubated without hagfish GTH (Nishiyama et al., 2015), indicating that gonadal *CYP11A* expression was induced by the pituitary GTH in the hagfish. However, as mentioned earlier, no relationship was obtained between the testicular development and plasma levels of estradiol or testosterone (Nishiyama et al., 2013).

Thus, there was a clear discrepancy between males and females in the relationship between the transcriptional levels of *CYP11A* and plasma steroid levels. These results clearly suggest a possibility that male hagfish uses other steroids than estradiol or testosterone as major androgens. In support of this possibility, recent studies in the lamprey have emphasized the importance of non-classical steroids, such as androstenedione and 15α-hydroxylated sex steroids (15α-hydroxytestosterone and 15α-hydroxyprogesterone) in serving as functional androgens (Lowartz et al., 2003; Young et al., 2007; Bryan et al., 2007, 2008). A receptor for androstenedione was recently described in the lamprey by Bryan et al. (2007). Since hagfish gonads also produce substantial amounts of unusual androgens, such as 6β-hydroxy testosterone and 5α-androstane-3β, 7α, 17β-triol, as well as androstenedione (Hirose et al., 1975; Kime et al., 1980; Kime and Hews, 1980), some of these steroids may act as functional androgens. Further study is required in order to clarify the role of these steroids in hagfish.

Localization of steroidogenic cells in the hagfish gonads

It is well established that sex steroid hormones are produced in the cells comprising the growing follicles (theca cells and granulosa cells) in the ovary of female gnathostomes. A two-cell type model has been proposed for the follicular steroidogenesis in the ovary (see Nagahama et al., 1994; Gore-Langton and Armstrong, 1998). In the model, steroid synthesis from cholesterol to androgen is performed in the theca cells, followed by the synthesis of estrogen from androgen in the granulosa cells.

By *in situ* hybridization, *CYP11A* mRNA signals were found in the theca cells of the ovary of *P. atami* (Nishiyama et al., 2015) (Figure 9.9), which is well accordance with those in gnathostomes. This result contrasted to the previous electron microscopy in the hagfish, in which cells, showing the characteristics associated with steroidogenesis, were not observed in the ovary of *M. glutinosa* (Fernholm, 1972) or *E. stoutii* (Tsuneki and Gorbman, 1977a). The reason for the difference of the results was not clear, since Tsuneki and Gorbman (1977a) studied various structures of the hagfish ovary including those of large oocytes. Possibly, the theca cells of the hagfish ovary may not show typical ultrastructural features of steroidogenic cells.

It is also well established that sex steroid hormones are produced in the interstitial cells (Leydig cells) of the testis in male gnathostomes. For example, steroidogenic enzymes such as CYP11A, CYP17, and 3β-HSD

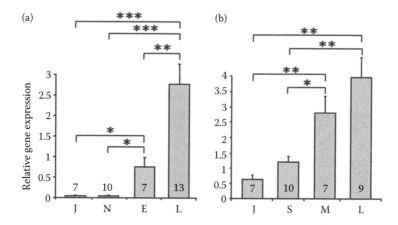

Figure 9.9 Relative *CYP11A* gene expressions in the gonads of female (a) and male (b) hagfish. The *CYP11A* mRNA levels were normalized by β-actin mRNA levels. Relative values are expressed as mean ±SE. Number in each column indicates number of animals studied. (a) J, juvenile ovary; N, non-vitellogenic adult ovary; E, early vitellogenic adult ovary; L, late vitellogenic adult ovary. (b) J, juvenile testis; S, adult testis with small GSI; M, adult testis with medium GSI; L, adult testis with large GSI. *$P < 0.05$; **$P < 0.01$; ***$P < 0.001$. (From Nishiyama, M. et al. 2015. *Gen Comp Endocrinol* 212:1–9.)

were reported in the Leydig cells of the rainbow trout testis (Kobayashi et al., 1998). However, in some teleost species, such as pike and char, urodele amphibians and turtles, typical interstitial cells are absent, and a ring of circumtubular cells (tubule-boundary cells) are considered to be the site of the production of sex steroid hormones (Gorbman and Bern, 1962). In the lamprey, Leydig cells in the testis are also suggested, in a histological study, to be sex steroid hormone-producing cells (Larsen, 1973, cited by Gorbman, 1983).

In the hagfish, Tsuneki and Gorbman (1977b) described the ultrastructure of the testis of *E. stoutii*: they found no apparent steroidogenic cells until a body length of about 40 cm is attained. At that time, cells with the features of Leydig cells (e.g., smooth ER and tubular cristae) appeared among the spermatogenic follicles. In well agreement with Tsuneki and Gorbman (1977b), expression levels of *CYP11A* mRNA were very low in juveniles with their total length of less than 40 cm, but increased significantly in relation to testicular development in the brown hagfish (Nishiyama et al., 2015) (Figure 9.9). Histological observations by *in situ* hybridization further revealed that *CYP11A* mRNA was expressed in both the Leydig cells and tubule-boundary cells of the developing testis (Nishiyama et al., 2015) (Figure 9.10). Thus, both types of cells, Leydig cells

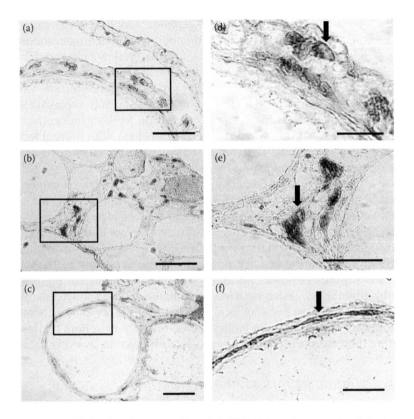

Figure 9.10 Cellular localization of hagfish *CYP11A* in the gonads of the brown hagfish. (a) *CYP11A* mRNA signals in the theca cells of late vitellogenic egg. Rectangular area in (a) is magnified in (d). Arrow indicates cells of the theca externa-expressing *CYP11A* mRNA. (b,c) *CYP11A* mRNA signals in the Leydig cells (b) and tubule-boundary cells (c) in the testis with large GSI. Rectangular areas in (b) and (c) are magnified in (e) and (f), respectively. Arrows indicate Leydig cells (e) and tubule-boundary cells (f) expressing *CYP11A* mRNA, respectively. Scale bars: a–c: 100 μm; d–f: 10 μm. (From Nishiyama, M. et al. 2015. *Gen Comp Endocrinol* 212:1–9.)

and tubule-boundary cells, are considered the steroid-producing cells in the hagfish testis. It seems most likely that one of these two types of cells is adopted as the steroidogenic cells of the testis in gnathostomes.

Functional corpus luteum in the hagfish

Follicular atresia is a common feature in the hagfish ovaries (Gorbman, 1983), and it appears to be the method by which some 100 oocytes are reduced to approximately 30 that are grown and are ovulated. Using *in vitro* organ-cultured ovaries of *M. glutinosa* supplemented with pregnenorone, Powell et al. (2006) recently demonstrated that larger amounts of progesterone were released from atretic follicles (yellow bodies only) than from normal follicles. They hypothesized that hagfish possessed functional corpora lutea-like structures that produced progesterone. On the other hand, Nishiyama et al. (2013) reported in *P. atami* that plasma levels of sex steroid hormones including progesterone were significantly lower in adult females that possessed atretic follicles along with normal follicles than they were in females that possessed only normal follicles (Figure 9.8), indicating reduced steroidogenic activity in females that possessed atretic follicles. Thus, a clear discrepancy is found on progesterone production of atretic hormones between Powell et al. (2006) and Nishiyama et al. (2013). Several possible explanations could be considered: (1) As Powell et al. (2006) pointed out that only specialized yellow bodies have potent steroidogenic activity, and thus most atretic follicles of *P. atami* do not have progesterone secreting activity. (2) Progesterone synthesis by the yellow bodies is suppressed under normal physiological conditions. (3) It is also possible that the progesterone released from the yellow bodies does not go to general circulation but acts locally at the gonadal levels. Further studies are needed to clarify the biological significance of yellow bodies and whether they represent "a proto corpus luteum" seen in jawed vertebrates.

Hypothalamic factors regulating the gonadotropic function of hagfish

GnRH

The synthesis and the secretion of GnRH are the key neuroendocrine function in the hypothalamic regulation of the HPG axis. To date, two to three isoforms of GnRH have been identified in representative species of all classes of gnathostomes and lampreys (Kavanaugh et al., 2008; Sower et al., 2009). GnRHs are also identified in tunicates (Adams et al., 2003), and several invertebrates belonging to lophotrochozoans (mollusk and annelid; Tsai and Zhang, 2008; Zhang et al., 2008), but not in the ecdysozoan lineages. On the other hand, adipokinetic hormone (AKH) has been identified as the ligand of the GnRH receptor of the insects, *Drosophia* and *Bombyx* (Staubli et al., 2002). An AKH–GnRH-like neuropeptide has been identified in the nematode *Caenorhabditis elegans* (Lindemans et al., 2009). A comparative and phylogenetic approach shows that the ecdysozoan AKHs, lophotrochozoan GnRHs, and chordate GnRHs are structurally related, and suggested that they all originate from a common ancestor (Lindemans et al., 2011).

In the hagfish, GnRH has not yet been identified, but previous chromatographic and immunohistochemical studies suggested the presence of a GnRH-like molecule in the hypothalamic–neurohypophysial area (Braun et al., 1995; Sower et al., 1995; Kavanaugh et al., 2005). Sower et al. (1995) showed using the techniques of immunocytochemistry (ICC), HPLC and radioimmunoassay (RIA) with a specific lamprey GnRH-III antisera localized a lamprey-III GnRH-like molecule in the hypothalamus, adenohypophysis, and the neurohypophysis of the Atlantic hagfish (*M. glutinosa*). In particular, a dense accumulation of GnRH-like immunoreaction was observed in the dorsal wall of the neurohypophysis with the use of antisera against chicken GnRH-II (GnRH2 type), salmon GnRH (GnRH3 type), lamprey GnRH-I (GnRH3 type) and lamprey GnRH-III (GnRH3 type) (Sower et al., 1995; Oshima et al., 2001) (Figure 9.2). Also reported in 1995, using six different antisera to GnRH (salmon PBL-49, lamprey 21–134, lamprey 1459, lamprey 1467, chicken-II and mammalian), ir-salmon GnRH-like molecule was shown to be present in the preoptic cells, hypothalamic infundibular nucleus, hypophysial stalk and distributed fibers in the brain of the Pacific hagfish, *E. stoutii* (Braun et al., 1995). The identity of an ir-salmon GnRH-like molecule supports that latest information that a-type GnRH3 arose early in the vertebrate lineage (Decatur et al., 2013). Based on synteny data, these authors proposed that there were four lineages of GnRH (GnRH1, 2, 3 and 4) that arose before the divergence of the ancestral agnathans and gnathostome lineages. GnRH4 was lost during these events. Lamprey GnRH-I and -III previously proposed to be part of a Group 4 are now considered to be part of the GnRH3 lineage (Decatur et al., 2013). Thus, it seems reasonable that hagfish may have retained a GnRH2 and/or GnRH3-like molecule.

In addition, Kavanaugh et al. (2005) reported seasonal changes in hypothalamic ir-GnRH in relation to gonadal reproductive stages in the Atlantic hagfish. Based on these investigations, it is likely hagfish indeed have a GnRH or GnRH-like peptide in the brain, although the primary amino acid structure of GnRH has not been identified in these ancient fish. The primary structure of the GnRH (-like) molecules needs to be determined to confirm the presence of specific GnRH(s) in the hagfish brain. The presence and the location of a GnRH-like substance in the brain of the hagfish have led the authors to hypothesize that GnRH has a neuroendocrine function acting on the pituitary (Braun et al., 1995; Sower et al., 1995).

In vertebrates, the neuroendocrine axis has a central role in the control of reproduction by integrating internal and external cues during key developmental reproductive stages. As stated previously, GnRH is considered the major hypothalamic hormone orchestrating reproduction in all vertebrates. There is evidence as described in this chapter that the hypothalamic–pituitary axis emerged in the early ancestral vertebrates and that hagfish have certain conserved aspects of this complex neuroendocrine axis in the coordination and the integration of environmental and hormonal cues in controlling reproduction.

RFamide peptides

RFamide peptides play various important roles in the central nervous system in both invertebrates and vertebrates (Ukena and Tsutsui, 2005; Tsutsui and Ukena, 2006). Among RFamide peptides, PQRFamide peptide group and LPXRFamide (X = L or Q) peptide group share a highly conserved C-terminal Pro-Gln-Arg-Phe-NH2 motif (PQRFa motif), which are considered to be important for the interaction with their receptors, as well as the structure of their receptors showed high-sequence similarities (for reviews, see Ukena and Tsutsui, 2005; Tsutsui, 2009; Tsutsui et al., 2010). PQRFa peptides are mainly expressed in the spinal cord and medulla oblongata (Vilim et al., 1999; Liu et al., 2001), and act as neurotransmitters or neuromodulators in the opioid system in mammals (Roumy et al., 2007). They are also expressed in the hypothalamus (Vilim et al., 1999; Kalliomäki and Panula, 2004; Goncharuk et al., 2006) and have other functions, such as cardiovascular regulation (Panula et al., 1996), neuroendocrine function(s) (Jhamandas and MacTavish, 2003; Jhamandas et al., 2006), and locomotor regulation (Kotlinska et al., 2007). LPXRFamide peptide group includes GTH-inhibitory hormone (GnIH) (see Tsutsui, 2009). GnIH was shown to be located in the hypothalamic–pituitary system and to decrease GTH secretion from the pituitary (Tsutsui, 2009; Tsutsui et al., 2010). However, studies on teleosts and amphibians have shown that functions of LPXRFa peptides were stimulatory or inhibitory (Koda et al., 2002; Ukena et al., 2003; Amano et al., 2006; Zhang et al., 2010; Shahjahan et al., 2011).

Recently, PQRFa peptides including LPQRFa were identified from the brains of sea lamprey (Osugi et al., 2006, 2012) and brown hagfish (Osugi et al., 2011). In the lamprey, LPQRFa peptide-positive neurons were localized in the hypothalamus, and their fibers were terminated in close proximity to GnRH-III neurons in addition to the neurohypophysis (Osugi et al., 2012). Moreover, intraperitoneal injection of LPQRFa stimulated the expression of lamprey GnRH-III in the hypothalamus and GTHβ mRNA expression in the pituitary (Osugi et al., 2012). Similarly, several PQRFamide peptides were identified in the brain of the brown hagfish (Osugi et al., 2011). Based on *in situ* hybridization and immunohistochemistry, hagfish PQRFamide peptide precursor mRNA and its translated peptides were localized in the infundibular nucleus of the hypothalamus. Dense immunoreactive fibers were found in the infundibular nucleus and some of them were terminated on blood vessels within the infundibular nucleus. Furthermore, LPQRFa peptide, one of the hagfish PQRFa peptides, significantly stimulated the expression of GTHβ mRNA in the cultured hagfish pituitary. The latter results clearly suggest that GTH functions of the hagfish pituitary are controlled by the hypothalamic factors.

Hypothalamic–pituitary system of the hagfish

Neither the hagfish nor lamprey has the anatomical equivalent of a median eminence to convey the neurohormones to the anterior pituitary. Most vertebrates except agnathans and teleost fish have a portal vascular system (median eminence) for transferring neurohormones from the hypothalamus to the adenohypophysis (Holmes and Ball, 1974; Gorbman et al., 1983). In teleosts, there is a direct innervation of the pars distalis in the anterior pituitary by neurosecretory neurons from the hypothalamus (Holmes and Ball, 1974; Gorbman et al., 1983). The agnathans do not have nervous or vascular communication between the brain and the pituitary (Holmes and Ball, 1974; Gorbman et al., 1983) (Figures 9.1a,b). It has been suggested the brain regulation of the pituitary in agnathans is achieved via diffusion. In lampreys, the authors concluded from their studies that neurosecretory peptides such as GnRH diffuse from the brain to the adenohypophysis and thus regulate its secretory activity (Nozaki et al., 1994). Similarly it is generally considered that in hagfish, the hypothalamic factors, such as GnRH, reach the adenohypophysis simply by diffusion (Nozaki et al., 1975; Tsukahara et al., 1986; Gorbman, 1995). However, the dorsal wall of the hagfish neurohypophysis, where ir-GnRH nerve fibers and ir-AVT nerve fibers are terminated (Nozaki and Gorbman, 1983; Braun et al., 1995; Oshima et al., 2001) (Figure 9.2), is far from the adenohypophysis by the presence of the neurohypophysis itself. On the other hand, the blood vessels are richly distributed on the surface of the dorsal wall, and make the posterior hypophysial vascular plexus (Gorbman et al., 1963; Kobayashi and Uemura, 1972). Although most blood in the posterior hypophysial vascular plexus enter the posterior hypophysial vein of the anterior cardinal system, several small vessels proceed from the dorsal wall to the adenohypophysis in *E. burgeri* (Kobayashi and Uemura, 1972). These small vessels may contribute the regulation of the adenohypophysial functions. A pair of small blood vessels from the hypothalamus also enters the posterior hypophysial vascular plexus (Gorbman et al., 1963). These identified small blood vessels, along with some PQRFamide neuronal fibers terminating on the blood vessels within the hypothalamus (Osugi et al., 2011), suggest that further studies are needed to understand the anatomical relationship between the hypothalamus and the pituitary in hagfish. In an evolutionary sense, there are three different types of brain regulation of the pituitary that have developed in the vertebrates: the agnathan diffusional type; the teleostean direct innervational type and the vascular type seen in all other vertebrates (Nozaki et al., 1994; Gorbman, 1995). During evolution of the vertebrates, structural features of the pituitary and hypothalamus also evolved that perhaps optimized the communication between these tissues as vertebrates became larger and more complicated in form and distance between the hypothalamus and pituitary increased significantly (Gorbman, 1995). Unlike the lamprey with a diffusional type of brain regulation of the pituitary, the hagfish may represent an intermediate stage in the hypothalamic–pituitary anatomical relationship in vertebrates and may have both a diffusional and the beginnings of a "pre-median eminence."

Conclusion

The pituitary gland and all major adenohypophysial hormones were major evolutionarily events that emerged prior to or during the differentiation of the ancestral jawless vertebrates (agnathans). The acquisition of the pituitary gland in vertebrates along with the differentiation of the hypothalamus has led to physiological divergence, including reproduction, growth, metabolism, stress and osmoregulation in subsequent evolution of jawed vertebrates. Since hagfish represent the most basal and primitive vertebrate that diverged over 550 million years ago (Janvier, 1996), they are of particular importance in understanding the evolution of the HPG axis related to vertebrate reproduction. Recent studies clearly show that the hagfish have a conserved, functional HPG axis similar to that of more advanced gnathostomes. However, there are distinct differences in both lamprey and hagfish HPG axes compared with later evolved differences suggesting an intermediate stage in development of the pituitary and its hormones. An understanding of the evolutionary events in these jawless vertebrates is critical in our understanding of the evolution of the HPG axis. We propose that this HPG system likely evolved from an ancestral, pre-vertebrate exclusively neuroendocrine mechanism by gradual emergence of components of a new control level, the pituitary gland and can provide important clues for understanding the organization of the hypothalamus and pituitary as essential regulatory systems in all vertebrates.

Acknowledgments

This work was supported in parts by Grant-in Aid for Basic Research and by Japanese Association for Marine Biology (JAMBIO) from Ministry of Education, Culture, Sports, Science and Technology (MEXT), Japan to M.S. Partial funding was provided by the National Science Foundation and New Hampshire Agricultural Experiment Station to S.A.S.

References

Adams, B.A., J.A. Tello, J. Erchegyi, C. Warby, D.J. Hong, K.O. Akinsanya, G.O. Mackie, W. Vale, J.E. Rivier and N.M. Sherwood. 2003. Six novel gonadotropin-releasing hormones are encoded as triplets on each of two genes in the protochordate, Ciona intestinalis. Endocrinology 114:1907–1919. doi:10.1210/en.2002-0216.

Amano, M., S. Moriyama, M. Iigo, S. Kitamura, N. Amiya, K. Yamamori, K. Ukenaand and K. Tsutsui. 2006. Novel fish hypothalamic neuropeptides stimulate the release of gonadotrophins and growth hormone from the pituitary of sockeye salmon. *J Endocrinol* 188:417–423.

Aroura, S., F.-A. Weltzien, N.L. Belle and S. Dufour. 2007. Development of real-time RT-PCR assays for eel gonadotropins and their application to the comparison of *in vivo* and *in vitro* effects of sex steroids. *Gen Comp Endocrinol* 153:333–343. doi:10.1016/j.ygcen.2007.02.027.

Banerjee, A. and I. Khan. 2008. Molecular cloning of FSH and LH b subunits and their regulation of estrogen in Atlantic croaker. *Gen Comp Endocrinol* 155:827–837. doi:10.1016/j.ygcen.2007.09.016.

Braun, C.B., H. Wichtand and R.G. Northcutt. 1995. Distribution of gonadotropin-releasing hormone immunoreactivity in the brain of the Pacific hagfish *Eptatretus stoutii* (Craniata, Myxinoidea). *J Comp Neurol* 353:464–476. doi:0.1002/cne.903530313.

Bryan, M.B., A.P. Scott and W. Li. 2007. The sea lamprey (*Petromyzon marinus*) has a receptor for androstenedione. *Biol Reprod* 77:688–696. doi: 10.1095/biolreprod.107.061093.

Bryan, M.B., A.P. Scott and W. Li. 2008. Sex steroids and their receptors in lampreys. *Steroids* 73:1–12. doi:10.1016/j.steroids.2007.08.011.

Conel, J.L. 1931. The genital system of the Myxinoidea: A study based on notes and drawings of these organs in Bdellostoma made by Bashford Dean. In: E.W. Gudger (ed.), *The Bashford Dean Memorial Volume: Archaic Fishes*. American Museum of Natural History, New York, pp. 67–102.

Dayhoff, M.O. 1976. *Atlas of Protein Sequence and Structure*. Vol. 5, Suppl. 2. National Biomedical Research Foundation, Silver Springs, MD, USA.

Decatur, W.A., J.A. Hall, J.J. Smith, W. Li and S.A. Sower. 2013. Insight from the lamprey genome: Glimpsing early vertebrate development via neuroendocrine-associated genes and shared synteny of gonadotropin-releasing hormone (GnRH). *Gen Comp Endocrinol* 192:237–245. doi:10.1016/j.ygcen.2013.05.020.

Dickey, J.T. and P. Swanson. 1998. Effects of sex steroids on gonadotropin (FSH and LH) regulation in coho salmon (*Oncorhynchus kisutch*). *J Mol Endocrinol* 21:291–306. doi:10.1677/jme.0.0210291.

Dos Santos, S., C. Bardet, S. Bertrand, H. Escriva, D. Habert and B. Querat. 2009. Distinct expression patterns of glycoprotein hormone-a2 (GPA2) and -b5 (GPB5) in a basal chordate suggest independent developmental functions. *Endocrinology* 150:3815–3822. doi: 10.1210/en.2008–1743.

Dos Santos, S., S. Mazan, B. Venkatesh, J. Cohen-Tannoudji and B. Querat. 2011. Emergence and evolution of the glycoprotein hormone and neurotrophin gene families in vertebrates. *BMC Evol Biol* 11:332. doi: 10.1186/1471-2148-11-332.

Evennett, P. J. and J. M. Dodd. 1963. Endocrinology of reproduction in the river lamprey. *Nature* 197:715–716.

Fernholm, B. 1972. Is there any steroid hormone formation in the ovary of the hagfish, *Myxine glutinosa*? *Acta Zool (Stockh)* 53:235–242.

Forey, P. and P. Janvier. 1993. Agnathans and the origin of jawed vertebrates. *Nature* 361:129–134.

Forey, P. and P. Janvier. 1994. Evolution of the early vertebrates. *Am Sci* 82:554–566.

Goncharuk, V.D., R.M. Buijs, D. Mactavish and J.H. Jhamandas. 2006. Neuropeptide FF distribution in the human and rat forebrain: A comparative immunohistochemical study. *J Comp Neurol* 496:572–593.

Gorbman, A. 1983. Reproduction in cyclostome fishes and its regulation. In: W.S. Hoar and D.J. Randall (eds.), *Fish Physiology*, Vol. IXA. Academic Press, New York, London, pp. 1–28.

Gorbman, A. 1990. Sex differentiation in the hagfish *Eptatretus stoutii*. *Gen Comp Endocrinol* 77:309–323.

Gorbman, A. 1995. Olfactory origins and evolution of the brain – pituitary endocrine system: Facts and speculation. *Gen Comp Endocrinol* 97:171–178.

Gorbman, A. and A.H. Bern. 1962. *A Textbook of Comparative Endocrinology*. Wiley, New York.

Gorbman, A. and W.W. Dickhoff. 1978. Endocrine control of reproduction in hagfish. In: J. Gaillard and H.H. Boer (eds.), *Comparative Endocrinology*. Elsevier, Amsterdam, pp. 49–54.

Gorbman, A., W.W. Dickhoff, S.R. Vigna, N.B. Clark and C.L. Ralph. 1983. *Comparative Endocrinology*. John Wiley & Sons, New York, 572 pp.

Gorbman, A., H. Kobayashi and H. Uemura. 1963. The vascularisation of the hypophysial structure of the hagfish. *Gen Comp Endocrinol* 3:505–514. doi:10.1016/0016-6480(63)90083-2.

Gore-Langton, R.E. and D.T. Armstrong. 1998. Follicular steroidogenesis and its control. In: E. Knobil, J.D. Neill, L.L. Ewing, G.S. Greenwald, C.L. Markert and D.W. Pfaff (eds.), *The Physiology of Reproduction*, Vol. 1. Raven Press, New York, pp. 331–385.

Habibi, H.R., R. de Leeuw, C.S. Nahorniak, H.J. Th. Goos and R.E. Peter 1989. Pituitary gonadotropin-releasing hormone (GnRH) receptor activity in goldfish and catfish: Seasonal and gonadal effects. *Fish Physiol Biochem* 7:109–118. doi:10.1007/BF00004696.

Hardisty, M.W. 1979. *Biology of Cyclostomes*. Chapman & Hall, London.

Hardisty, M.W. and B.I. Baker. 1982. Endocrinology of lampreys. In: M.W. Hardisty and I.C. Potter (eds.), *The Biology of Lampreys*, Vol. 4B. Academic Press, London, pp. 1–115.

Hirose, K., B. Tamaoki, B. Fernholm and H. Kobayashi. 1975. *in vitro* bioconversions of steroids in the mature ovary of the hagfish, *Eptatretus burgeri*. *Comp Biochem Physiol* 51B:403–408. doi:10.1016/0305-0491(75)90029-2.

Holmes, R.L. and J.N. Ball. 1974. *The Pituitary Gland, A Comparative Account*. Cambridge University Press, London.

Holland, L.Z., R. Albalat, K. Azumi, È. Benito-Gutiérrez, M.J. Blow, M. Bronner-Fraser, F. Brunet et al. 2008. The amphioxus genome illustrates vertebrate origins and cephalochordate biology. *Genome Res* 18:1100–1111. doi:10.1101/gr.073676.107.

Huggard, D., Z. Khakoo, G. Kassam, S.S. Mahmoud and H.R. Habibi. 1996. Effect of testosterone on maturational gonadotropin subunit messenger ribonucleic acid levels in the goldfish pituitary. *Biol Reprod* 54:1184–1191. doi:10.1095/biolreprod54.6.1184.

Huggard-Nelson, D.L., P.S. Nathwani, A. Kermouni and H.R. Habibi. 2002. Molecular characterization of LH-beta and FSH-beta subunits and their regulation by estrogen in the goldfish pituitary. *Mol Cell Endocrinol* 188:171–193. doi:10.1016/S0303-7207(01)00716-X.

Ichikawa, T., H. Kobayashi and M. Nozaki. 2000. Seasonal migration of the hagfish, *Eptatretus burgeri*, Girard. *Zool Sci* 17:217–223.

Jhamandas, J.H. and D. MacTavish. 2003. Central administration of neuropeptide FF causes activation of oxytocin paraventricular hypothalamic neurones that project to the brainstem. *J Neuroendocrinol* 15:24–32.

Jhamandas, J.H., D. MacTavish and K.H. Harris. 2006. Neuropeptide FF (NPFF) control of magnocellular neurosecretory cells of the rat hypothalamic paraventricular nucleus (PVN). *Peptides* 27:973–979.

Janvier, P. 1996. *Early Vertebrates*. Clarendon Press, Oxford.

Kalliomäki, M.L. and P. Panula. 2004. Neuropeptide FF, but not prolactin-releasing peptide, mRNA is differentially regulated in the hypothalamic and medullary neurons after salt loading. *Neuroscience* 124:81–87.

Kavanaugh, S.I., M. Nozaki and S.A. Sower. 2008. Origins of gonadotropin-releasing hormone (GnRH) in vertebrates: Identification of a novel GnRH in a basal vertebrate, the sea lamprey. *Endocrinology* 149:3860–3869. doi: 10.1210/en.2008-0184.

Kavanaugh, S.I., M.L. Powell and S.A. Sower. 2005. Seasonal changes of gonadotropin-releasing hormone in the Atlantic hagfish *Myxine glutinosa*. *Gen Comp Endocrinol* 140:136–143. doi:10.1016/j.ygcen.2004.10.015.

Kawauchi, H. and S.A. Sower. 2006. The dawn and evolution of hormones in the adenohypophysis. *Gen Comp Endocrinol* 148:3–14. doi:10.1016/j.ygcen.2005.10.011.

Kawauchi, H., K. Suzuki, H. Itoh, P. Swanson, N. Naito, Y. Nagahama, M. Nozaki, Y. Nakai and S. Itho. 1989. The duality of teleost gonadotropins. *Fish Physiol Biochem* 7:29–38. doi:10.1007/BF00004687.

Kawauchi, H., K. Suzuki, T. Yamazaki, S. Moriyama, M. Nozaki, K. Yamaguchi, A. Takahashi, J. Youson and S.A. Sower. 2002. Identification of growth hormone in the sea lamprey, an extant representative of a group of the most ancient vertebrates. *Endocrinology* 143:4916–4921. doi:10.1210/en.2002-220810.

Kazeto, Y., S. Ijiri, S. Adachi and K. Yamauchi. 2006. Cloning and characterization of a cDNA encoding cholesterol side-chain cleavage cytochrome P450 (CYP11A1): Tissue-distribution and changes in the transcript abundance in ovarian tissue of Japanese eel, *Anguilla japonica*, during artificially induced sexual development. *J Steroid Biochem Mol Biol* 99:121–128.

Kime, D.E. and E.A. Hews. 1980. Steroid biosynthesis by the ovary of the hagfish *Myxine glutinosa*. *Gen Comp Endocrinol* 42:71–75. doi:10.1016/0016-6480(80)90258-0.

Kime, D.E., E.A. Hews and J. Gafter. 1980. Steroid biosynthesis by testes of the hagfish *Myxine glutinosa*. *Gen Comp Endocrinol* 41:8–13.

Kobayashi, H., T. Ichikawa, H. Suzuki and M. Sekimoto. 1972. Seasonal migration of the hagfish, *Eptatretus burgeri*. *Jpn J Ichthyol* 19:191–194.

Kobayashi, T., M. Nakamura, H. Kajiura-Kobayashi, G. Young and Y. Nagahama. 1998. Immunolocalization of steroidogenic enzymes (P450scc, P450c17, P450arom, and 3beta-HSD) in immature and mature testes of rainbow trout (*Oncorhynchus mykiss*). *Cell Tissue Res* 292:573–577.

Kobayashi, M., Y.C. Sohn, Y. Yoshiura and K. Aida. 2000. Effects of sex steroids on the mRNA levels of gonadotropin subunits in juvenile and ovariectomized goldfish *Carassius auratus*. *Fish Sci* 66:223–231. doi:10.1046/j.1444-2906.2000.00038.x.

Kobayashi, M. and N.E. Stacey. 1990. Effects of ovariectomy and steroid hormone implantation on serum gonadotropin levels in female goldfish. *Zool Sci* 7:715–721.

Kobayashi, H. and H. Uemura. 1972. The neurohypophysis of the hagfish, *Eptatretus burgeri* (Girard). *Gen Comp Endocrinol Suppl* 3:114–124.

Koda, A., K. Ukena, H. Teranishi, S. Ohta, K. Yamamoto, S. Kikuyama and K. Tsutsui. 2002. A novel amphibian hypothalamic neuropeptide: Isolation, localization, and biological activity. *Endocrinology* 143:411–419.

Kotlinska, J., A. Pachuta, T. Dylag and J. Silberring. 2007. The role of neuropeptide FF (NPFF) in the expression of sensitization to hyperlocomotor effect of morphine and ethanol. *Neuropeptides* 41:51–58.

Kusakabe, M., I. Nakamura, J. Evans, P. Swanson and G.H. Young. 2006. Changes in mRNAs encoding steroidogenic acute regulatory protein, steroidogenic enzymes and receptors for gonadotropins during spermatogenesis in rainbow trout testes. *J Endocrinol* 189:541–554.

Larsen, L.O. 1973. Development in adult, freshwater river lampreys and its hormonal control. Starvation, sexual maturation and natural death. Thesis, University of Copenhagen.

Larsen, L.O. and B. Rothwell. 1972. Adenohypophysis. In: M.W. Hardisty and I.C. Potter (eds.), *The Biology of Lampreys*, Vol. 2. Academic Press, London, pp. 1–67.

Larsen, D. and P. Swanson. 1997. Effects of gonadectomy on plasma gonadotropins I and II in coho salmon, *Oncorhynchus kisutch*. *Gen Comp Endocrinol* 108:152–160. doi:10.1006/gcen.1997.6958.

Levavi-Sivan, B., J. Biran and E. Fireman. 2006. Sex steroids are involved in the regulation of gonadotropin-releasing hormone and dopamine D2 receptors in female tilapia pituitary. *Biol Reprod* 75:642–650. doi:10.1095/biolreprod.106.051540.

Lindemans, M., T. Jannsen, I. Beets, L. Temmerman, E. Meelkop and L. Schoofs. 2011. *Frontiers in Endocrinology* 12 July, 2011. doi:10.3389/fendo.2011.00016.

Lindemans, M., F. Liu, T. Janssen, S.J. Husson, I. Mertens, G. Gade and L. Schoofs. 2009. Adipokinetic hormone signaling through the gonadotropin-releasing hormone receptor modulates egg-laying in *Caenorhabditis elegans*. *Proc Natl Acad Sci USA* 106:1642–1647. doi:10.1073/pnas.0809881106.

Liu, Q., X.M. Guan, W.J. Martin, T.P. McDonald, M.K. Clements, Q. Jiang, Z. Zeng et al. 2001. Identification and characterization of novel mammalian neuropeptide FF-like peptides that attenuate morphine-induced antinociception. *J Biol Chem* 276:36,961–36,969.

Lowartz, S., R. Petkam, R. Renaud, F.W.H. Beamish, D.E. Kime, J. Raeside and J.F. Leatherland. 2003. Blood steroid profile and *in vitro* steroidogenesis by ovarian follicles and testis fragments of adult sea lamprey, *Petromyzon marinus*. *Comp Biochem Physiol A* 134:365–376. doi:10.1016/S1095-6433(02)00285-4.

Matty, A.J., K. Tsuneki, W.W. Dickhoff and A. Gorbman. 1976. Thyroid and gonadal function in hypophysectomized hagfish, *Eptatretus stoutii*. *Gen Comp Endocrinol* 30:500–516. doi:10.1016/0016-6480(76)90120-9.

Miki, M., T. Shimotani, K. Uchida, S. Hirano and M. Nozaki. 2006. Immunohistochemical detection of gonadotropin-like material in the pituitary of brown hagfish (*Paramyxine atami*) correlated with their gonadal functions and effect of estrogen treatment. *Gen Comp Endocrinol* 148:15–21. doi:10.1016/j.ygcen.2006.01.018.

Nagahama, Y., M. Yoshikuni, M. Yamashita and M. Tanaka. 1994. Regulation of oocyte maturation in fish. In: N.M. Sherwood and C.L. Hew (eds.), *Fish Physiology, Vol. XIII, Molecular Endocrinology of Fish*. Academic Press, San Diego, pp. 393–439. Nakabayashi, K., H. Matsumi, A. Bhalla, J. Bae, S. Mosselman, S.Y. Hsu and A.J.W. Hsueh. 2002. Thyrostimulin, a heterodimer of two new human glycoprotein hormone subunits, activates the thyroid-stimulating hormone receptor. *J Clin Invest* 109:1445–1452. doi:10/1172/JCI200214340.

Nakamura, I., J. Evans, M. Kusakabe, Y. Nagahama and G.H. Young. 2005. Changes in steroidogenic enzyme and steroidogenic acute regulatory protein messenger RNAs in ovarian follicles during ovarian development of rainbow trout (*Oncorhynchus mykiss*). *Gen Comp Endocrinol* 144:224–231.

Nishiyama, M., H. Chiba, K. Uchida, T. Shimotani and M. Nozaki. 2013. Relationships between plasma concentrations of sex steroid hormones and gonadal development in the brown hagfish, *Paramyxine atami*. *Zool Sci* 30:967–974.

Nishiyama, M., K. Uchida, N. Abe and M. Nozaki. 2015. Molecular cloning of cytochrome P450 side-chain cleavage and changes in its mRNA expression during gonadal development of brown hagfish, *Paramyxine atami*. *Gen Comp Endocrinol* 212:1–9. doi:org/10.1016/j.ygcen.2015.01.014.

Nozaki, M. 2008. The hagfish pituitary gland and its putative adenohypophysial hormones. *Zool Sci* 25:1028–1036. doi:10.2108/zsj.25.1028

Nozaki, M. 2013. Hypothalamic–pituitary–gonadal endocrine system in the hagfish. Front *Endocrinol* 4:200. doi:10.3389/fendo. 2013.00200.

Nozaki, M., B. Fernholm and H. Kobayashi. 1975. Ependymal absorption of peroxidase into the third ventricle of the hagfish *Eptatretus burgeri* (Girard). *Acta Zool* 56:265–269.

Nozaki, M. and A. Gorbman. 1983. Immunocytochemical localization of somatostatin and vasotocin in the brain of the Pacific hagfish, *Eptatretus stoutii*. *Cell Tissue Res* 229:541–550.

Nozaki, M., A. Gorbman and S.A. Sower. 1994. Diffusion between the neurohypophysis and the adenohypophysis of lampreys, *Petromyzon marinus*. *Gen Comp Endocrinol* 96:385–391.

Nozaki, M., T. Ichikawa, K. Tsuneki and H. Kobayashi.2000. Seasonal development of gonads of the hagfish, *Eptatretus burgeri*, correlated with their seasonal migration. *Zool Sci* 17:225–232.

Nozaki, M., K. Ominato, T. Shimotani, H. Kawauchi, J.H. Youson and S.A. Sower. 2008. Identity and distribution of immunoreactive adenohypophysial cells in the pituitary during the life cycle of sea lampreys, *Petromyzon marinus*. *Gen Comp Endocrinol* 155:403–412. doi:10.1016/j.ygcen.2007.07.012.

Nozaki, M., Y. Oshima, M. Miki, T. Shimotani, H. Kawauchi and S.A. Sower. 2005. Distribution of immunoreactive adenohypophysial cell types in the pituitaries of the Atlantic and the Pacific hagfish, *Myxine glutinosa* and *Eptatretus burgeri*. *Gen Comp Endocrinol* 143:142–150. doi:10.1016/j.ygcen.2005.03.002

Nozaki, M., T. Shimotani and K. Uchida. 2007. Gonadotropin-like and adrenocorticotropin-like cells in the pituitary gland of hagfish, *Paramyxine atami*: Immunohistochemistry in combination with lectin histochemistry. *Cell Tissue Res* 328:563–572. doi:10.1007/s00441-006-0349-3

Nozaki, M., K. Uchida, K. Honda, T. Shimotani and M. Nishiyama. 2013. Effects of Estradiol or testosterone treatment on expression of gonadotropin subunit mRNAs and proteins in the pituitary of juvenile brown hagfish, *Paramyxine atami. Gen Comp Endocrinol* 189:142–150. doi:10.1016/j.ygcen.2013.04.034.

Oshima, Y., K. Ominato and M. Nozaki. 2001. *Distribution of GnRH-like immunoreactivity in the brain of lampreys and hagfish*. Annual Activity Reports of the Sado Marine Biological Station, Niigata University, 31:4–5.

Osugi, T., K. Uchida, M. Nozaki and K. Tsutsui 2011. Characterization of novel RFamide peptides in the central nervous system of the brow hagfish: Isolation, localization, and functional analysis. *Endocrinology* 152:4252–4264. doi:10.1210/en.2011-1375.

Osugi, T., K. Ukena, S.A. Sower, H. Kawauchi and K. Tsutsui. 2006. Evolutionary origin and divergence of PQRFamide peptides and LPXRFamide peptides in the RFamide peptide family: Insights from novellamprey RFamide peptides. *FEBS J* 273:1731–1743.

Panula, P., A.A. Aarnisalo and K. Wasowicz. 1996. Neuropeptide FF, a mammalian neuropeptide with multiple functions. *Prog Neurobiol* 48:461–487.

Park, J.-I.I., J. Semeyonov, C.L. Cheng and S.Y.T. Hsu. 2005. Conservation of the heterodimeric glycoprotein hormone subunit family proteins and the LGR signaling system from nematodes to humans. *Endocrine* 26:267–276. doi:10.1385/ENDO:26:3:267.

Patzner, R.A. 1998. Gonads and reproduction in hagfishes. In: J.M. Jørgensen, J.P. Lomholt, R.E. Weber and H. Malte (eds.), *The Biology of Hagfishes*. Chapman & Hall, London, pp. 378–395.

Patzner, R.A. and T. Ichikawa.1977. Effect of hypophysectomy on the testis of the hagfish, *Eptatretus burgeri*, Girard (Cyclostomata). *Zool Anz* 199:371–380.

Powell, M.L., S. Kavanaugh and S.A. Sower. 2004. Seasonal concentrations of reproductive steroids in the gonads of the Atlantic hagfish, *Myxine glutinosa*. *J Exp Zool* 301A, 352–360. doi:10.1002/jez.a.20043.

Powell, M.L., S.I. Kavanaugh and S.A. Sower. 2005. Current knowledge of hagfish reproduction: Implications for fisheries management. *Integr Comp Biol* 45:158–165.

Powell, M.L., S. Kavanaugh and S.A. Sower. 2006. Identification of a functional corpus luteum in the Atlantic hagfish, *Myxine glutinosa*. *Gen Comp Endocrinol* 148:95–101.

Quérat, B., Y. Arai, A. Henry, Y. Akama, T.J. Longhurst and J.M. Joss.2004. Pituitary glycoprotein hormone beta subunits in the Australian lungfish and estimation of the relative evolution rate of these subunits within vertebrates. *Biol Reprod* 70:356–363. doi:10.1095/biolreprod.103.022004.

Quérat, B., A. Sellouk and C. Salmon. 2000. Phylogenetic analysis of the vertebrate glycoprotein hormone family including new sequences of sturgeon *(Acipenser baeri)* beta subunits of the two gonadotropins and the thyroid-stimulating hormone. *Biol Reprod* 63:222–228. doi:10.1095/biolreprod63.1.222.

Quérat, B., C. Tonnerre-Donicarli, F. Géniès and C. Salmon. 2001. Duality of gonadotropins in gnathostomes. *Gen Comp Endocrinol* 124:308–314. doi:10.1006/gcen.2001.7715.

Roumy, M., C. Lorenzo, S. Mazères, S. Bouchet, J.M. Zajac and C. Mollereau.2007. Physical association between neuropeptide FF and m-opioid receptors as a possible molecular basis for anti-opioid activity. *J Biol Chem* 282:8332–8342.

Saligaut, C., B. Linard, E.L. Mananos, O. Kah, B. Breton and M. Govoroun. 1998. Release of pituitary gonadotropins GtHI and GtHII in the rainbow trout *(Oncorhynchus mykiss)*: Modulation by estradiol and catecholamines. *Gen Comp Endocrinol* 109:302–309.

Schützinger, S., H.S. Choi, R.A. Patzner and H. Adam. 1987. Estrogens in plasma of the hagfish, *Myxine glutinosa* (Cyclostomata). *Acta Zool (Stockh)* 68:263–266. doi:10.1111/j.1463-6395.1987.tb00893.x.

Shahjahan, M., T. Ikegami, T. Osugi, K. Ukena, H. Doi, A. Hattori, K. Tsutsui and H. Ando. 2011. Synchronised expressions of LPXRFamide peptide and its receptor genes: Seasonal, diurnal and circadian changes during spawning period in grass puffer. *J Neuroendocrinol* 23:39–51.

Smith, J.J., S. Kuraku, C. Holt, T. Sauka-Spengler, N. Jiang, M.S. Campbell, M.D. Yandell et al. 2013. Sequencing of the sea lamprey (*Petromyzon marinus*) genome provides insights into vertebrate evolution. *Nat Genet* 45(4):415–421.

Sower, S.A. 1998. Brain and pituitary hormones of lampreys, recent findings and their evolutionary significance. *Am Zool* 38:15–38. doi:10.1093/icb/38.1.15.

Sower, S.A., M. Freamat and S.I. Kavanaugh. 2009. The origins of the vertebrate hypothalamic-pituitary-gonadal (HPG) and hypothalamic–pituitary–thyroid (HPT) endocrine systems: New insights from lampreys. *Gen Comp Endocrinol* 161:20–29. doi:10.1016/j.gcen.2008.11.023.

Sower, S.A., W.A. Decatur, K.N. Hausken, T.J. Marquis, S.L. Barton, J. Gargan, M. Freamat et al. 2015. Emergence of an ancestral glycoprotein hormone in the pituitary of the sea lamprey, a basal vertebrate. *Endocrinology* [Epub ahead of print].

Sower, S.A., S. Moriyama, M. Kasahara, A. Takahashi, M. Nozaki, K. Uchida, J.M. Dahlstrom and H. Kawauchi. 2006. Identification of sea lamprey GTHbeta-like cDNA and its evolutionary implications. *Gen Comp Endocrinol* 148:22–32. doi:10.1016/j.ygcen.2005.11.009.

Sower S.A., M. Nozaki, C.J. Knox and A. Gorbman. 1995. The occurrence and distribution of GnRH in the brain of Atlantic hagfish, an Agnathan, determined by chromatography and immunocytochemistry. *Gen Comp Endocrinol* 97: 300–307. doi:10.1006/gcen.1995.1030.

Staubli, F., T.J. Jorgensen, G. Cazzamali, M. Williamson, C. Lenz, L. Sondergarrd, P. Rooepstorff and C.J. Grimmelikhuijzen. 2002. Molecular identification of the insect adipokinetic hormone receptors. *Proc Natl Acad Sci USA* 99:3446–3451. doi:10.1073/pnas.052556499.

Sudo, S., Y. Kuwabara, J.I. Park, S.Y. Hsu and A.J. Hsueh. 2005. Heterodimeric fly glycoprotein hormone-alpha2 (GPA2) and glycoprotein hormone-beta5 (GPB5) activate fly leucine-rich repeat-containing G protein-coupled receptor-1 (DLGR1) and stimulation of human thyrotropin receptors by chimeric fly GPA2 and human GPB5. *Endocrinology* 146:3596–3604. doi:10.1210/en.2005-0317.

Suzuki, M., M. Kubokawa, K. Nagasawa and A. Urano. 1995. Sequence analysis of vasotocin cDNAs of the lamprey, *Lampetra japonica* and the hagfish *Eptatretus burgeri*—Evolution of cyclostome vasotocin precursors. *J Mol Endocrinol* 14:67–77. doi:10.1677/jme.0.0140067.

Tando, Y. and K. Kubokawa. 2009a. Expression of the gene for ancestral glycoprotein hormone beta subunit in the nerve cord of amphioxus. *Gen Comp Endocrinol* 162:329–339. doi:10.1016/j.ygcen.2009.04.015.

Tando, Y. and K. Kubokawa. 2009b. A homolog of the vertebrate thyrostimulin glycoprotein hormone a subunit (GPA2) is expressed in amphioxus neurons. *Zool Sci* 26:409–414. doi:10.2108/zsj.26.409.

Tsai, P.S. and L. Zhang. 2008. The emergence and loss of gonadotropin-releasing hormone in protostomes: Orthology, phylogeny, structure, and function. *Biol Reprod* 79:798–805. doi:10.1095/biolreprod.108.070185.

Tsukahara, T., A. Gorbman and H. Kobayashi. 1986. Median eminence equivalence of the neurohypophysis of the hagfish, *Eptatretus burgeri*. *Gen Comp Endocrinol* 61:348–354. doi:org/10.1016/0016-6480(86)90220-0.

Tsuneki, K. and A. Gorbman. 1977a. Ultrastructure of the ovary of the hagfish *Eptatretus stoutii*. *Acta Zool (Stockh)* 58:27–40.

Tsuneki, K. and A. Gorbman. 1977b. Ultrastructure of the testicular interstitial tissue of the hagfish *Eptatretus stoutii. Acta Zool (Stockh)* 58:17–25.

Tsuneki, K., M. Ouji and H. Saito. 1983. Seasonal migration and gonadal changes in the hagfish *Eptatretus burgeri. Jpn J Ichthyol* 29:429–440.

Tsutsui, K. 2009. A new key neurohormone controlling reproduction, gonadotropin-inhibitory hormone (GnIH): Biosynthesis, mode of action and functional significance. *Prog Neurobiol* 88:76–88.

Tsutsui, K. and K. Ukena. 2006. Hypothalamic LPXRF-amide peptides in vertebrates: Identification, localization and hypophysiotropic activity. *Peptides* 27:1121–1129.

Tsutsui, K., G.E. Bentley, G. Bedecarrats, T. Osugi, T. Ubukaand and L.J. Kriegsfeld. 2010. Review: Gonadotropin-inhibitory hormone (GnIH) and its control of central and peripheral reproductive function. *Front Neuroendocrinol* 31:284–295.

Uchida, K., S. Moriyama, H. Chiba, T. Shimotani, K. Honda, M. Miki, A. Takahashi, S.A. Sower and M. Nozaki. 2010. Evolutionary origin of a functional gonadotropin in the pituitary of the most primitive vertebrate, hagfish. *Proc Natl Acad Sci USA* 107:15,832–15,837. doi:10.1073/pnas.100220810.

Ukena, K., A. Koda, K. Yamamoto, T. Kobayashi, E. Iwakoshi-Ukena, H. Minakata, S. Kikuyama and K. Tsutsui. 2003. Novel neuropeptides related to frog growth hormone-releasing peptide: Isolation, sequence, and functional analysis. *Endocrinology* 144:3879–3884.

Ukena, K. and K. Tsutsui. 2005. A new member of the hypothalamic RF amide peptide family, LPXRF-amide peptides: Structure, localization, and function. *Mass Spectrom Rev* 24:469–486.

Vacher, C., F. Ferrière, M.H. Marmignon, E. Pellegrini and C. Saligaut. C. 2002. Dopamine D2 receptors and secretion of FSH and LH: Role of sexual steroids on the pituitary of the female rainbow trout. *Gen Comp Endocrinol* 127:198–206. doi:10.1016/S0016-6480(02)00046-1.

Vilim, F.S., A.A. Aarnisalo, M.L. Nieminen, M. Lintunen, K. Karlstedt, V.K. Kontinen, E. Kalso, B. States, P. Panula and E. Ziff. 1999. Gene for pain modulatory neuropeptide NPFF: Induction in spinal cord by noxious stimuli. *Mol Pharmacol* 55:804–811.

Weisbart, M., W.W. Dickhoff, A. Gorbman and D.R. Idler. 1980. The presence of steroids in the sera of the Pacific hagfish, *Eptatretus stoutii*, and the sea lamprey, *Petromyzon marinus. Gen Comp Endocrinol* 41:506–519.

Young, B.A., M.B. Bryan, J.R. Glenn, S.S. Yun, A.P. Scott and W. Li. 2007. Dose–response relationship of 15a-hydroxylated sex steroids to gonadotropin-releasing hormones and pituitary extract in male sea lampreys (*Petromyzon marinus*). *Gen Comp Endocrinol* 151:108–115. doi:10.1016/j.ygcen.2006.12.005.

Yu, J.Y.L., W.W. Dickhoff, P. Swanson and A. Gorbman. 1981. Vitellogenesis and its hormonal regulation in the Pacific hagfish, *Eptatretus stoutii* L. *Gen Comp Endocrinol* 43:492–502. doi:10.1016/0016-6480(81)90234-3.

Zhang, Y., S. Li, Y. Liu, D. Lu, H. Chen, X. Huang, X. Liu, Z. Meng, H. Lin and C.H. Cheng. 2010. Structural diversity of the GnIH/GnIH receptor system in teleost: Its involvement in early development and the negative control of LH release. *Peptides* 31:1034–1043.

Zhang, L., J.A. Tello, W. Zhang and P.S. Tsai. 2008. Molecular cloning, expression pattern, and immunocytochemical localization of a gonadotropin-releasing hormone-like molecule in the gastropod mollusk, *Aplysia californica. Gen Comp Endocrinol* 156:201–209. doi:10.1016/j.ygcen.2007.11.015.

chapter ten

Corticosteroid signaling pathways in hagfish

Nic R. Bury, Alexander M. Clifford, and Gregory G. Goss

Contents

Introduction

There are seven nuclear hormone receptor (NR) families, termed NR0–7. NR1 is a large group containing the thyroid hormone receptors, retinoic acid receptors, peroxisome proliferator-activated receptors, vitamin D, and ecdysone receptors, as well as a number of orphan receptors. NR2 consists of the retinoid-X-receptors as well as Coup-TF, HNF4 TR2, and four orphan receptors. NR3 consists of the steroid receptors (SRs), whereas NR4 are the NGFIB orphan receptors. NR5 FTZ-F1 and NR6 GCNF1 comprise orphan receptors, whereas NR0 contains gene with sequences that resemble fragments of other receptors, but whose function is unclear. The focus of this chapter is the corticosteroid hormone receptors that belong to the NR3 family, which in terrestrial vertebrates includes the sex steroid receptors—estrogen (ER), androgen (AR) and progesterone (PR), and the adrenal gland hormone receptors—the corticosteroid receptors (CRs), which are further divided into the glucocorticoid

257

(GR) and mineralocorticoid (MR) receptors, as well as a group known as the estrogen-related receptors (ERRs). The primary role of the steroid receptors is to act as ligand-dependent transcription factors that enable the cell to increase the expression of target genes. This is an oversimplification and some SRs act via a nongenomic pathway or influence other hormone-signaling pathways via protein–protein interactions (e.g., Glass and Rosenfeld, 2000; Borski et al., 2001; Lee et al., 2012). The SR ligands are small lipophilic compounds that readily cross the cell membrane and include hormones such as estradiol, testosterone, and cortisol. The array of ligands and receptors present in vertebrates forms a complex system that has evolved to regulate or influence a vast number of physiological processes (e.g., reproduction, energy metabolism, the immune system, and development) and this system is critical for interpreting the brain's response to internal and external stimuli. Chapter 9 covers a more detailed review of the sex steroid hormone synthesis pathway in hagfish. This chapter will take a comparative approach to track the evolution of the steroid-signaling pathway, including the steroid synthesis pathway, as well as the steroid receptors, but will focus primarily on the CRs.

Chordate phylogeny and steroid receptors

The two invertebrate groups, urochordates and cephalochordates, along with the vertebrates make up the chordates. The cephalochordates were long believed to be the closest relatives to the vertebrates due to the morphological similarities to the vertebrates; however, complete genome sequencing of a number of urochordatese (*Ciona intestinalis*, *C. savigny*, and *Oikopleura dioica*) reveals that they are a sister clade to the vertebrates (Delsuc et al., 2006). Thus, it appears that urochordates have lost a significant number of features found in cephalochordates and retained in early vertebrates. This includes members of the steroid receptor, with the cephalochordates possessing SRs, which are absent in the urochordates, but present in the vertebrates (Figure 10.1).

There is ongoing debate on how to class the two extant groups of cyclostomes, the lamprey and hagfish. Cladistic-based approaches suggest paraphyly, with the lamprey being more closely related to the jawed vertebrates, gnathostomes, than the hagfish (Maisey, 1986). However, molecular evidence, such as ribosomal RNA, mitochondrial DNA, and encoded gene sequencing analysis, supports monophyly (Mallatt and Sullivan, 1998; Kuraka et al., 1999; Delarbre et al., 2002). Heimberg et al. (2010) provide further evidence for a monophyletic Cyclostomata. Deep sequencing of small RNA of hagfish, lamprey, tunicates, and jawed fish enables a comparison of the micro-RNA (miRNA) between the various groups. Hagfish shared 44 out of the 46 specific miRNA and possessed four unique miRNA, only shared with lamprey. In addition, the

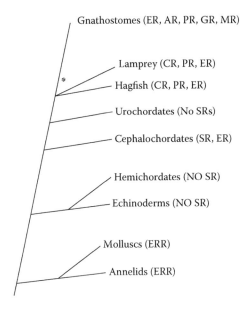

Figure 10.1 The emergence of steroid receptors in the deuterostomes. ERRs are present in annelids and mollusks, but are absent in other invertebrates. A proposed whole gene duplication event occurred early in the chordate lineage and gave rise to duplicate receptors termed an ER and SR that are present in the cephalochordates, but have been lost in the urochordates. Further duplication events (Smith and Keinart, 2014) have given rise to three of the potential four duplicates of ER and SR being retained in hagfish. In addition to the ER, two new receptors appear, the PR and the CR. A further round of duplications sees the suite of steroid receptor present in the gnathostomes that now includes the AR and the GRs and MRs, the latter being products derived from CR duplication. The asterisks denoted the current debate on whether hagfish and lamprey are monophyletic or paraphyletic (see "Chordate phylogeny and steroid receptors").

hagfish and lamprey possessed 15 unique paralogues and within 18 of the shared miRNA there were 22 unique substitutions (Heimberg et al., 2010). An assumption for the use of miRNA for phylogenetic analysis is that because of their important role in regulating development, the rate of mutations to form novel miRNA would be exceedingly rare. However, Thomson et al. (2014) have recently questioned this assumption and the use of some miRNA information for such analyses. Thomson et al. (2014) identified that there are heterogeneous rates of miRNA gain and loss as well as secondary loss of these molecules. These observations, coupled with potential sampling errors, means that caution should be taken when inferring phylogeny from these data sets. A re-examination of Heimberg et al. (2010) work drew the conclusion that the miRNA data were equivocal in resolving whether lamprey and hagfish are sister clades. If the cyclostomes were monophyletic, it would suggest either that the ancestral vertebrate resembled the hagfish and the lamprey have undergone astonishing convergent evolution to obtain many features of the gnathostomes, or alternatively, the ancestral vertebrate was more complex than originally thought with the lamprey retaining a number of features that are present in gnathostomes. In this scenario, the hagfish has undergone a remarkable and unprecedented loss of features in the vertebrate lineage, suggesting that the hagfish and lamprey diverged early on, following the emergence of the cyclostome clade around 525MYA (Near, 2009). The SRs are an example of a group of proteins that hagfish have retained and which are also present in the lamprey (Figure 10.1; Bridgham et al., 2006). The retention of these molecules in extant members of hagfish suggests that when they emerged in an ancestral vertebrate many millions of years ago they played an extremely important physiological and/or developmental role. These SR protein further evolved into the suite of SRs present in gnathostomes (Figure 10.1; see "**Steroid receptor evolution**").

Hormone synthesis pathways

Pituitary gland and associated peptides

The pituitary gland of vertebrates is divided into two regions: the adenohypophysis and neurohypophysis. The adenohypophysis is further divided into rostral pars distalis, proximal distalis, and pars intermedia, and the neurohypophysis into the anterior and posterior neurohypophysis. In fish, mammals, and birds, the signaling cascade that leads to the synthesis of corticosteroids starts with the hypothalamus-releasing corticotrophin-releasing factor, which stimulates the adenohypophysis to produce the glycoprotein pro-opiomeanocortin (POMC). POMC is a precursor protein that is further fragmented into the functional peptides adrenocorticotrophic hormone (ACTH), melanotropins (MSH), and β-lipotropin. ACTH and β-lipotropin can be further cleaved to form α-MSH and corticotrophin-like intermediate lobe peptide, and γ-lipoprotein and β-endorphin, respectively (Malagoli et al., 2011). ACTH is released into the circulation and binds to membrane G protein-coupled ACTH receptors of the adrenal cortex of land vertebrate and interrenal tissue of the head kidney of fish to stimulate corticosteroid synthesis. Thus, the control of corticosteroid release is via the hypothalamus–pituitary–adrenal (HPA) or –interrenal (HPI) axis. The sex steroid synthesis is under the control of the hypothalamus–pituitary–gonadal (HPG) axis, where the hypothalamus releases gonadotropin-releasing hormone (GnRH) that results in the pituitary-secreting gonadotropins (GTH). In the gnathostomes, there are two GTHs, luteinizing hormone (LH) and follicle-stimulating hormone (FSH), which in conjunction with the thyroid-stimulating hormone form part of the pituitaries glycoprotein hormone (GPH) family. LH and FSH stimulate the testis and ovary to synthesize testosterone and estradiol, respectively.

In the hagfish, the adenohypophysis of the pituitary, which is formed from the endoderm, but interestingly from the ectoderm in all other vertebrates (Gorbman, 1983), consists of a series of cell clusters embedded in connective tissue below the neurohypophysis (Ball and Baker, 1969). These two regions are separated by a layer of connective tissue (Uchida et al., 2013) and there is no distinction between the pars distalis and pars intermedia, which is seen in the pituitary of other vertebrates (Holmes and Ball, 1974; Nozaki et al., 2008). In contrast, the lamprey pituitary is far better defined with clear adenohypophysis and neurohypophysis with the adenohypophysis divided further into the three regions rostral pars distalis, proximal distalis, and pars intermedia, similar to the structure observed in teleost fish and other vertebrates (Kawauchi and Sower, 2006).

Buckingham et al. (1985) used hagfish pituitary extract to show that it had ACTH-like activity in a mammalian-based assay system; this activity was considerably less than the standard assay using mammalian ACTH. The cloning of the lamprey POMC gene revealed two POMC-related genes (Takahashi et al., 1995b), one containing ACTH and β-endorphin, termed pro-opiocortin (POC) and the other encoding two MSHs, termed pro-opiomelanotropin (POM) (Heinig et al., 1995). Both POC and POM products are glycoproteins (Takahashi et al., 1995b). The ACTH of lamprey is larger (60 amino acids) compared with other vertebrates (39 amino acids), but the first 22 amino acids include MSH-like sequences followed by four basic amino acids, characteristic of vertebrate ACTH (Takahashi et al., 1995a). Nozaki et al. (2007) used antibodies raised to the lamprey ACTH along with lectin histochemistry (lectins will bind to glycoconjugates) to identify a group of cells within the hagfish pituitary that are ACTH reactive, but this signal was extremely weak. This study, however, also identified a separate group of cells that respond to ovine LH antibodies with the intensity of staining altering during different stages of gonadal development, indicating a division in function between cell types in the hagfish pituitary; one involved in reproduction, whereas the others in mediating the response to corticotrophin stimulus. The precise molecules involved in the steroid synthesis cascade have not been determined in hagfish, but they are likely to be glycoproteins derived from a POMC-like precursor protein. This hypothesis is based on discovery of similar peptides in lamprey (Takahashi et al., 1995a), knowledge that the origin of POMC extends back into the bilaterian lineage and that POMC and peptides derived from this precursor protein showing high sequence similarity to their mammalian counterparts have been detected in protostomes (annelids: Salzet et al., 1997; mollusks: Stefano et al., 1999). The role of ACTH in invertebrates is unclear; they are not involved in steroid synthesis, but may well modulate the immune response because ACTH-receptor-like proteins have been identified in mussel immunocytes (Ottaviani et al., 1998) and human ACTH influence motility of these cells (Sassi et al., 1998).

Knowledge on the control of corticosteroid release in agnathans has been further complicated by a recent study by Roberts et al. (2014) in lamprey. This study built on previous work by Close et al. (2010) that identified that the active putative corticosteroid in lamprey plasma was 11-deoxycortisol and not cortisol or corticosterone. The mesonephros of the lamprey kidney was shown to convert 11α hydroxyprogesterone to 11-deoxycortisol (Close et al., 2010; Roberts et al., 2014). However, treatment with the four lamprey ACTHs previously identified (Takahashi et al., 2005) did not stimulate 11-deoxycortisol synthesis. In contrast, other peptides involved in stress or osmoregulation, such as CRH, arginine-vasotocin (AVT), or reproduction, such as GnRH had a corticotrophic effect. AVT acts via vasotocin receptors and has previously been shown to induce corticosteroid synthesis in fish (Wendelaar Bonga, 1997; Gilchriest et al., 2001; Sangiao-Alvarellos et al., 2006). These results suggest that the stimulus for corticosteroid synthesis in lamprey may occur via different pathways, and whether this is also seen in hagfish awaits further study.

Steroid synthesis

The steroids consist of a core of 17 carbon atoms arranged as three cyclohexane rings and a fourth cyclopentane ring (Figure 10.2a). Each carbon is numbered 1–17 and the three cyclohexane rings are labeled A–C and the cyclopentane ring termed D (Figure 10.2a). The steroid synthesis pathway involves reactions catalyzed by a number of cytochrome P450 (CYP) and hydroxysteroid dehydrogenase (HSD) enzymes and it is the subtle changes in positioning of hydroxyl and hydrogen moieties on the carbon atoms that alter the steroid structure and determine ligand/receptor specificity (Figure 10.2b). The pathway is dependent on the activity of CYP11A1 (also known as desmolase of P450scc), which converts cholesterol into pregnenolone. Pregnenolone is the precursor of progesterone that is formed following the removal of hydrogen from the hydroxyl group on C3 of the molecule due to the activity of 3β HSD. CYP21 converts progesterone into 11-deoxycorticosterone, which in turn is converted either into corticosterone via CYP11B1 or into aldosterone via CYP11B2. However, CYP11B2 is only present in land vertebrates, and no aldosterone is present in fish. An alternative synthesis pathway sees pregnenolone being converted into 17-OH-pregnenolone due to the activity of CYP17. 3β HSD plays an important role in this pathway converting 17-OH-pregnenolone into 17-OH-progesterone, and subsequently, CYP21 adds a hydroxyl group to C21 to form 11-deoxycortisol. CYB11B1 transforms 11-deoxycortisol into cortisol (Figure 10.3b). In the testosterone/estradiol synthesis pathway, the precursor 17-OH-progesteone is converted into intermediate androstenedione via the activity of CYP17.

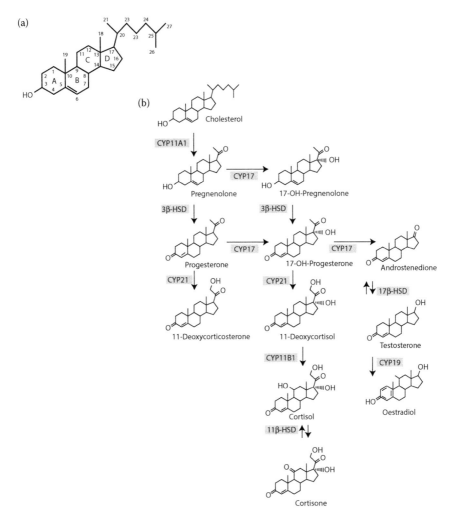

Figure 10.2 The vertebrate steroid synthesis pathway. Cholesterol (a) is the precursor for steroids; its structure consists of 17 carbon atoms arranged as three cyclohexane rings, labeled A–C and a fourth cyclopentane ring labeled D, with an additional 10 carbons labeled 18–27. (b) Members of the cytochrome P450 (CYP11A1, CYP17, CYP19, CYP21) and HSD (3β HSD, 11β HSD, and 17β HSD) family of enzymes convert cholesterol into cortisol and estradiol via a number of active steroid intermediaries (e.g., testosterone and progesterone). In terrestrial vertebrates, 11-deoxycorticosteorne is converted into the mineralocorticoid aldosterone (not shown), via the action CYP11B1, an enzyme not present in fish. (Adapted from Bury, N.R., Sturm, A. 2007. *Gen Comp Endocrinol* 153: 47–56.)

(a)

AmphioxusSR	pcavchcpstglhygvyac**eg**cksffhrahkrahpyvcpannncvidrrlkkncpacrlkkcllmgm
HumanGR	klclvcsdeasgchygvltc**gs**ckvffkravegqhnylcagrndciidkirrkncpacryrkclqagm
HumanMR	kiclvgdeasgchygvvtc**gs**ckvffkravegqhnylcagrndciidkirrkncpacrlqkclqagm
HagfishCR	kaclicrdeasgchygvltc**gs**ckgffkraiegqhnylcagrndciidkirrkncpacrlrkciqagm
LampeyCR	kaclicsdeasgchygvltc**gs**ckvffkravegqhnylcagrndciidkirrkncpacrlrkciqagm

(b)

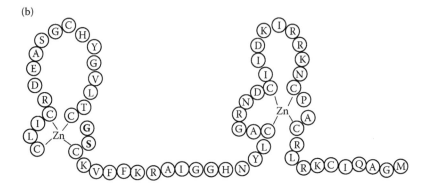

Figure 10.3 (a) Sequence alignment of the human GRs and MRs, hagfish and lamprey CR, and amphioxus steroid receptor (CR) DNA-binding region or C-domain. (b) A diagrammatic representation of the formation of the two zinc fingers by this region. The letters in bold, **gs**, are conserved in GRs, MRs, and CRs, and form part of the P-box that recognizes the GREs upstream of target genes. In contrast, the corresponding amphioxus sequence is **eg,** a receptor signature recognizing the ERE.

Two further enzymes, 17β HSD and CYP19 (also known as aromatase), produce testosterone and the aromatized A-ring characteristic of the natural estrogens, respectively (Figure 10.2b). The corticosterone/aldosterone synthesis pathway is not isolated, and CYP17 can convert progesterone into 17-OH-progesterone, which then enters the cortisol and estradiol synthesis pathway (Figure 10.2b).

A number of studies (reviewed in Janer and Porte, 2007) demonstrate the presence of immune-like estradiol and testosterone in mollusks, crustacean, and testosterone in echinoderms, indicating the synthesis pathway for these steroids emerged early on in metazoan evolution. However, the genome of *Ciona* lacks the enzymes necessary for steroid synthesis. It is only in the cephalochordate that we see CYP11, CYP17, CYP19, and 17β HSD gene transcripts present in the ovaries of amphioxus necessary for the synthesis of the sex steroid (Mizuta and Kubokawa, 2007). Holland et al. (2008) confirmed these observations via whole genome sequencing of *Branchiostoma floridae*; however, the genome lacked CYP21 or CYP11A1 enzymes necessary for the formation of the corticosteroids. Thus, it would appear that the ability to synthesize corticosteroids emerged in the vertebrates.

Evidence for corticosteroid synthesis in hagfish is still debatable. Sparsely distributed cells within the pronephroi have been identified in hagfish that in response to heterologous ACTH increase in number and undergo hypertrophy (Idler and Burton, 1976). Idler and Burton (1976) termed these as "presumptive" interrenal cells (PICs) because they were unable to confirm that these cells were able to convert cholesterol into corticosteroids, and the site of corticosteroid synthesis in hagfish remains to be clearly identified. A double isotopic technique for identifying steroids identified very low submicrogram levels of cortisol in the plasma of Atlantic hagfish (Weisbart and Idler, 1970), which could be partially, but significantly elevated, in response to ACTH injection. Similarly, Wales (1988) could not detect cortisol in control Atlantic hagfish using a radio-immunoassay approach, but found elevated plasma levels following arachidonic acid, prostaglandin E2, and porcine ACTH injection. Weisbart et al. (1980) also reported cortisol, 11-deoxycortisol, and corticosterone in plasma of Pacific hagfish but repeated injection of heterologous ACTH was necessary to elevate the plasma concentrations of these steroids and levels that were extremely low compared with the elevation in cortisol seem in fish in response to a stress (Wendelaar Bonga, 1997). But, to fully understand the control of corticosteroid synthesis in hagfish, it will be necessary to identify the homologs to ACTH in these fish and repeat similar experiment as those described above.

The presence of corticosteroids in the plasma of hagfish, however, indicates the presence of the full set of CYP and HSD enzymes necessary for corticosteroid synthesis. To date, no CYP21 or CYP11B1, critical enzymes in the corticosteroid synthesis pathway (Figure 10.2b), has been identified in any hagfish species. In this context, recent next-generation Illumina sequencing of transcripts from the slime gland and gill tissue of Pacific hagfish identified the presence of orthologs of the lamprey CYP11A1, CYP17, and 3β HSD (Clifford and Goss, unpublished), but were unable to identify CYP21 and CYP11B1. The lack of evidence for the presence of the transcripts does not necessarily mean they are not present, as they may be expressed in other tissues or at much lower concentrations. Regardless, the presence of the Lamprey orthologs (critical for sex steroid hormone synthesis) in tissue other than ovary or testis in hagfish is interesting. Clearly, further study is necessary to identify the key enzymes involved in steroid synthesis and their location in hagfishes.

Steroid receptors

Steroid receptor structure and function

The steroid receptor proteins are divided into a number of domains, for example, four domains in the vertebrate GR. For each receptor, the LBD and DBD possess conserved amino acids conferring hormone specificity in the ligand-binding pocket and the recognition of specific palindromic DNA sequences in the response elements upstream of the target genes (Figures 10.3 and 10.4). In mammals, the GRs reside in the cytoplasm where they remain inactive. Following ligand–receptor binding, the proteins that are bound to the receptor to maintain its inactive state are released and the receptors form homodimers, which migrate to the nucleus. The homodimers attach to response elements upstream of target genes and initiate transcription. Upon completion of transcription, the receptors detach, the ligand dissociates, and the receptor returns to the cytoplasm. This is a simplified description and throughout the SR family there are variations of this theme. For example, a proportion of the mammalian MR and the teleost fish (Rainbow trout) GR are present in the nucleus in the absence of hormone (Becker et al., 2008) and an isoform of the mammalian GR that lacks the last 52 amino acids of the N-terminal region, termed GR-beta, can bind cortisol but is transcriptionally inactive, acting as a negative regulator of GR-alpha activity (Schaaf et al., 2008).

Steroid receptor evolution

Thornton (2001) proposed a ligand exploitation hypothesis as an explanation for the emergence of the complex hormone steroid/receptor family in vertebrates. In this scenario, the ancestral steroid receptor would have estrogen-like properties. Gene duplication of the ancestral estrogen-like receptor gave rise to the genetic material where mutations within the steroid-binding domain enabled the receptor to exploit the other steroids that are precursors for estradiol (e.g., testosterone and progesterone) in steroid synthesis pathway (Figure 10.2) and develop neofunctionalization. The hypothesis that the ancestral receptor was an ER is supported by the isolation of an ER gene ortholog in the opisthobranch *Aplysia californica* (Thornton et al., 2003). Genes for other invertebrate ERs have been identified (Keay et al., 2006), but a number of them appear to be constitutively active and their precise function awaits elucidation.

Bridgham et al. (2008) identified orthologs of two SRs in the cephalochordate *B. floridae* genome termed bfER and bfSR. The full-length clones of these receptors were inserted into a mammalian expression vector and were used to transfect Chinese Hamster Ovary cells (CHOK-1) along with a reporter plasmid containing either vertebrate estrogen response elements (EREs) or glucocorticoid response elements (GREs) upstream of the luciferase gene. In this system, the production of luciferase indicates that a ligand has bound to the receptor and induces transcription. bfSR was transcriptionally activated in the presence of estrogen but not the corticosteroids, and it only recognized the mammalian ERE and not the GRE. The bfER was transcriptionally unresponsive in this assay system, but acted as a negative regulator of bfSR, via the formations of heterodimers. Despite the assay being performed out of context, for example, in mammalian cell lines and using a reporter construct containing mammalian response elements, it demonstrates that the SR is able to bind ligands, recognize the conserved vertebrate EREs, and is capable of recruiting the necessary mammalian co-activators for gene expression. The amino acids in the P-box of the DBD domain that confer ERE specificity are conserved in the bfSR (Figure 10.3). Both the bfER and bfSR are expressed in the amphioxus ovary and testis, suggesting that they are both involved in germ cell development. However, after duplication of an ancestral SR, the bfSR retained the estrogen-sensitive properties, whereas the bfER lost these traits, but was retained as a transcription repressor. The loss of transcriptional functionality in the bfER is attributed to two amino acid substitutions, R394C and F404L (human ER numbering), in the ligand-binding domains (Bridgham et al., 2008). The bfSR amino acid residues (R392 and F404) are conserved in other vertebrate ERs and are involved in hydrogen bond formation and stabilization of the ligand–receptor complex (Tremblay et al., 1999).

Two rounds of whole-genome duplication (WGD), often referred to as 1R and 2R, are thought to have occurred in the vertebrate lineage (Figure 10.1) (Kuraku et al., 2009; Smith et al., 2013), and a further WGD occurred in the teleostei (Hoegg et al., 2004). However, recent high-density meiotic and comparative mapping of the sea lamprey genome supports only 1R and not the 2R initially proposed (Smith and Keinart, 2014). The potential misinterpretation of another WGD event is due to the occurrence of several evolutionarily independent segmental duplications in the chordates (Smith and Keinart, 2014). The debate on the number of WGD continues, but whatever the mechanism by which genetic material was duplicated in the early vertebrate lineage it gave rise to three distinct hormone receptors, the CR, ER, and PR in cyclostomes (Bridgham et al., 2006). The cloning of the Atlantic hagfish CR (Bridgham et al., 2006) reveals that the P-box of the DNA binding domain contains the conserved amino acids that characterize a receptor capable of recognizing a GRE (Figure 10.3). This was functionally confirmed in transfection studies where the hagfish CR recognizes GRE (Bridgham et al., 2006). The Atlantic hagfish CR is promiscuous, being activated by a number of different corticosteroids such as cortisol, corticosterone, 11-deocycortisol, 11-deocycorticosteorne, and aldosterone, marginally by the progestins, but not the androgens or estrogens (Bridgham et al., 2006). This profile of steroid-induced transactivation is similar to that seen with the vertebrate MR and suggests that the ancestral CR was "MR-like"; there are two amino acid substitutions in the steroid-binding pocket of the vertebrate GR, which confer glucocorticoid specificity (Bridgham et al., 2006). However, sequence alignment of the steroid-binding domain of the human MR, GR, and hagfish and lamprey CR does not provide convincing evidence as to whether the hagfish CR is MR- or GR-like (Figure 10.4). Of the 22 amino acids identified to be important for recognition of the synthetic gluco-corticoid dexamethasone in the rat GR (Bledsoe et al., 2002), only 14 are conserved between the four receptors and there is a 77.3% and 72.7% similarity between the hagfish CR, and the hGR and hMR, respectively (Figure 10.5). A 3D model of the lamprey CR (Baker et al., 2011) may be useful for predicting the structure of active CR in hagfish. Leucine and methionine at positions 35 and 121 (number for Figure 10.5) are impor-tant for interactions between the receptor and hydroxyl group on C17 of 11-deoxycortisol (Baker et al., 2011), the proposed active corticoste-roid in lamprey (Close et al., 2010). These amino acids are also present in hagfish and two further significant amino acids, cysteine at 42 and methionine at 79, form van der Waals contact, with the steroids are also conserved in the agnathans but absent in the hGR.

Functional role of corticosteroids in hagfish

Very little is known about the physiological processes that corticosteroid controls in hagfish. Hagfish differ to lamprey and other gnathostomes in their ionoregulatory strategy because they are osmoconformers, how-ever they do regulate plasma sulfate and magnesium ion concentrations (Bellamy and Chester, 1961). To try and identify the role of corticoste-roids in hagfish, we administered hormones via the intraperitoneal cav-ity within a bolus of coconut oil containing 0, 20, 100, or 200 mg/kg of cortisol, corticosterone, or 11-deoxycorticosterone (DOC). Plasma sam-ples were taken on days 4 and 7 for glucose measurements to assess any potential changes in glucoconeogenesis, a cortisol-regulated process in other vertebrates. Gill total ATPase activity measurements via a modified method from McCormick (1993) were also measured on day 7 to assess if there was any stimulus in potential ATP-dependent ion transport pro-cess. Results showed a significant increase in plasma glucose in those fish treated with 200 mg DOC/kg only at 4 days postadministration (Figure 10.6a), but no effect of hormone treatment on gill ATPase activity was noted (Figure 10.6b). These preliminary studies provided some evidence that DOC may be an active hormone in hagfish, but the evidence was not overwhelming, with only moderate changes to plasma glucose lev-els. These experiments were performed prior to the discovery by Close et al. (2010) that 11-deoxycortisol putatively regulates ion homeostasis in lamprey, and the 3D model of the lamprey CR which shows the key amino acids that interact with 11-deocycortisol (Baker et al., 2011) is also conserved in the hagfish CR: consequently, it will be of interest to assess whether 11-deoxycortisol is the active corticosteroid in hagfish.

(a)

```
                  *  * *** ***** *  * *           **  ** *
HumanGR    pqltptlvslleviepevlyagydssvpdstwrimttlnmlggrqviaavkwakaipgfrnlhlddqmtllqyswmflmafalg  84
HumanMR    raltpspvmvleniepeivyagydsskpdtaenllstinrlagkmiqvvkwakvlpgfknlpledqitlliqyswmclssfals  84
hagfishCR  pvlspplvstlqviepdiisagfdnsramtttyllssintlcekqlvflvkwakampgfrslhiddqmvliqyswmgimafamg  84
LampeyCR   pifsptliailqaiepevmsgydntrsqttaymlsslnrlcdkqlsivkwakslpgfrnlhiddqmvliqyswmglmsfams    84

                         *
HumanGR    wrsyrqssanllcfapdliineqrmtlpcmydqckhmlyvsselhrlqvsyeeylcmktllllssvpkdglksqelfdeirmty  168
HumanMR    wrsykhtnsqflyfapdlvfneekmhqsamyelcqgmhqislqfvrlqtfeeytimkvlllstipkdglksqaafeemrtny    168
hagfishCR  wrsyintncellyfapdlifneqrmkqsamydlclgmrnigeemmrmtmspdefrcmkavlllstipkeglkcqtsfeelrmty  168
LampeyCR   wrsfqhtnskllyfapdlvfdetrmqgsamyqlcvemrqvsedfmklqvtseeflcmkaillllstvpqeglksqgcfeemrisy 168

                 *         *    **     ** *
HumanGR    ikelgkaivkregnssqnwqrfyqltklldsmhevvenllnycfqtfldk-tmsiefpemlaeiitnqipkysngnikkllfhqk  252
HumanMR    Ikelrkmvtkcpmnsgqswqrfyqltklldsmhdlvsdllefcfytfreshalkvefpamlveiisdqlpkvesgnakplyfhrk  253
hagfishCR  Irelhravgqqtsspvqcwkrfyqltrlldsmhnlvggllefcfmtftqselwsvefpenmseiitaqlphvlaghahalrfhkk  253
LampeyCR   Irelnrtiarteknavqcwqrfyqltklldcmqdlvskllefcfatftqcqwsvefpdnmaeiisaqlashhgrearalhfhkk   253
```

(b)

	Amino add number																					
	35	38	39	41	42	45	75	76	79	80	83	86	98	117	121	207	210	211	214	222/3	224/5	228/9
hGR	M	L	N	L	G	Q	W	M	M	A	L	R	F	Q	M	L	Y	C	T	I	F	L
hMR	L	L	N	L	G	Q	W	M	S	S	L	R	F	L	M	L	F	C	T	V	F	L
HagCR	L	L	N	L	C	Q	W	M	M	A	M	R	F	L	M	L	F	C	T	V	F	M
LamCR	L	L	N	L	C	Q	W	M	M	S	M	R	F	L	M	L	F	C	T	V	F	M
	*	**	**	**	+	**	**	**	#	~	+	**	**	*	**	**	*	**	**	*	**	+

Figure 10.4 (a) Sequence alignment of the human GRs and MRs, hagfish, and lamprey CR hormone-binding region or E-domain. Asterisks indicate the key amino acids identified by Bledsoe et al. (2002) that interact with the synthetic glucocorticoid, dexamethasone. (b) Alignment of these key amino acids. **Those amino acid conserved between all receptors. *Similarity between the two agnathan CRs and the hGR and the hMR. # A similarity between the two agnathan CRs and the hGR, a ~ indicates the hagfish CR is the same as the hGR, whereas the lamprey CR is similar to the hMR, and + indicates the agnathan CRs share that amino acid, but this differs to that of the hGR and hMR. The numbers refer to the position of the corresponding hGR amino acid in (a).

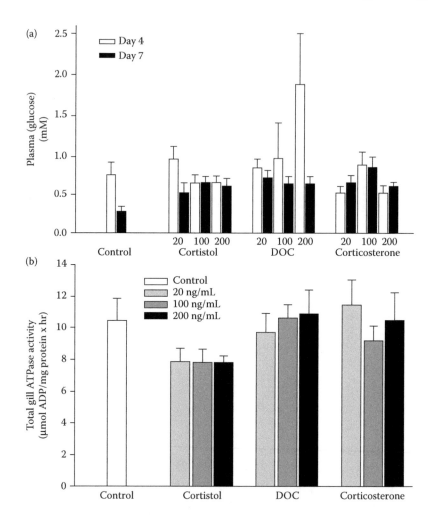

Figure 10.5 (a) Plasma glucose concentrations and (b) gill ATPase activity in Pacific hagfish (*Eptatretus stoutii*) given 20, 100, 200 mg/kg of either cortisol 11-deoxycorticosteorne (DOC) or corticosterone as a coconut oil bolus injected into the intraperitoneal cavity, after 4 (white bars) and 7 days (black bars), for glucose measurements and 7 days, for gill ATPase activity. Asterisks indicate significant difference between the treatment and control hagfish (two-way analysis of variance followed by a Tukey's post-hoc test, $p < 0.05$.)

Conclusion

Very little is currently known about the corticosteroid signaling pathway and the physiological roles it may influence in hagfish. There is firm molecular evidence that there are CRs present in hagfish and that this protein when expressed in mammalian expression systems is activated by corticosteroids (Bridgham et al., 2006). The control and site of corticosteroid synthesis in hagfish has not been ascertained. There is no distinction between the pars distalis and pars intermedia of the pituitary (Holmes and Ball, 1974; Nozaki et al., 2008), and no definitive ACTH-like peptide has been identified. However, a lamprey ACTH antibody was able to detect a weak signal for ACTH reactive cells in the hagfish pituitary (Nozaki et al., 2007). A number of PICs, based on increased hypertrophy following ACTH injection in the Atlantic hagfish, have been identified (Idler and Burton, 1976), but it was not possible to verify that these cells are able to synthesize corticosteroids. Though corticosteroids have been measured in the plasma of hagfish, often these concentrations are a lot lower than those measured in other vertebrates. For example, Weisbart et al. (1980) could not detect cortisol in control or saline-injected hagfish and only a small rise to 0.22 ng/mL was observed in those organisms injected mammalian ACTH. In contrast, in response to a stressor teleost fish show a plasma cortisol rise in the region of 10 ng/mL (Bury et al., 1995; Wendelaar Bonga, 1997). Thus, though there is evidence for the presence of an HPI axis in hagfish, similar to that in lamprey and teleost fish, further studies to verify this are required. However, the retention of the CR in extant hagfish species would strongly suggest it is functionally active and plays a significant role in hagfish physiology, but its precise role requires elucidation.

Acknowledgments

This research was supported by an NSERC Discovery Grant to G.G.G. N.B. was supported by the Royal Society International outgoing short visit scheme and in part by the Biotechnology and Biological Science Research Council of the United Kingdom (BB/E0016337/1). A.M.C. was supported by an NSERC Post-Graduate Doctoral Scholarship, the Alberta Innovates Technology Futures Graduate Student Scholarship and the Donald M. Ross Scholarship. Research findings reported herein were conducted at Bamfield Marine Sciences Centre, Bamfield, BC, Canada.

References

Baker, M. E., K. Y. Uk and P. Asnaashari. 2011. 3D models of lamprey corticoid receptor complexed with 11-deoxycortisol and deoxycorticosterone. *Steroids* 76:1451–1457.

Ball, J. N. and B. I. Baker. 1969. The pituitary gland: Anatomy and histophysiology. In W. S. Hoar and D. J. Tandall (eds.), *Fish Physiology*, Vol. 2. Academic Press, New York.

Becker, H., A. Sturm, J. E. Bron, K. Schirmer and N. R. Bury. 2008. The A/B domain of the teleost glucocorticoid receptor influences partial nuclear localisation in the absence of hormone. *Endocrinology* 149:4567–4576.

Bellamy, D. and I. C. Jones. 1961. Studies on *Myxine glutinosa*—I. The chemical composition of the tissues. *Comp Biochem Physiol A* 3:175–183.

Bledsoe, R. K., V. G. Montana, T. B. Stanley, C. J. Delves, C. J. Apolito, D. D. McKee, T. G. Consler et al. 2002. Crystal structure of the glucocorticoid receptor ligand binding domain reveals a novel mode of receptor dimerization and coactivator recognition. *Cell* 110:93–105.

Borski, R. J., G. N. Hyde, S. Fruchtman and W. S. Tsai. 2001. Cortisol suppresses prolactin release through a non-genomic mechanism involving interactions with the plasma membrane. *Comp Biochem Physiol B Biochem Mol Biol* 129:533–541

Bridgham, J. T., J. E. Brown, A. Rodríguez-Marí, J. M. Catchen and J. W. Thornton. 2008. Evolution of a new function by degenerative mutation in cephalochordate steroid receptors. *PLoS Genet* 4:e1000191.

Bridgham, J. T., S. M. Caroll and J. W. Thornton. 2006. Evolution of hormone–receptor complexity by molecular exploitation. *Science* 312:97–101.

Buckingham, J. C., J. H. Leach, E. Plisetskaya, S. A. Sower and A. Gorbman. 1985. Corticotrophin-like bioactivity in the pituitary gland and brain of the Pacific hagfish, *Eptatretus stoutii*. *Gen Comp Endocrinol* 57:434–437.

Bury, N. R., F. B. Eddy and G. A. Codd. 1995. The effects of the cyanobacterium *Microcystis aeruginosa*, the cyanobacterial hepatotoxin microcystin-LR and ammonia on growth rate and ionic regulation of brown trout (*Salmo trutta*). *J Fish Biol* 46:1042–1054.

Bury, N.R., Sturm, A. 2007. Evolution of the corticosteroid receptors in teleost fish. *Gen Comp Endocrinol* 153: 47–56.

Close, D. A., S.-S. Yun, S. D. McCormick, A. J. Wildbill and W. Li. 2010. 11-Deoxycortisol is a corticosteroid hormone in the lamprey. *Proc Natl Acad Sci USA* 107:13,942–13,947.

Delarbre, C., C. Gallut, V. Barriel, P. Janvier and G. Gachelin. 2002. Complete mitochondrial DNA of the hagfish, *Eptatretus burgeri*: The comparative analysis of mitochondrial DNA sequences strongly supports the cyclostome monophyly. *Mol Phylogenet Evol* 22:184–192.

Delsuc, F., H. Brinkmann, D. Chourrout and H. Philippe. 2006. Tunicates and not cephalochordates are the closest living relatives of vertebrates. *Nature* 439:965–968.

Gilchriest, B. J., D. J. Tipping, L. Hake, A. Levy and B. I. Baker. 2001. Differences in arginine vasotocin gene transcripts and cortisol secretion in trout with high or low endogenous melanin-concentrating hormone secretion. *J Neuroendocrinol* 13:407–411.

Glass, C. K. and M. G. Rosenfeld. 2000. The coregulator exchange in transcriptional functions of nuclear receptors. *Gene Dev* 14:121–141.

Gorbman, A. 1983. Early development of the hagfish pituitary gland—Evidence for the endodermal origin of the adenohypophysis. *Am Zool* 23: 639–654.

Heimberg, A. M., R. Cowper Sallari, M. Sémon, P. C. J. Donoghue and K. J. Peterson. 2010. MicroRNAs reveal the interrelationships of hagfish, lampreys and gnathostomes and the nature of the ancestral vertebrate. *Proc Natl Acad Sci USA* 107:19,379–19,383.

Heinig, J. A., F. W. Keeley, P. Robson, S. A. Sower and J. H. Youson. 1995. The appearance of proopiomelanocortin early in vertebrate evolution: Cloning and sequencing of POMC from a Lamprey pituitary cDNA library. *Gen Comp Endocrinol* 99:137–144.

Hoegg, S., H. Brinkmann, J. S. Taylor and A. Meyer 2004. Phylogenetic timing of the fish-specific genome duplication correlates with the diversification of teleost fish. *J Mol Evol* 59:190–203.

Holland, L. Z., R. Albalat, K. Azumi, E. Benito-Gutiérrez, M. J. Blow, M. Bronner-Fraser, F. Brunet et al. 2008. The amphioxus genome illuminates chordate origins and cephalochordate biology. *Genome Res* 18:1100–1111.

Holmes, R. L. and J. N. Ball. 1974. *The Pituitary Gland, A Comparative Account*. Cambridge University Press, London.

Idler, D. R. and M. P. M. Burton. 1976. The pronephroi as the site of presumptive interregnal cells in the hagfish *Myxine glutinosa* L. *Comp Biochem Physiol* 53A:73–77.

Janer, G. and C. Porte. 2007. Sex steroids and potential mechanisms of non-genomic endocrine disruption in invertebrates. *Ecotoxicology* 16:145–160.

Kawauchi, H. and S. A. Sower. 2006. The dawn and evolution of hormones and the adenohypophysis. *Gen Comp Endocrinol* 148:3–14.

Keay, J., J. T. Bridgham and J. W. Thornton. 2006. The *Octopus vulgaris* estrogen receptor is a constitutive transcriptional activator: Evolutionary and functional implications. *Endocrinology* 147:3861–3869.

Kuraka, S., D. Hoshiyama, K. Katoh, H. Suga and T. Miyata. 1999. Monophyly of lamprey and hagfish supported by nuclear DNA-coded genes. *J Mol Evol* 49:729–735.

Kuraku, S., A. Meyer and S. Kuratani. 2009. Timing of genome duplications relative to the origin of the vertebrates: Did cyclostomes diverge before or after? *Mol Biol Evol* 26:47–59.

Lee, S. R., H. K. Kim, J. B. Youm, L. A. Dizon, I. S. Song, S. H. Jeong, D. Y. Seo et al. 2012. Non-genomic effect of glucocorticoids on cardiovascular system. *Pflügers Arch* 464:549–559.

Maisey, J. G. 1986. Heads and tails: A chordate phylogeny. *Cladistics* 2:201–256.

Malagoli, D., A. Accorsi and E. Ottaviani. 2011. The evolution of pro-opiomelano-cortin: Looking for the invertebrate fingerprints. *Peptide* 32:2137–2140.

Mallatt, J. and J. Sullivan. 1998. 28S and 18S rDNA sequences support the monophyly of hagfish and lamprey. *Mol Biol Evol* 15:1706–1718.

McCormick, S. D. 1993. Methods for nonlethal gill biopsy and measurement of Na$^+$K$^+$-ATPase activity. *Can. J. Fish. Aquat. Sci.* 50: 656–658.

Mizuta, T. and K. Kubokawa. 2007. Presence of sex steroids and cytochrome P450 genes in amphioxus. *Endocrinology* 148:3554–3565.

Near, T. J. 2009. Conflict and resolution between phylogenies inferred from molecular and phenotypic data sets for hagfish, lamprey and gnathostomes. *J Exp Zool* 312B:749–761.

Nozaki, M., Ominato, K., Shimotani, T., Kawauchi, H., Youson, J.H., Sower, S.A. 2008. Identity and distribution of immunoreactive adenohypophysial cells in the pituitary during the life cycle of sea lampreys, *Petromyzon marinus*. *Gen Comp Endocrinol* 155: 403–412.

Nozaki, M., T. Simotani and K. Uchida. 2007. Gonadotropin-like and adrenocorticotrophin-like cells in the pituitary gland of hagfish, *Paramyxine atami*: Immunohistochemistry in combination with lectin histochemistry. *Cell Tissue Res* 328:563–572.

Ottaviani E., A. Franchini and I. Hanukoglu. 1998. *in situ* localization of ACTH receptor-like mRNA in molluscan and human immunocytes. *Cell Mol Life Sci* 54:139–142.

Roberts, B. W., W. Didier, S. Rai, N. S. Johnson, S. Libants, S. S. Yun and D. A. Close. 2014. Regulation of a putative corticosteroid, 17,21-dihydroxypregn-4-ene,3,20-one, in sea lamprey, *Petromyzon marinus*. *Gen Comp Endocrinol* 196:17–25.

Salzet, M., B. Salzet-Raveillon, C. Cocquerelle, M. Verger-Bocquet, S. C. Pryor, C. M. Rialas, V. Laurent and G. B. Stefano. 1997. Leech immunocytes contain proopiomelanocortin: Nitric oxide mediates hemolymph proopiomelano-cortin processing. *J Immunol* 159:5400–5411.

Sangiao-Alvarellos, S., S. Polakof, F. J. Arjona, A. Kleszczynska, M. P. Martín Del Río, J. M. Míguez, J. L. Soengas and J. M. Mancera. 2006. Osmoregulatory and metabolic changes in the gilthead sea bream *Sparus auratus* after arginine vasotocin (AVT) treatment. *Gen Comp Endocrinol* 148:348–358.

Sassi, D., D. Kletsas and E. Ottaviani. 1998. Interactions of signaling pathways in ACTH (1–24)-induced cell shape changes in invertebrate immunocytes. *Peptides* 19:1105–1110.

Schaaf, M. J., D. Champagne, I. H. van Laanen, D. C. van Wijk, A. H. Meijer, O. C. Meijer, H. P. Spaink and M. K. Richardson. 2008. Discovery of a functional glucocorticoid receptor beta-isoform in zebrafish. *Endocrinology* 149:1591–1599.

Smith, J. J. and M. C. Keinart. 2014. The sea lamprey meiotic map resolves ancient vertebrate genome duplications. *BioRvix*. doi: http://dx.doi.org/10.1101/008953.

Smith, J. J., S. Kuraka, C. Holt, T. Sauka-Spenjler, N. Jiang, M. S. Campbell, M. D. Yandell et al. 2013. Sequencing of the sea lamprey (*Petromyzom marinus*) genome provides insights into vertebrate evolution. *Nat Genet* 45:415–423.

Stefano, G. B., B. Salzet-Raveillon and M. Salzet 1999. *Mytilus edulis* hemolymph contains pro-opiomelanocortin: LPS and morphine stimulate differential processing. *Brain Res Mol Brain Res* 63:340–350.

Takahashi, A., Y. Amemiya, M. Nozaki, S. A. Sower, J. Joss, A. Gorbman and H. Kawauchi. 1995a. Isolation and characterization of melanotropins from lamprey pituitary lands. *Int J Pept Protein Res* 46:197–204.

Takahashi, A., Y. Amemiya, M. Sarashi, S. A. Sower, and H. Kawauchi. 1995b. Melanotropin and corticotrophin are encoded on two distinct genes in the lamprey, the earliest evolved extant vertebrate. *Biochem Biophys Res Commun* 213:490–498.

Takahashi, A., O. Nakata, M. Kasahara, S. A. Sower, and H. Kawauchi. 2005. Structures for the proopiomelanocortin family genes proopiocortin and proopiomelanotropin in the sea lamprey Petromyzon marinus. *Gen Comp Endocrinol* 144:174–181.

Thomson, R. C., D. C. Plachetzki, D. L. Mahler and B. R. Moore. 2014. A critical appraisal of the use of microRNA data in phylogenetics. *Proc Natl Acad Sci USA* 111:E3659–E3668.

Thornton, J. W. 2001. Evolution of vertebrate steroid receptors from an ancestral estrogen receptor by ligand exploitation and serial genome expansions. *Proc. Natl Acad Sci USA* 98:5671–5676.

Thornton, J. W., E. Need and D. Crews. 2003. Resurrecting the ancestral steroid receptor: Ancient origin of estrogen signaling. *Science* 301:1714–1717.

Tremblay, A., G. B. Tremblay, F. Labrie and V. Giguère. 1999. Ligand-independent recruitment of SRC-1 to estrogen receptor beta through phosphorylation of activation function AF-1. *Mol Cell* 3:513–519.

Uchida, K., S. Moriyama, H. Chiba, T. Shimotani, K. Honda, M. Miki, A. Takahashi, S. A. Sower and M. Nozaki. 2010. Evolutionary origin of a functional gonadotropin in the pituitary of the most primitive vertebrate, hagfish. *Proc Natl Acad Sci USA* 107:15,832–15,837.

Uchida, K., S. Moriyama, S. A. Sower and M. Nozaki. 2013. Glycoprotein hormone in the pituitary of hagfish and its evolutionary implications. *Fish Physiol Biochem* 39:75–83.

Wales, N. A. M. 1988. Hormone studies in *Myxine glutinosa*: Effects of the eicosanoids arachidonic acid, prostaglandin E_1, E_2, A_2 and $F_{2\alpha}$, thromboxane B_2 and of indomethacin on plasma cortisol, blood pressure, urine flow and electrolyte balance. *J Comp Physiol B* 158:621–626.

Weisbart, M., W. W. Dickhoff, A. Gorbman and D. R. Idler. 1980. The presence of steroids in the sera of the Pacific hagfish *Eptatretus stoutii*, and the Sea lamprey, *Petromyzon marinus*. *Gen Comp Endocrinol* 41:506–519.

Weisbart, M. and D. R. Idler. 1970. Re-examination of the presence of corticosteroids in two cyclostomes, the Atlantic hagfish (*Myxine glutinosa* L.) and the sea lamprey (*Petromyzon marinus* L.). *J Endocrinol* 46:29–43.

Wendelaar Bonga, S. E. 1997. The stress response in fish. *Physiol Rev* 77:591–625.

chapter eleven

Acid/base and ionic regulation in hagfish

Alexander M. Clifford, Gregory G. Goss,
Jinae N. Roa, and Martin Tresguerres

Contents

Introduction

Hagfish are the living representative of the most basal craniate and share an ionic profile similar to that seen in marine invertebrates, representing the evolutionary transition between invertebrate and vertebrate organisms. It is these unique physiological and evolutionary positions that make hagfish an excellent model to study the evolution of acid/base (A/B) and ionic regulation. Hagfishes diverged from the main vertebrate lineage more than 500 million years ago with fossil evidence supporting that their morphology has remained unchanged for ~450 million years (Holland and Chen, 2001). Hagfishes are exclusively osmoconforming marine animals with most living at considerable depths and are the only living chordate group to maintain their plasma [Na$^+$] to [Cl$^-$] almost iso-osmotic to that of seawater (Hardisty, 1979). Coupling the unique osmoregulatory characteristics of hagfish along with the fact that fossil evidence comes solely from marine-based sediments, it may be inferred that hagfish never entered a freshwater environment. Considering that their morphology has remained relatively unchanged, it can be hypothesized that their current physiology may reflect the physiology of vertebrate ancestors prior to their invasion of estuaries, freshwater, and land (Hardisty, 1979; Holland and Chen, 2001).

Hagfish are stenohaline osmoconformers who maintain their blood osmolarity slightly hypertonic in comparison to seawater. They have no ability or need to regulate plasma [Na$^+$] and [Cl$^-$], but they have been shown to regulate [Ca^{2+}], [Mg^{2+}], and [SO$_4^{2-}$], which are hypo-regulated (Bellamy and Jones, 1961). Despite the inability to regulate [Na$^+$] and [Cl$^-$], a number of studies have identified genes associated with ion transport and acid/base regulation (Karnaky, 1998; Choe et al., 1999, 2002; Edwards et al., 2001; Tresguerres et al., 2006a).

Acid/base (A/B) regulation

All fishes including hagfish must maintain internal pH homeostasis, as the concentration of H$^+$ (pH) affects the kinetics of chemical reactions and the structure and function of proteins in all organisms (Whitten et al., 2005; Srivastava et al., 2007). Fish, similar to terrestrial vertebrates, use a strategy of buffering and excretion to defend against both intracellular and extracellular pH alterations. The processes of intracellular metabolism generate free H$^+$ and HCO$_3^-$ from CO$_2$, following the Henderson–Hasselbalch equation (reviewed by Claiborne et al., 2002). Although the intracellular buffering capacity in hagfishes is not well characterized, in higher fishes both inorganic and organic phosphates, proteins, and amino acid residues have been shown to limit free H$^+$. However, to maintain intracellular pH, critical for proper cellular function, cells must additionally actively import and export H$^+$ and HCO$_3^-$ via transport proteins located in the cell membrane (Roos and Boron, 1981). The ability of fishes to compensate for metabolic and environmental disturbances is limited by the aquatic environment in which they reside, and therefore most fishes regulate blood pH by secreting and absorbing excess H$^+$ and HCO$_3^-$ across specialized epithelia such as gills and skin (reviewed by Perry, 1997). Other factors that help maintain or affect extracellular fluid (ECF) and blood A/B homeostasis are buffer systems, such as CO$_2$/HCO$_3^-$, NH$_3$/NH$_4^+$, phosphates, and proteins (Heisler, 1986; Boron, 2004).

Hagfish habitat and behavior: Naturally acidotic environments

Hagfish are demersal and either burrow into the sediment (genus *Myxini*) or reside on top of the ocean bed (genus *Eptatretus*; Martini, 1998). These environments are usually quite hypoxic or even anoxic (McInerney and Evans, 1970; Smith and Hessler, 1974) and likely determine the low metabolic rates observed in hagfish (Munz and Morris, 1965; Forster, 1990). Hagfish also have a high capacity for anaerobic respiration (Sidell et al., 1984), which can lead to metabolic acidosis.

Hagfish have traditionally been considered to be scavengers; however, recently they have been observed to also actively hunt live prey, dragging the victim back into their burrows (Zintzen et al., 2011). Hagfish consume a wide variety of prey types including polychetes, squid, shrimp, crab, and teleosts (Shelton, 1978), sometimes robbing food captured by other fauna (Auster and Barber, 2006). Active hunting behavior is likely associated with burst swim activity potentially causing acute blood acidosis (Ruben and Bennett, 1980). However, hagfish are able to readily recover from such energetic bouts, with little disturbance to their blood A/B status (Davison et al., 1990).

Hagfish are best known for their opportunistic scavenging behavior, as they have been regularly reported to feed on dead and decaying carrion that falls to the ocean floor (Smith, 1985; Martini, 1998; Collins et al., 1999; Auster and Barber, 2006). Sources of carrion may include by-catch from commercial fishing (Collins et al., 1999) or expired large mammals (Smith and Baco, 2003), but the most likely source is leftovers from other predator–prey interactions. Hagfish burrow into the decomposing carcass (Jensen, 1966; Martini, 1998), which presents both opportunities and challenges. As the carcass is being consumed, decomposition of the tissues by bacteria is believed to generate an environment that is rich in free amino acids, which hagfish may absorb across gills and skin (Glover et al., 2011; Chapter 13). However, the metabolism of bacteria, the hagfish itself, and other animals feeding inside carcasses likely also results in elevations in PCO_2, PNH_3, $[H^+]$, and weak acids such as butyric acid, putrescine, and formic acid, which are known to cause acute acidification of tissues.

A/B regulatory organs

The gill is the main blood A/B regulatory organ in adult fish (Evans et al., 2005), and certain areas of the skin fulfill this role in larval fish before gills develop (Wells and Pinder, 1996). Similarly, gills were considered as the main A/B organ in hagfish (Evans, 1984; Edwards et al., 2001); however, recent research has demonstrated that the skin, especially in the posterior body, also plays an important role in blood A/B homeostasis (Clifford et al., 2014). Although the renal system in more derived aquatic species (e.g., Chondrichthyes, Osteichthyes) plays a distinctive role in A/B regulation (Claiborne et al., 2002), the hagfish renal output is too low (0.1–1.0 mL 100/g/day; Hardisty, 1979) to significantly contribute to acute A/B regulation. Nonetheless, hagfish kidneys have been shown to resorb Na$^+$, which may play a significant role in nutrient uptake from the filtrate (McInerney, 1974).

Gills

The gill pouches of the hagfish are morphologically different compared with gills from lampreys, Chondrichthyan fishes, and Osteichthyan fishes (Figure 11.1). Hagfish pouches are found along the length of the anterior half of the body and range from 6 to 13 pairs in number displaying both intra- and interspecific variability (Rauther, 1937; Chapter 1). All gill pouches are morphologically similar and consist of a muscular wall lined up with an epithelium known as the "primary filament," from which multiple "lamella"

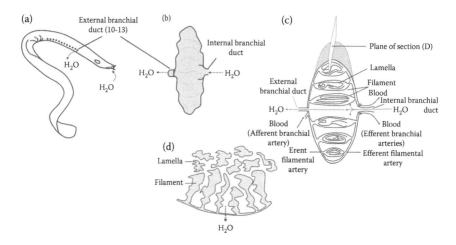

Figure 11.1 Schematic of the hagfish gill pouch gross anatomy and fine structure. (a) Water enters the hagfish through the nasopharyngeal duct and leaves across the 6–13 pairs of gill pouches located at each side of the body. (b) External morphology of a gill pouch. (c) Cross-section of a gill pouch. (Modified from Mallatt, J. and C. Paulsen. 1986. *Am J Anat* 177:243–269.) Water and blood flow in counter-current fashion. The surface area of filaments and lamella is enlarged by radial infoldings, as opposed to the typical structures found in lampreys, elasmobranchs, and teleosts. The plane of the section used in the immunohistochemistry study is illustrated on top of the pouch, which is also shown in more detail in (d). (d) Section across the gill pouch in the orientation indicated in (c). The gill filaments and lamella are shown in the same view as the immunohistochemistry sections of Figure 11.3. Notice how the radial infolding of the lamella results in an apparent separation from the filaments in this plane of section. (Adapted from Tresguerres et al. 2006a. *Comp Biochem Physiol A* 145:312–321.)

radiate toward the pouch lumen (Mallatt and Paulsen, 1986). Whether or not all gill pouches are also functionally identical has not yet been discerned.

Ultrastructural studies on hagfish gills have revealed the presence of pavement, basal and intermediate cells, and ionocytes (Elger and Hentschel, 1983; Bartels, 1984, 1985; Mallatt and Paulsen, 1986; reviewed in Chapter 1). Ionocytes, cells characterized as possessing large number of mitochondria, typically support transepithelial ATPase-driven ion-transporting processes. The ionocytes in hagfish gills have been shown to express ion transporter mRNAs and proteins associated with the excretion and absorption of H^+ and HCO_3^- for A/B regulation (Evans, 1984; Mallatt et al., 1987; Choe, 1999; Edwards et al., 2001; Tresguerres et al., 2007a; Parks et al., 2007b) and have also been demonstrated to uptake inorganic phosphate (Schultz et al., 2014) and amino acids (Glover et al., 2011).

Cellular mechanisms for gill A/B regulation

In all organisms, transepithelial A/B regulation is achieved by the combined action of various ion-transporting proteins differentially expressed in the basolateral or apical membranes, or in the cytoplasm; the overall result is excretion of excess H^+ and absorption of HCO_3^- to correct blood acidosis, and excretion of excess HCO_3^- and absorption of H^+ to correct blood alkalosis. In marine organisms, the most typical ion-transporting proteins involved in A/B regulation are Na^+-K^+-ATPase (NKA, located in the basolateral membrane), carbonic anhydrase (CA, cytoplasmic and membrane-bound), Na^+/H^+ exchangers (NHEs; HUGO gene family SLC9), Na^+/HCO_3^- cotransporters (NBCs; HUGO gene family SLC4), and Cl^-/HCO_3^- exchangers (CBEs; HUGO gene family SLC26) (with the latter three located in either apical or basolateral membrane, depending on the specific isoform and cell type). Another relevant protein is the V-type H^+-ATPase (VHA), in which intracellular localization and specific role varies depending on the species, cell type, and A/B condition (for review, see Evans et al., 2005).

In a generalized model, CO_2 enters the ionocytes from blood plasma (through a mechanism that may depend on plasma membrane-bound CA) and is also produced in mitochondria. Once in the ionocyte, cytoplasmic CA catalyzes its hydration into H^+ and HCO_3^-, which are subsequently differentially excreted to seawater or absorbed into the blood, depending on the A/B regulatory needs. To compensate for blood acidosis, H^+ are excreted to the surrounding seawater in exchange for Na^+ through apical NHEs; this transport is secondarily energized by basolateral NKA (Evans et al., 2005). The proteins responsible for HCO_3^- absorption are not as well characterized, although suitable candidates are basolateral CBEs and NBCs (Parks et al., 2007a). To compensate for blood alkalosis, HCO_3^- is excreted in exchange for Cl^- through apical CBEs. However, the rest of the mechanism is much less understood and may vary considerably from species to species. The driving force may be provided by NKA, for example, by creating a cell membrane potential that favors a low intracellular [Cl^-], or by linkage of apical Cl^-/HCO_3^- exchange with Na^+ uptake (e.g., Na^+-dependent CBEs) (Parks et al., 2009). In this case, H^+ transport from the ionocyte into the blood most likely takes place by basolateral NHEs, as described in the intestine of marine fish (Grosell and Genz, 2006) and crab gills (Tresguerres et al., 2008).

Another option involves basolateral VHA as described in shark gill ionocytes; in this mechanism, VHA "pumps" H^+ into the blood, which drives apical Cl^-/HCO_3^- exchange by removing H^+ from the cytosol (thus preventing CA inhibition and reversal by law of mass action) and by promoting Cl^- absorption into the blood across channels in the basolateral membrane (due to the membrane potential generated by H^+ transport) (Tresguerres et al., 2006b). Regardless of the specific mechanisms, lampreys, elasmobranchs, and bony fish have developed gill ionocyte subtypes specialized for acid or base secretion (reviewed in Evans et al., 2005). Hagfish follow the general model of Na^+/H^+ and Cl^-/HCO_3^- exchange across the gill epithelium (Evans, 1984: Figure 11.2a). However, histochemical and immunohistochemical evidence suggests that hagfish gills have a single type of ionocyte that alternatively performs H^+ or HCO_3^- secretion depending on the blood A/B condition (Tresguerres et al., 2006a). The vast majority of experiments on hagfish A/B regulation have been conducted on either the Atlantic hagfish (*Myxini glutinosa*) or the Pacific hagfish (*Eptatretus stoutii*) with little information on the other species (e.g., *E. burgeri*). Based on the published results, this chapter assumes that the cellular mechanisms for A/B regulation are similar for all hagfish species; for details about the experimental species used in each study, the reader is directed to the cited literature.

Hagfish decrease net H^+ or HCO_3^- secretion when placed in Na^+- or Cl^--free seawater, respectively (Evans, 1984). Given that hagfish represent the only chordate group assumed to have never entered freshwaters (Hardisty, 1979), the presence of these transport mechanisms has led to the hypothesis that these mechanisms evolved for pH balance prior to their use in ionoregulation and that these mechanisms were co-opted for Na^+ and Cl^- absorption for ionoregulation by the chordates that originally invaded estuarine and freshwater environments (Evans, 1984).

The first suggestion that hagfish gill ionocytes are involved in blood A/B regulation was biochemical evidence of NKA and CA activities in histological sections of hagfish gill pouches (Mallatt et al., 1987). Two subsequent studies used immunohistochemistry to localize NKA and NHEs in hagfish gill epithelium (Choe et al., 1999, 2002), and a third study suggested co-localization of NKA, V-H$^+$ATPase, and an NHE2-like in a single gill cell type (Tresguerres et al., 2006a). The anti-NKA and anti-VHA antibodies used in those studies were raised against amino

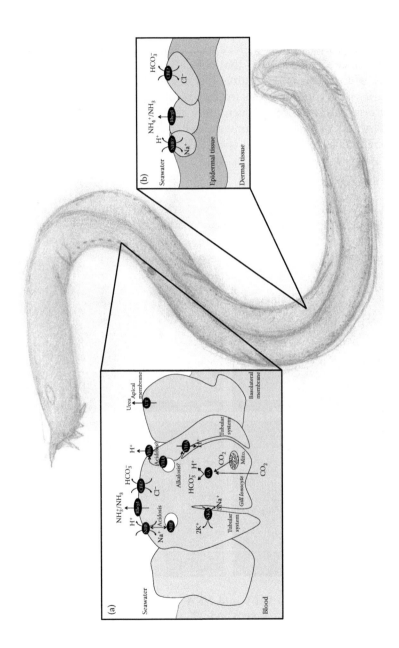

Figure 11.2 Molecular components of hagfish acid–base and ionoregulation in (a) gill and (b) skin. NKA, Na⁺/K⁺-ATPase; VHA, V-H⁺-ATPase; NHE, Na⁺/H⁺ exchanger; Rhcg1, Rh glycoprotein c1; UT, urea transporter; CBE, chloride bicarbonate exchanger; CA, carbonic anhydrase.

acid regions that are highly conserved in all animals and therefore there is little doubt they specifically recognize hagfish NKA and VHA proteins. However, the anti-NHE antibodies used were against mammalian NHE isoforms that are not as well conserved in other animals, so these results should be interpreted with caution. Another potential limitation of Tresguerres et al. (2006a) was that colocalization of NKA, VHA, and NHE2 was inferred from consecutive histological gill sections, which do not always contain the same cells (in fact, colocalization was not evident in all sections examined). However, more recent dual immunolocalization of NKA and VHA on the same gill section confirms that these two proteins are indeed present in the same cells (Figure 11.3).

Responses to metabolic acidosis

Hagfish have an outstanding capacity for recovering from metabolic acidosis; this has been determined from experiments that induced blood

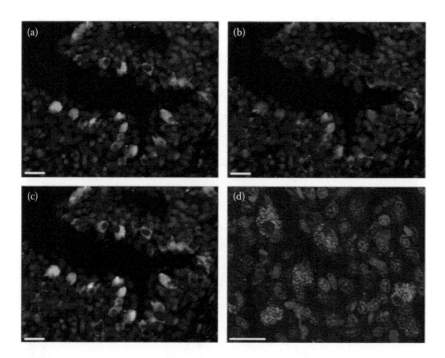

Figure 11.3 (**See color insert.**) VHA- and NKA-rich cells in hagfish gills. Dual immunolocalization of VHA (a, green) and NKA (b, red), with VHA and NKA immunoreactivity found in the same gill cells (c). At higher magnification with optical sectioning, NKA immunoreactivity was visible as punctate staining throughout the edge of cell, indicative of NKA localization along the basolateral membrane (d). Nuclei in blue. Scale bar: 20 μm.

acidosis by injections of H_2SO_4 (3 mequiv. H^+ kg^{-1}; McDonald et al., 1991; Edwards et al., 2001) or HCl (6 mequiv. H^+ kg^{-1}; Parks et al., 2007b) into the posterior caudal blood sinus. Blood pH suffered initial sharp reductions between ~0.6 (McDonald et al., 1991) and up to ~1.6 pH units (Parks et al., 2007b). However, blood pH was fully (McDonald et al., 1991) or near fully (Parks et al., 2007b) rectified within 6 h, demonstrating a significant capacity for rapid A/B regulation. In the latter study, subsequent injections caused significantly smaller reductions in blood pH compared with the initial injection, indicating increased capacity for acid secretion upon sustained acidosis. Metabolic acidosis resulted in upregulation in gill NHE2-like mRNA (Edwards et al., 2001) and protein abundance (Parks et al., 2007b), as well as protein insertion in the apical membrane of gill ionocytes (Parks et al., 2007b; Figure 11.2a). On the other hand, the protein abundance of NKA and VHA remained unchanged (Parks et al., 2007b). These results indicate that apical NHEs are involved in acid secretion and suggest that the basal levels of NKA are sufficient to sustain upregulated acid secretion. However, other NKA-activating mechanisms (NKA isoforms, phosphorylation) cannot be ruled out.

Responses to metabolic alkalosis

Little research has been conducted evaluating hagfish responses to metabolic alkalosis. It is also less clear which conditions would result in blood alkalosis in the wild. One possibility could arise following feeding in a hypercapnic environment (e.g. decaying carcass) where the resultant blood acidosis is compensated by increases in plasma [HCO_3^-] as has been described in other fishes (Brauner and Baker, 2009). Another potential source of blood alkalosis is a postprandial alkaline tide as observed in elasmobranchs (Wood et al., 2005; Tresguerres et al., 2007b), and both freshwater (Bucking and Wood, 2008) and seawater-acclimated trout (*Oncorhynchus mykiss*; Bucking et al., 2009). However, postprandial digestive strategies of hagfishes remain to be identified.

Nonetheless, it is clear that hagfish possess efficient mechanisms to rectify blood metabolic alkalosis (Tresguerres et al., 2007a). Injection of base (NaHCO$_3$; 6 mequiv. HCO_3^- kg^{-1}) caused sharp increases in blood pH and plasma [HCO_3^-]; both of which were rectified within 6 h. Similar to the acid injections described above, the impact of subsequent base injections on blood pH was diminished, indicating capacity for upregulation of base-secretory mechanisms. After four $NaHCO_3^-$ injections at 6 h intervals (24 h), there was an increase in VHA protein abundance both in gill homogenates and in enriched gill cell membrane fractions, estimated from Western blots. As accumulation of VHA in the apical membrane was not evident in immunostained gill sections, these results suggest a translocation of VHA to the basolateral membrane to counteract blood alkalosis as described in dogfish sharks (Tresguerres et al., 2005). However, the tubulovesicular system in hagfish gill ionocytes complicates differentiating between cytoplasmic and basolateral membrane localization using immunohistochemistry techniques; unlike shark ionocytes that lack a tubulovesicular system (e.g., see Tresguerres et al., 2005). Although Western blot analysis demonstrated no significant changes in NKA abundance in whole-gill homogenates, a reduction in NKA abundance was detected in gill cell membrane enriched fractions. As hagfish gills possess only one ionocyte subtype that is rich in both NKA and VHA, it was postulated that the differential activation of acid and base secretion occurs via differential insertion of NKA and VHA into the basolateral membrane and of NHE and an as of yet unidentified CBE into the apical membrane (Tresguerres et al., 2006a, 2007a; Parks et al., 2007b). In addition, a recent study has suggested that, at least at shorter time points (4 h), extrabranchial mechanisms are more important than gills for recovering from blood alkalosis (Clifford et al., 2014; see "Extrabranchial contributions to acid/base and ionoregulation").

In all other chordates studied, there exist two or more ionocyte subtypes (Doyle and Gorecki, 1961; Lin et al., 2006), with distinct utility for regulating acid/base status (Reid et al., 2003; Goss et al., 2011). In the case of lampreys and elasmobranchs, acid and base secretion is likely split between two subtypes: NKA-containing and V-H$^+$ATPase-containing (Piermarini and Evans, 2001; Piermarini et al., 2002; Choe et al., 2004; Tresguerres et al., 2005, 2006b). The same is true for teleost fish, where at least two or more different subtypes have been identified with at least one subtype capable of acid secretion and another for base secretion (for review, see Hwang and Lee, 2007; Hwang et al., 2011). The presence of a single ionocyte type in hagfish suggests that it is an ancestral character and that more derived groups evolved multiple ionocyte subtypes in relation to novel environmental and metabolic challenges (Tresguerres et al., 2007a).

Future studies on base-secreting mechanisms in hagfish should explore the relative contributions of branchial and extrabranchial mechanisms upon acute and chronic blood alkalosis, identify the putative apical CBE(s) involved, and ascertain the regulatory mechanisms that allow switching from acid to base secretion. To the latter aim, optimization of techniques such as isolated gill pouches and primary cultures of isolated gill cells would be beneficial, as they allow measuring H$^+$ and HCO$_3^-$ fluxes and intracellular pH dynamics, respectively.

Nitrogen balance in the hagfish

Build-up of ammonia (NH$_3$) and ammonia-related compounds within organisms is toxic, causing a variety of impairments including disruption of biochemical pathways, interference of ionic movement across transmembrane surfaces via ion substitution of other monovalent ions such as Na^+ and K^+, and disruption of extracellular and intracellular pH via the strongly basic NH_3 (Walsh, 1998). The high protein diet of hagfishes (Smith, 1985; Martini, 1998; Collins et al., 1999; Auster and Barber, 2006) results in uptake of a large amount of amino acids. Catabolism of these ingested amino acids not only supplies the carbon backbone necessary for conversion into fatty acids, and carbohydrates (Nelson et al., 2008), but also contributes to the net nitrogen load of the fish, which must be excreted (for excellent reviews, see Wood, 2001; Evans et al., 2005). Other sources of nitrogen include the deamination of protein during fasting periods or, perhaps more relevant to the scavenging behavior of the hagfishes, from a decaying food source. No data exists on the physicochemical character- istics inside or surrounding decaying carrion in aquatic environments, however Clifford et al. (2015) argue that such an environment would be similar to decomposition in terrestrial environments where putrefactive gases (hydrogen sulfide and ammonia) generated by anaerobic microflora lead to high deposits (~30 mmol/L) of nitrogenous compounds in the sur- rounding soil (Carter et al., 2007). Such deposits in an aquatic environ- ment would present high inwardly-directed ammonia gradients to an encapsulated animal during a prandial event.

Nitrogen handling organs

Gills

The gills are a primary site of NH_3 exchange for many aquatic vertebrates (Wood, 1993; Wright et al., 1995) and most aquatic invertebrates (Henry et al., 2012). Additionally, various strategies exist to combat high ammo- nia such as ureotelosis as in the lake magadi tilapia (*Oreochromis alcali- cus grahami*; Randall and Wright, 1989) and Gulf toadfish (*Opsanus beta*; Walsh et al., 1990), glutamine production in the swamp eel (*Monopterus albus*; Ip et al., 2010), and active NH_4^+ excretion used by the mudskipper (*Periophthalmodon schlosseri*) while burrowed (Randall et al., 1999; Wilson et al., 2000). Recently, hagfish have been shown to have low basal rates of NH_3/NH_4^+ and urea excretion, with both the gills and the skin being hypothesized as sites of excretion (Braun and Perry, 2010; Clifford et al., 2014, 2015).

Cellular mechanisms for nitrogen excretion

Despite reported low basal rates of nitrogenous waste excretion, hagfish have remarkable ability to excrete excess nitrogen taken up by either exposure to high environmental ammonia (HEA; Clifford et al., 2015) or via direct infusion of NH_4Cl (McDonald et al., 1991; Edwards et al., 2015) or urea (Braun and Perry, 2010). Transcellular transport of ammo- nia and urea may be, respectively, facilitated by Rh glycoproteins (Marini and Urrestarazu, 1997) and urea transporter (UT) proteins (Levine et al., 1973a, b; You et al., 1993). The Rh glycoproteins are homologs of the Amt superfamily and have multiple isoforms (Huang and Peng, 2005), however the isoforms of interest to fish physiology and ammonia excretion are Rhag, Rhbg, and Rhcg (Hung et al., 2007; Nakada et al., 2007; Nawata et al., 2007; Tsui et al., 2009). In gills from the marine teleost *Takifugu rubripes*, Rhag localizes to the apical and basolateral membranes of the pillar cells and mediates ammonia transport from red blood cells through pillar cells (Nakada et al., 2007). Rhbg localizes to the basolateral membrane of pavement cells, whereas Rhcg2 localizes to the apical membrane. Finally, Rhcg1 localizes on ionocytes on apical, gill/water interface. However, teleost fishes are highly derived compared with the hagfishes, so the roles and localizations of any hagfish-specific Rh glycoproteins may differ.

To date, two Rh glycoprotein isoforms from hagfish have been partially cloned, Rhbg and Rhcg, and are expressed in hagfish gill and skin (Edwards et al., 2015) with the presence of Rhag so far remaining elusive. Hagfish-specific Rhcg protein and mRNA expression have been demonstrated in both gill and skin tissues. In gill, Rhcg is localized to the basal aspect of the filament epithelium along the region of the filament closest to the blood margin and in the skin hRhcg immunoreactive cells were localized to regions surrounding mucous glands (Edwards et al. 2015). In addition, using heterologous antibodies and immunohistochemistry, Braun and Perry (2010) found Rhbg localized to the lining of larger blood vessels (i.e., noncapillary vessels), whereas Rhcg1 colocalized apically to NKA-enriched gill cells. Localization of Rhbg to the blood vessels points to a different role for this isoform in hagfish, likely mediating transcellular transport of ammonia from tissues to the blood (Braun and Perry, 2010).

Responses to nitrogen challenges

Basal ammonia excretion rates for hagfish are about 40–60 µmol kg^{-1} h^{-1} (Braun and Perry, 2010; Clifford et al., 2014, 2015), half that of most ammonotelic fishes (100–350 µmol kg^{-1} h^{-1}; Wood, 2001; Evans et al., 2005) and on par with larval and parasitic sea lamprey (*Peteromyzon marinus*; 20–50 µmol kg^{-1} h^{-1}; Wilkie et al., 1999, 2004) and the ureotelic elasmobranchs (~50 µmol kg^{-1} h^{-1}; reviewed by Wood, 2001). Multiple studies have described the ability of the hagfish to survive high inwardly-directed ammonia gradients with survival reported following 9 h exposure to 100 mM NH$_4$Cl in seawater (Braun and Perry, 2010) and 48 h exposure to 20 mM NH$_4$Cl (Clifford et al., 2015). These studies have also highlighted the capacity of hagfish to tolerate extremely high plasma ammonia loads with the highest reported plasma ammonia load peaking at 5450 µM following 24 h exposure to 20 mM HEA (Clifford et al., 2015). Remarkably, during continued exposure to HEA, hagfish are able to excrete ammonia against massively inwardly-directed gradients, thus stabilizing rates of uptake (Clifford et al., 2015). During recovery from HEA exposure, hagfish are capable of excreting ammonia/ammonium at high rates (~3000 µmol/kg/h; Clifford et al., 2015). The high rate of ammonia offloading is comprised of both branchial and extrabranchial transport with the branchial contribution comprising approximately 70%–80% and excretion through the skin comprising the remainder (Clifford et al., Unpublished data; see "Extrabranchial contributions to acid/base and ionoregulation").

A recent study where hagfish were injected with 3 mmol kg^{-1} NH_4Cl demonstrated that elevated plasma ammonia concentrations paralleled elevated ammonia excretion rates and coincided with initial significant upregulation of Rhcg mRNA expression in the gill and the skin, suggesting that the transcriptional regulation of Rh glycoproteins may respond to elevated plasma ammonia. The study also used homologous antibodies to demonstrate the subsequent significant upregulation of hagfish Rhcg protein, suggesting that Rh glycoproteins are involved in the regulation of ammonia excretion in hagfish (Edwards et al., 2015).

Basal excretion of urea in hagfish has been reported to be ~30 µmol/kg^{-1} h^{-1} (Braun and Perry, 2010), again on par with excretion rates observed in parasitic lamprey (Wilkie et al., 2004). In *E. stoutii*, urea excretion did not change during exposure to HEA (10–100 mM) and during recovery in nominally ammonia-free seawater (Braun and Perry, 2010; Clifford et al., 2015). However, following infusion of NH_4Cl that raised plasma ammonia to 10 mM, threefold increases in urea excretion were observed following a 6-h period. Direct infusion of urea (to raise plasma urea to ~100 mM) caused an immediate (0–3 h) 1000-fold surge in urea excretion which tapered to ~250-fold compared with control animals (Braun and Perry, 2010).

Ionoregulation

Similar to A/B balance, the maintenance of stable plasma ionic composition is crucial for normal physiological function. Ionic regulation is traditionally associated with Na^+ and Cl^- transport because these are the two major constituents of internal fluids and therefore are the major contributors to total osmolarity. However, plasma ions such as K^+, Ca^{2+}, Mg^{2+}, and SO_4^{2-} are also maintained at significantly lower concentrations compared with the environment; these are especially relevant for hagfish plasma as they are the only ions whose concentrations are significantly lower than seawater (Currie and Edwards, 2010). Ions that contribute to A/B balance (e.g., H^+, HCO_3^-, NH_4^+) are discussed in previous sections.

Ionic gradients between internal fluids and the marine environment originated in the first unicellular organisms; these gradients were essential for processes such as energy generation, uptake of molecules from the environment, and counteraction of negatively charged proteins (Yancey, 2005). As an ionic gradient is associated with water fluxes that can disrupt cells by swelling and shrinking, cells also accumulate solutes such as proteins, amino acids, and other osmolytes to bridge the osmotic gap with the environment.

The osmolarity of hagfish plasma is essentially the same as in seawater (Smith, 1932) and therefore hagfish do not experience major water fluxes with the environment; this strategy is called "osmoconforming." Later works demonstrated that hagfish plasma is also similar to seawater in ionic composition (Robertson, 1954; Bellamy and Jones, 1961), with the exception of Ca^{2+}, Mg^{2+}, and SO_4^{2-} (Bellamy and Chester Jones, 1961). Due to the similarity of osmotic strategies, hagfish have been considered to have more in common with invertebrates than the vertebrates (Currie and Edwards, 2010). Indeed, Na^+ and Cl^- are the major osmolytes in hagfish plasma which lack the capacity to regulate plasma osmolality, [Na^+], and [Cl^-] (Sardella et al., 2009), similar to most marine invertebrates. The strict osmoconforming strategy likely reflects an exclusive evolutionary history of hagfish in seawater and it can also explain why there are no fossil records of hagfish in sediments other than of marine origin and why there are not extant estuarine or freshwater hagfishes.

Although plasma [Na^+] and [Cl^-] are similar to seawater, plasma [Ca^{2+}], [Mg^{2+}], and [SO_4^{2-}] are regulated to levels one-half to one-third that of seawater (Robertson, 1954, 1976; Bellamy and Jones, 1961). For example, unlike [Na^+] and [Cl^-] (Sardella et al., 2009), hagfish can actively regulate [SO_4^{2-}] in the absence of any changes in glomerular filtration rate (Clifford et al., unpublished). Similarly, information about mechanisms for active regulation of plasma [Ca^{2+}] and [Mg^{2+}] is still lacking. Given the phylogenetic position of the hagfish and its ability to regulate these divalent ions but not plasma [Na^+] and [Cl^-], it seems that ionoregulation of divalent ions evolved prior to regulation of monovalent ions. However, validation of this hypothesis requires further study examining the ionoregulatory capacities of organisms from major clades ancestral and derived in relation to the agnathans.

Extrabranchial contributions to acid/ base and ionoregulation

Recent developments in hagfish physiology have begun to highlight the extrabranchial aspects of acid/base and ionoregulatory status, with specific regard to the skin performing nutrient acquisition. Glover et al. (2011) proposed that the feeding environment (i.e., putrefied carrion) of the hagfish would be high in amino acid content and that the hagfish may have adapted to take up nutrients across multiple epithelia. The group hypothesized that while burrowing into a carcass to feed, hagfish optimize nutritional uptake via amino acid transport through both the skin and gill. The findings from this study did indeed show *in vitro* uptake of L-alanine and L-glycine through the intestine and more importantly in the gills and the skin (Glover et al., 2011; see Chapter 12), thus highlighting the adaptations that hagfish have acquired to maximize nutrient uptake while burrowing into food sources. The intestine, skin, and gills of hagfish have also recently been shown to facilitate uptake of inorganic phosphate, a limiting nutrient in the demersal habitat of the hagfish (Schultz et al., 2014; reviewed in Chapter 12).

In addition to being nutrient-rich, the environment in and around the putrefying carrion is also likely high in PCO_2, PNH_3, and acidic pH (see "Hagfish habitat and behavior: Naturally acidotic environments"). Presumably, hagfish have developed similar extrabranchial adaptations in ionoregulatory capacity to cope with these environmental stressors when hagfish burrow into carrion (Jensen, 1966; Martini, 1998). Anecdotally, we have noticed that while feeding, the posterior portion of the animal often remains outside the carrion. Thus, the caudal region is immersed in seawater that is nominally free from putrefactive gases and would therefore present favorable gradients to offload toxicants (excess NH_4^+, NH_3, H^+, HCO_3^-, CO_2). We are currently investigating the potential role of extrabranchial tissues (specifically the skin, intestine, and kidney) in ionoregulation, A/B regulation, and nitrogen handling as unique adaptations to the hagfish feeding behavior/strategy.

Concluding remarks

Although the description of the unique ionoregulatory strategy of hagfishes has been known for more than 85 years (Smith, 1932), the molecular and cellular mechanisms have only recently begun to be described. The recent development of advanced sequencing approaches combined with strategies/techniques for understanding cellular mechanisms of ion and acid–base regulation have resulted in a resurgence of interest in hagfish as a key evolutionary species. Future developments in our understanding should provide valuable new information on the physiology of this unique organism and will aid in our understanding of the evolution of ionoregulatory and acid–base regulatory systems.

Acknowledgments

A.M.C. was supported by an NSERC Post-Graduate Doctoral Scholarship, the Alberta Innovates Technology Futures Graduate Student Scholarship, the Andrew Stewart Memorial Graduate Prize, the University of Alberta

Doctoral Prize of Distinction, the Richard E. (Dick) Peter Memorial Graduate Scholarship and the Donald M. Ross Scholarship and was supported by a Discovery Grant Awarded to G.G.G. J.N.R. was supported by an NIH Training Grant in Marine Biotechnology (GM067550) and a National Science Foundation grant (#IOS1354181) to M.T., who was also supported by an Alfred P. Sloan Foundation Research Fellowship (grant #BR2013-103). The authors wish to thank Alyssa Weinrauch for illustrations, Dr. Susan Edwards for critical review and improvements to the M.S., and the staff at Bamfield Marine Sciences Centre, Bamfield, BC, Canada, for continued support for hagfish research.

References

Auster, P. J. and K. Barber. 2006. Atlantic hagfish exploit prey captured by other taxa. *J Fish Biol* 68:618–621.

Bartels, D. H. 1984. Orthogonal arrays of particles in the gill epithelium of the Atlantic hagfish, *Myxine glutinosa*. *Cell Tissue Res* 238:657–659.

Bartels, H. 1985. Assemblies of linear arrays of particles in the apical plasma membrane of mitochondria-rich cells in the gill epithelium of the Atlantic hagfish (*Myxine glutinosa*). *Anat Rec* 211:229–238.

Bellamy, D. and I. C. Jones. 1961. Studies on *Myxine glutinosa*—I. The chemical composition of the tissues. *Comp Biochem Physiol A* 3:175–183.

Boron, W. F. 2004. Regulation of intracellular pH. *Adv Physiol Educ* 28:160–179.

Braun, M. H. and S. F. Perry. 2010. Ammonia and urea excretion in the Pacific hagfish *Eptatretus stoutii*: Evidence for the involvement of Rh and UT proteins. *Comp Biochem Physiol A* 157:405–415.

Brauner, C. J. and D. W. Baker. 2009. Patterns of acid–base regulation during exposure to hypercarbia in fishes. In M. L. Glass and S. C. Wood (eds.), *Cardio-Respiratory Control in Vertebrates*. Springer, Berlin, Heidelberg, pp. 43–63.

Bucking, C., J. L. Fitzpatrick, S. R. Nadella and C. M. Wood. 2009. Post-prandial metabolic alkalosis in the seawater-acclimated trout: The alkaline tide comes in. *J Exp Biol* 212:2159–2166.

Bucking, C. and C. W. Wood. 2008. The alkaline tide and ammonia excretion after voluntary feeding in freshwater rainbow trout. *J Exp Biol* 211:2533–2541.

Carter, D. O., D. Yellowlees and M. Tibbett. 2007. Cadaver decomposition in terrestrial ecosystems. *Naturwissenschaften* 94:12–24.

Choe, K. P., S. Edwards, A. I. Morrison-Shetlar, T. Toop and J. B. Claiborne. 1999. Immunolocalization of Na$^+$/K$^+$-ATPase in mitochondrion-rich cells of the Atlantic hagfish (*Myxine glutinosa*) gill. *Comp Biochem Physiol A* 124:161–168.

Choe, K. P., A. I. Morrison-Shetlar, B. P. Wall and J. B. Claiborne. 2002. Immunological detection of Na$^+$/H$^+$ exchangers in the gills of a hagfish, *Myxine glutinosa*, an elasmobranch, *Raja erinacea*, and a teleost, *Fundulus heteroclitus*. *Comp Biochem Physiol A* 131:375–385.

Choe, K. P., S. O'Brien, D. H. Evans, T. Toop and S. L. Edwards. 2004. Immunolocalization of Na+/K+-ATPase, carbonic anhydrase II, and vacuolar H+-ATPase in the gills of freshwater adult lampreys, *Geotria australis. J Exp Zool A Comp Exp Biol* 301A:654–665.

Claiborne, J. B., S. L. Edwards and A. I. Morrison-Shetlar. 2002. Acid–base regulation in fishes: Cellular and molecular mechanisms. *J Exp Zool* 293:302–319.

Clifford, A. M., G. G. Goss and M. P. Wilkie. 2015. Adaptations of a deep sea scavenger: Extreme ammonia tolerance and active NH_4^+ excretion by the Pacific hagfish (*Eptatretus stoutii*). *Comp Biochem Physiol A* 182:64–74.

Clifford, A. M., S. C. Guffey and G. G. Goss. 2014. Extrabranchial mechanisms of systemic pH recovery in hagfish (*Eptatretus stoutii*). *Comp Biochem Physiol A* 168:82–89.

Collins, M. A., C. Yau, C. P. Nolan, P. M. Bagley and I. G. Priede. 1999. Behavioural observations on the scavenging fauna of the Patagonian slope. *J Mar Biol Assoc UK* 79:963–970.

Currie, S. and S. L. Edwards. 2010. The curious case of the chemical composition of hagfish tissues—50 years on. *Comp Biochem Physiol A* 157:111–115.

Davison, W., J. Baldwin, P. S. Davie, M. E. Forster and G. H. Satchell. 1990. Exhausting exercise in the hagfish, *Eptatretus cirrhatus*: The anaerobic potential and the appearance of lactic acid in the blood. *Comp Biochem Physiol A* 95:585–589.

Doyle, W. L. and D. Gorecki. 1961. The so-called chloride cell of the fish gill. *Physiol Zool* 34:81–85.

Edwards, S. L., J. Arnold, S. D. Blair, M. Pray, R. Bradley, O. Erikson and P. J. Walsh. 2015. Ammonia excretion in the Atlantic hagfish (*Myxine glutinosa*) and responses of an Rhc glycoprotein. *Am J Phys Reg Int* 308:R769–778.

Edwards, S. L., J. B. Claiborne, A. I. Morrison-Shetlar and T. Toop. 2001. Expression of Na^+/H^+ exchanger mRNA in the gills of the Atlantic hagfish (*Myxine glutinosa*) in response to metabolic acidosis. *Comp Biochem Physiol A* 130:81–91.

Elger, M. and H. Hentschel. 1983. Morphological evidence for ionocytes in the gill epithelium of the hagfish *Myxine glutinosa*. *Bull Mt Desert Island Biol Lab* 23:4–8.

Evans, D. H. 1984. Gill Na^+/H^+ and Cl^-/HCO_3^- exchange systems evolved before the vertebrates entered fresh water. *J Exp Biol* 113:465–469.

Evans, D. H., P. M. Piermarini and K. P. Choe. 2005. The multifunctional fish gill: Dominant site of gas exchange, osmoregulation, acid–base regulation, and excretion of nitrogenous waste. *Physiol Rev* 85:97–177.

Forster, M. E. 1990. Confirmation of the low metabolic rate of hagfish. *Comp Biochem Physiol A* 96:113–116.

Glover, C., C. Bucking and C. Wood. 2011. Adaptations to *in situ* feeding: Novel nutrient acquisition pathways in an ancient vertebrate. *Proc R Soc B* 278:3096–3101.

Goss, G., K. Gilmour, G. Hawkings, J. H. Brumbach, M. Huynh and F. Galvez. 2011. Mechanism of sodium uptake in PNA negative MR cells from rainbow trout, *Oncorhynchus mykiss* as revealed by silver and copper inhibition. *Comp Biochem Physiol A* 159:234–241.

Grosell, M. and J. Genz. 2006. Ouabain-sensitive bicarbonate secretion and acid absorption by the marine teleost fish intestine play a role in osmoregulation. *Am J Physiol Regul Integr Comp Physiol* 291:R1145–R1156.

Hardisty, M. W. 1979. *Biology of Cyclostomes*. Chapman & Hall, London.

Heisler, N. 1986. Buffering and transmembrane ion transfer processes. In N. Heisler (ed.), *Acid–Base Regulation in Animals*. Elsevier, Amsterdam, pp. 3–47.

Henry, R. P., Č. Lucu, H. Onken and D. Weihrauch. 2012. Multiple functions of the crustacean gill: Osmotic/ionic regulation, acid–base balance, ammonia excretion, and bioaccumulation of toxic metals. *Front Physiol* 3:431.

Holland, N. D. and J. Chen. 2001. Origin and early evolution of the vertebrates: New insights from advances in molecular biology, anatomy, and paleontology. *Bioessays* 23:142–151.

Huang, C.-H. and J. Peng. 2005. Evolutionary conservation and diversification of Rh family genes and proteins. *Proc Natl Acad Sci USA* 102:15,512–15,517.

Hung, C. Y. C., K. N. T. Tsui, J. M. Wilson, C. M. Nawata, C. M. Wood and P. A. Wright. 2007. Rhesus glycoprotein gene expression in the mangrove killifish *Kryptolebias marmoratus* exposed to elevated environmental ammonia levels and air. *J Exp Biol* 210:2419–2429.

Hwang, P. P. and T. H. Lee. 2007. New insights into fish ion regulation and mito-chondrion-rich cells. *Comp Biochem Physiol A* 148:479–497.

Hwang, P.-P., T.-H. Lee and L.-Y. Lin. 2011. Ion regulation in fish gills: Recent progress in the cellular and molecular mechanisms. *Am J Physiol Regul Integr Comp Physiol* 301:R28–R47.

Ip, Y. K., A. S. L. Tay, K. H. Lee and S. F. Chew. 2010. Strategies for surviving high concentrations of environmental ammonia in the swamp eel *Monopterus albus*. *Physiol Biochem Zool* 77:390–405.

Jensen, D. 1966. The hagfish. *Sci Am* 214:82–90.

Karnaky, K. J. Jr. 1998. Osmotic and ionic regulation. In D. H. Evans (ed.), *The Physiology of Fishes*. CRC Press, Boca Raton, pp. 157–176.

Levine, S., N. Franki and R. M. Hays. 1973a. A saturable, vasopressin-sensitive carrier for urea and acetamide in the toad bladder epithelial cell. *J Clin Invest* 52:2083–2086.

Levine, S., N. Franki and R. M. Hays. 1973b. Effect of phloretin on water and solute movement in the toad bladder. *Journal of Clinical Investigation* 52:1435–1442.

Lin, L.-Y., J.-L. Horng, J. G. Kunkel and P.-P. Hwang. 2006. Proton pump-rich cell secretes acid in skin of zebrafish larvae. *Am J Physiol Cell Physiol* 290:C371–C378.

Mallatt, J., D. M. Conley and R. L. Ridgway. 1987. Why do hagfish have gill "chloride cells" when they need not regulate plasma NaCl concentration? *Can J Zool* 65:1956–1965.

Mallatt, J. and C. Paulsen. 1986. Gill ultrastructure of the Pacific hagfish *Eptatretus stoutii*. *Am J Anat* 177:243–269.

Marini, A. M. and A. Urrestarazu. 1997. The Rh (Rhesus) blood group polypeptides are related to NH_4^+ transporters. *Trends in Biochem Sci* 22:460–461.

Martini, F. H. 1998. The ecology of hagfishes. In J. M. Jørgensen, J. P. Lomholt, R. E. Weber, and H. Malte (eds.), *The Biology of Hagfishes*. Springer, Dordrecht, The Netherlands, pp. 57–77.

McDonald, D. G., V. Cavdek, L. Calvert and C. L. Milligan. 1991. Acid–base regulation in the Atlantic hagfish *Myxine glutinosa*. *J Exp Biol* 161:201–215.

McInerney, J. E. 1974. Renal sodium reabsorption in the hagfish, *Eptatretus stoutii*. *Comp Biochem Physiol A* 49:273–280.

McInerney, J. E. and D. O. Evans. 1970. Habitat characteristics of the Pacific hagfish, *Polistotrema stoutii*. *J Fish Res Bd Can* 27:966–968.

Munz, F. W. and R. W. Morris. 1965. Metabolic rate of the hagfish, *Eptatretus stoutii* (Lockington) 1878. *Comp Biochem Physiol A* 16:1–6.

Nakada, T., C. M. Westhoff, A. Kato and S. Hirose. 2007. Ammonia secretion from fish gill depends on a set of Rh glycoproteins. *FASEB J* 21:1067–1074.

Nawata, C. M., C. C. Y. Hung, T. K. N. Tsui, J. M. Wilson, P. A. Wright and C. M. Wood. 2007. Ammonia excretion in rainbow trout (*Oncorhynchus mykiss*): Evidence for Rh glycoprotein and H^+-ATPase involvement. *Physiol Genom* 31:463–474.

Nelson, D. L., A. L. Lehninger and M. M. Cox. 2008. *Lehninger Principles of Biochemistry*, 5th ed. W. H. Freeman, New York.

Parks, S. K., M. Tresguerres and G. G. Goss. 2007a. Interactions between Na^+ channels and Na^+-HCO_3^- cotransporters in the freshwater fish gill MR cell: A model for transepithelial Na^+ uptake. *Am J Physiol Cell Physiol* 292:C935–C944.

Parks, S. K., M. Tresguerres and G. G. Goss. 2007b. Blood and gill responses to HCl infusions in the Pacific hagfish (*Eptatretus stoutii*). *Can J Zool* 85:855–862.

Parks, S. K., M. Tresguerres and G. G. Goss. 2009. Cellular mechanisms of Cl^- transport in trout gill mitochondrion-rich cells. *Am J Physiol Regul Integr Comp Physiol* 296:R1161–R1169.

Perry, S. F. 1997. The chloride cell: Structure and function in the gills of freshwater fishes. *Ann Rev Physiol* 59: 325–347.

Piermarini, P. M. and D. H. Evans. 2001. Immunochemical analysis of the vacuolar proton-ATPase B-subunit in the gills of a euryhaline stingray (*Dasyatis sabina*): Effects of salinity and relation to Na^+/K^+-ATPase. *J Exp Biol* 204:3251–3259.

Piermarini, P. M., J. W. Verlander, I. E. Royaux and D. H. Evans. 2002. Pendrin immunoreactivity in the gill epithelium of a euryhaline elasmobranch. *Am J Physiol Regul Integr Comp Physiol* 283:R983–R992.

Randall, D. J., J. M. Wilson, K. W. Peng, T. W. K. Kok, S. S. L. Kuah, S. F. Chew, T. J. Lam et al. 1999. The mudskipper, *Periophthalmodon schlosseri*, actively transports NH_4^+ against a concentration gradient. *Am J Physiol Regul Integr Comp Physiol* 277:R1562–R1567.

Randall, D. J. and P. A. Wright. 1989. The interaction between carbon dioxide and ammonia excretion and water pH in fish. *Can J Zool* 67:2936–2942.

Rauther, M. 1937. Kiemen der anamnier-kiemendarm-derivate der Cyclostomen und fische. die driisenartigen derivate des kiemendarmes und kiemengange bei den Cyclostomen und fischen. In *Handbuch der Vergleichenden Anatomie der Wirbeltiere*. Urban & Schwarzenberg, Berlin, pp. 211–278.

Reid, S. D., G. S. Hawkings, F. Galvez and G. G. Goss. 2003. Localization and characterization of phenamil-sensitive Na^+ influx in isolated rainbow trout gill epithelial cells. *J Exp Biol* 206:551–559.

Robertson, J. D. 1954. The chemical composition of the blood of some aquatic chordates, including members of the tunicata, cyclostomata and osteichthyes. *J Exp Biol* 31:424–442.

Robertson, J. D. 1976. Chemical composition of the body fluids and muscle of the hagfish *Myxine glutinosa* and the rabbit-fish *Chimaera monstrosa*. *J Zool* 178:261–277.

Roos, A. and W. F. Boron. 1981. Intracellular pH. *Physiol Rev* 61:296–434.

Ruben, J. A. and A. F. Bennett. 1980. Antiquity of the vertebrate pattern of activity metabolism and its possible relation to vertebrate origins. *Nature* 286:886–888.

Sardella, B. A., D. W. Baker and C. J. Brauner. 2009. The effects of variable water salinity and ionic composition on the plasma status of the Pacific Hagfish (*Eptatretus stoutii*). *J Comp Physiol B* 179:721–728.

Schultz, A. G., S. C. Guffey, A. M. Clifford and G. G. Goss. 2014. Phosphate absorption across multiple epithelia in the Pacific hagfish (*Eptatretus stoutii*). *Am J Physiol Regul Integr Comp Physiol* 307:R643–R652.

Shelton, R. G. J. 1978. On the feeding of the hagfish *Myxine Glutinosa* in the North Sea. *J Mar Biol Assoc UK* 58:81–86.

Sidell, B. D., D. B. Stowe and C. A. Hansen. 1984. Carbohydrate is the preferred metabolic fuel of the hagfish (*Myxinie glutinosa*) heart. *Physiol Zool* 57:266–273.

Smith, C. R. 1985. Food for the deep sea: Utilization, dispersal, and flux of nekton falls at the Santa catalina basin floor. *Deep Sea Res A Oceanogr Res Pap* 32:417–442.

Smith, C. R. and A. R. Baco. 2003. Ecology of whale falls at the deep-sea floor. *Oceanogr Mar Biol* 41:311–354.

Smith, H. W. 1932. Water regulation and its evolution in the fishes. *Quart Rev Biol* 7:1–26.

Smith, K. L. Jr. and R. R. Hessler. 1974. Respiration of benthopelagic fishes: *In situ* measurements at 1230 meters. *Science* 184:72–73.

Srivastava, J., D. L. Barber and M. P. Jacobson. 2007. Intracellular pH sensors: Design principles and functional significance. *Physiology (Bethesda)* 22:30–39.

Tresguerres, M., F. Katoh, H. Fenton, E. Jasinska and G. G. Goss. 2005. Regulation of branchial V-H +-ATPase, Na$^+$/K$^+$-ATPase and NHE2 in response to acid and base infusions in the Pacific spiny dogfish (*Squalus acanthias*). *J Exp Biol* 208:345–354.

Tresguerres, M., S. K. Parks and G. G. Goss. 2006a. V-H$^+$-ATPase, Na$^+$/K$^+$-ATPase and NHE2 immunoreactivity in the gill epithelium of the Pacific hagfish (*Epatretus stoutii*). *Comp Biochem Physiol A* 145:312–321.

Tresguerres, M., S. K. Parks and G. G. Goss. 2007a. Recovery from blood alkalosis in the Pacific hagfish (*Eptatretus stoutii*): Involvement of gill V-H$^+$-ATPase and Na$^+$/K$^+$-ATPase. *Comp Biochem Physiol A* 148:133–141.

Tresguerres, M., S. K. Parks, F. Katoh and G. G. Goss. 2006b. Microtubule-dependent relocation of branchial V-H +-ATPase to the basolateral membrane in the Pacific spiny dogfish (*Squalus acanthias*): A role in base secretion. *J Exp Biol* 209:599–609.

Tresguerres, M., S. K. Parks, S. E. Sabatini, G. G. Goss and C. M. Luquet. 2008. Regulation of ion transport by pH and [HCO$_3^-$] in isolated gills of the crab Neohelice (Chasmagnathus) granulata. *Am J Physiol Regul Integr Comp Physiol* 294:R1033–R1043.

Tresguerres, M., S. K. Parks, C. M. Wood and G. G. Goss. 2007b. V-H$^+$-ATPase translocation during blood alkalosis in dogfish gills: Interaction with carbonic anhydrase and involvement in the postfeeding alkaline tide. *Am J Physiol Regul Integr Comp Physiol* 292:R2012–R2019.

Tsui, T. K. N., C. Y. C. Hung, C. M. Nawata, J. M. Wilson, P. A. Wright and C. M. Wood. 2009. Ammonia transport in cultured gill epithelium of freshwater rainbow trout: The importance of Rhesus glycoproteins and the presence of an apical Na$^+$/NH$_4^+$ exchange complex. *J Exp Biol* 212:878–892.

Walsh, P. J. 1998. Nitrogen excretion and metabolism. In D. H. Evans (ed.), *The Physiology of Fishes*, 2nd edn. CRC Press, Boca Raton, FL, pp. 199–214.

Walsh, P. J., E. Danulat and T. P. Mommsen. 1990. Variation in urea excretion in the gulf toadfish *Opsanus beta*. *Mar Biol* 106:323–328.

Wells, P. and A. Pinder. 1996. The respiratory development of Atlantic salmon. I. Morphometry of gills, yolk sac and body surface. *J Exp Biol* 199:2725–2736.

Whitten, S. T., B. García-Moreno and V. J. Hilser. 2005. Local conformational fluctuations can modulate the coupling between proton binding and global structural transitions in proteins. *Proc Natl Acad Sci USA* 102:4282–4287.

Wilkie, M. P., S. Turnbull, J. Bird, Y. S. Wang, J. F. Claude and J. H. Youson. 2004. Lamprey parasitism of sharks and teleosts: High capacity urea excretion in an extant vertebrate relic. *Comp Biochem Physiol A* 138:485–492.

Wilkie, M. P., Y. Wang, P. J. Walsh and J. H. Youson. 1999. Nitrogenous waste excretion by the larvae of a phylogenetically ancient vertebrate: The sea lamprey (*Petromyzon marinus*). *Can J Zool* 77:707–715.

Wilson, J. M., D. J. Randall, M. Donowitz, A. W. Vogl and A. K. Ip. 2000. Immunolocalization of ion-transport proteins to branchial epithelium mitochondria-rich cells in the mudskipper (*Periophthalmodon schlosseri*). *J Exp Biol* 203:2297–2310.

Wood, C. M. 1993. Ammonia and urea metabolism and excretion. In D. H. Evans (ed.), *The Physiology of Fishes*. CRC Press, Boca Raton, FL, pp. 379–425.

Wood, C. M. 2001. Influence of feeding, exercise, and temperature on nitrogen metabolism and excretion. In P. A. Anderson and P. A. Wright (eds.), *Nitrogen Excretion, Fish Physiology*. Elsevier, San Diego, pp. 201–238.

Wood, C. M., M. Kajimura, T. P. Mommsen and P. J. Walsh. 2005. Alkaline tide and nitrogen conservation after feeding in an elasmobranch (*Squalus acanthias*). *J Exp Biol* 208:2693–2705.

Wright, P. A., P. Part and C. M. Wood. 1995. Ammonia and urea excretion in the tidepool sculpin (*Oligocottus maculosus*): Sites of excretion, effects of reduced salinity and mechanisms of urea transport. *Fish Physiol Biochem* 14:111–123.

Yancey, P. H. 2005. Organic osmolytes as compatible, metabolic and counteracting cytoprotectants in high osmolarity and other stresses. *J Exp Biol* 208:2819–2830.

You, G., C. P. Smith, Y. Kanai, W. S. Lee, M. Stelzner and M. A. Hediger. 1993. Cloning and characterization of the vasopressin-regulated urea transporter. *Nature* 365:844–847.

Zintzen, V., C. D. Roberts, M. J. Anderson, A. L. Stewart, C. D. Struthers and E. S. Harvey. 2011. Hagfish predatory behaviour and slime defence mechanism. *Sci Rep* 1:131.

chapter twelve

Feeding, digestion, and nutrient absorption in hagfish

Chris N. Glover and Carol Bucking

Contents

Introduction

The acquisition, processing, and assimilation of food are fundamental functions in all animals. Feeding provides organisms with the building blocks and energy required for homeostasis, growth, reproduction, and development and as such could be considered the most essential of all physiological functions. Despite this, our knowledge of feeding and the related processes of digestion and nutrient absorption in the hagfishes is surprisingly poor.

An understanding of feeding and feeding-related processes provides insight into the critical roles that hagfishes have in benthic ecosystems. For example, the large biomass of these animals in deep sea settings is likely essential for nutrient recycling and ecosystem engineering (Martini et al., 1997; Knapp et al., 2011), whereas knowledge of hagfish diets and energy requirements is vital to the sustainability of hagfish fisheries (Knapp et al., 2011). Furthermore, their phylogenetic positioning at the branch point of vertebrate evolution (Heimberg et al., 2010) means that hagfish feeding behaviors and digestive functions could offer insight into the evolution of a variety of phenomena such as the chemosensory detection of food items, gut complexity, and nutrient transport.

Although our knowledge of hagfish feeding, digestion, and nutrient absorption remains rudimentary, in recent years there have been some significant advances. These include an improved understanding of hagfish feeding niches and behaviors (e.g., Zintzen et al., 2011); a more detailed appreciation of feeding biomechanics (Clark and Summers, 2007); and studies that have characterized unusual nutrient absorption pathways (Glover et al., 2011a). As with many other aspects of its biology, knowledge of feeding, digestion, and nutrient absorption in the hagfishes is challenging existing paradigms.

Feeding

Diet and feeding modes

The initial studies that informed our understanding of hagfish feeding, digestion, and nutrient absorption focussed on the analysis of hagfish gut contents (Table 12.1). These identified a wide variety of dietary components and included material that could have only been sourced from terrestrial carrion (e.g., insect and bird material; Shelton, 1978; Martini, 1998) and items most likely attained by predation (e.g., high quantities of shrimp and polychaetes; Shelton, 1978; Johnson, 1994). Recent combined analysis of stable isotopes and fatty acids has revealed that *Eptatretus cirrhatus* obtains between 40% and 50% of its organic matter from chemoautotrophs (McLeod and Wing, 2007). Additionally, the individual variation in isotopic signatures revealed a wide-ranging generalist diet, exploiting a number of trophic levels (McLeod and Wing, 2007). Together, these findings have led to the hypothesis that hagfishes consume a variety of prey (Table 12.1) and display two main modes of feeding: opportunistic scavenging and active predation.

As opportunistic scavengers, hagfishes are known to exploit seafloor carcasses (Smith and Baco, 2003) and to take advantage of fishery discards (Catchpole et al., 2006). This characteristic has been utilized for scientific study and the commercial exploitation of the species, in that baited traps

Table 12.1 Diets of hagfishes

Hagfish species	Diet components	References
E. atami	Teleost fishes	Strahan and Honma (1960)
E. burgeri	Cartilaginous fish; teleost fish	Dean (1904), Fernholm (1974)
E. cirrhatus	Mammal	Strahan (1962)
E. deani	Polychaetes; cephalopods; crustaceans; echinoderms; teleost fishes	Wakefield (1990), Johnson (1994)
E. longipinnis	Decapod crustacean	Strahan (1975)
E. stouti	Polychaetes; cephalopods; crustaceans; hagfish eggs; teleost fishes	Worthington (1905), Johnson (1994)
M. glutinosa	Nemerteans; priapuloids; gastropods; cephalopods; polychaetes; insects; crustaceans; hagfish eggs; cartilaginous fish; teleost fish; birds; mammals	Cole (1913), Gustafson (1935), Strahan (1963), Shelton (1978), Martini (1998), Auster and Barber (2006)
Neomyxine sp.	Teleost fish	Zintzen et al. (2011)

can be used to attract hagfish for harvest or for observation by underwater cameras (e.g., Davies et al., 2006). However, hagfish are known to occur in very high densities, and such large biomasses are unlikely to be supported purely by scavenging (Martini, 1998).

Evidence from gut content analysis and estimates of hagfish energetic demands predict that predation is an important feeding mode in hagfish. Traditionally, hagfish have been considered to focus predation on small invertebrates such as polychaetes and shrimp (Shelton, 1978; Martini, 1998). However, there is also evidence for active predation on larger animals. For example, *Eptatretus longipinnis* was first discovered after being extracted from the abdomen of a living lobster (Strahan, 1975). Baited camera traps have also led to observation of active predation in a *Neomyxine* species (Zintzen et al., 2011). In this recording, a single individual was observed to scour muddy sand in an area of substrate proximal to fish burrows. Upon prey detection, the hagfish entered the burrows and, at that point, was thought to exude slime to immobilize and potentially suffocate their prey, before extracting it from the burrow with its mouth (Zintzen et al., 2011). It is worth noting that this behavior was observed within close proximity to the bait, a more readily available food source.

A third mode of feeding, encounter competition, has also been observed in hagfish. Utilizing remotely operated vehicles with cameras, Auster and Barber (2006) noted two distinct instances of *Myxine glutinosa* feeding on prey items that had been previously captured, and were in the process of being consumed, by other benthic species (in this case an echinoderm and a crab). This opportunistic approach to acquiring food underlies the significant flexibility of hagfish feeding behaviors, a trait that likely contributes toward their success in habitats in which competition for feed resources may be intense.

More recently, a fourth feeding mode, passive, has been detailed. In order to access soft tissues, hagfish will enter decaying carcasses (Martini, 1998), an environment enriched in dissolved nutrients. Evidence suggests that hagfish can absorb these dissolved nutrients directly across gill and skin epithelia (Glover et al., 2011a; see section "Nutrient absorption across the gill and skin epithelia").

Plasticity in feeding behaviors

Hagfishes feed primarily via active predation or opportunistic scavenging, but knowledge of how these modes differ as a function of species, developmental stage, and habitat is critical for understanding the ecology of hagfish and benthic ecosystems. Some insight into species differences in feeding behaviors has been provided by studies exploiting stable isotopes to infer positioning of hagfish within food webs. For example, off the coast of New Zealand, there are two *Eptatretus* species (*E. cirrhatus* and a species that has yet to be taxonomically described; Zintzen et al., 2013) that share near identical habitats and body forms. However, despite these similarities, these two species occupy different trophic niches, with *E. cirrhatus* more reliant on lower trophic levels than its co-occurring species (Zintzen et al., 2013). Both the *Eptatretus* species feed at higher trophic levels than the significantly smaller *Neomyxine biplincata*, which is found in similar geographical locations, but at shallower depths. It was proposed that *N. biplincata* subsists mostly on predation of small invertebrates, whereas the *Eptatretus* species rely to varying degrees on carrion scavenging (Zintzen et al., 2013).

Habitat depth is likely to be an important factor driving different feeding niches among sympatric species. It is generally considered that scavenging increases with depth, a function of decreased food densities (Yeh and Drazen, 2011). In the study of Zintzen et al. (2013), the hagfish species known to be most regularly found the deepest (*Eptatretus* sp.) had the stable isotope profile that reflected the highest trophic level, and the hagfish species which was found most regularly in shallower waters (*N. biplincata*) fed at the lowest levels. This suggests that feeding mode is shaped by environmental factors.

On the basis of these findings, it would be predicted that hagfishes more dependent on episodic scavenging at depth would display lower metabolic rates than those species more dependent on active predation. However, measures of oxygen consumption rates of hagfish do not correspond with this hypothesis (Drazen et al., 2011). This is more likely a reflection of experimental difficulties in attaining true resting metabolic rates for hagfish than evidence arguing against feeding niche differentiation with depth.

The plasticity of hagfish feeding modes extends to their ability to go long periods without feeding. Kench (1989) reported that feeding once every 2 weeks was appropriate for long-term maintenance of hagfish (*Eptatretus hexatrema*) in the laboratory, whereas *Eptatretus stouti* survived starvation periods of up to, but not greatly exceeding, 9 months (Tamburri and Barry, 1999). A major contributor to this ability is likely to be the very low metabolic rates of hagfish (Forster, 1990; Lesser et al., 1996; Drazen et al., 2011). This would greatly reduce energy demands and facilitate survival on a temporally and spatially limited food source.

Individual variation in feeding modes

Complicating the understanding of how feeding modes may differ between hagfish species are the large inter-individual differences in feeding modes within species. In two studies of hagfishes (*Eptatretus* sp.) from either near shore (McLeod and Wing, 2007) or off shore environments (MacAvoy et al., 2002), a large variation in stable isotope signatures has been noted. This indicates hagfish caught from similar habitats, with presumably similar access to food, may have quite distinct diets. More research is clearly required to delineate the factors that drive hagfish feeding mode.

Ontogenetic changes in hagfish diets and feeding modes have not been specifically examined, but by analogy with teleost fish, significant differences in diet between juvenile and adult hagfish might be expected (Mittelbach and Persson, 1998). However, Clark and Summers (2012) showed that, on the basis of feeding apparatus morphometrics, juvenile and adult *E. stouti* were likely to subsist on similar prey items. Differences in how juvenile and adult hagfish skin handle dissolved nutrient uptake have, however, been observed. The skin of juvenile *E. stouti* (<30 g) displayed uptake rates for L-alanine approximately twice those of adult skin (Glover et al., 2011a), which may suggest a greater reliance on passive feeding in early life stages.

Feeding and the role of slime

One of the defining features of hagfishes is their production of slime. In addition to being a potential tool for subduing prey (discussed earlier), slime may have other key roles in feeding. Scavenging hagfish are reliant on an episodic food source, and there is significant selection pressure in being able to maximize utilization of feeding opportunities. Consequently, noting the presence of slime on carrion, it has been suggested that slime functions as a repellent, discouraging competition for food resources by other scavengers (Martini, 1998; Zintzen et al., 2011). This is supported by both field and laboratory observations. In baited trap studies, hagfish slime clogged the gills of a scavenging ratfish leading to its death by suffocation (Davies et al., 2006). In laboratory experiments, carrion-feeding gastropods avoided consumption of fish flesh covered in hagfish slime (Tamburri and Barry, 1999). However, in the same study, it was noted that the amphipod *Orchomene obtusus* was not deterred and indeed fed on the slime itself (Tamburri and Barry, 1999). This may still have the desired effect, as the exploitation of slime by competing scavengers would still leave the carrion intact for hagfish consumption.

Feeding behavior: Sensing food

In benthic habitats, food may be spatially and temporally dispersed, and therefore the ability to detect carrion falls is critical. Hagfishes have degenerate eyes and rely mainly on tactile and chemosensory cues for identifying food. To this end, hagfish appear to have well-developed chemosensory abilities (Braun, 1996).

Once food is detected by a hagfish, a characteristic search behavior is initiated (Tamburri and Barry, 1999). The hagfish lifts its head into the water column moving it from side to side, presumably to isolate the direction of the chemosensory stimulus. Thereafter, the animal swims toward the food source, with head down and the barbels framing the mouth in contact with the substrate. Periodically, feeding-like eversions of the dental plate occur (see "Feeding mechanism"). At least in the case of *E. stouti* held in laboratory conditions, the initiation of feeding behavior occurs only in response to a carcass, with an "aged carcass" initiating a greater proportion of responding hagfish. This is hypothesized to be a consequence of the relatively greater chemical signal that results, rather than a specific cue related to death or decomposition (Tamburri and Barry, 1999). Amino acids are reported to be the main food attractant molecules in fish, although other chemicals associated with biological breakdown may also have an important role (Derby and Sorensen, 2008).

Supporting a role for amino acids in initiating hagfish feeding, Døving and Holmberg (1974) have shown a neurological response of the *M. glutinosa* olfactory system to amino acids, although this was induced only at relatively high concentrations (100 µM), suggesting that this might be a cue effective only in close proximity to a carrion fall. It remains to be determined whether the olfactory or gustatory senses are more important in mediating food-seeking behavior. However, it is intriguing to note that hagfishes possess a unique sensory modality, the Schreiner organs, which are hypothesized to have a similar function to vertebrate taste buds

(Braun, 1998; Kirino et al., 2013). Given their distribution over the entire skin surface of the hagfish, it could be that these play a role in detecting nutrient signals and promoting scavenging behavior.

Feeding mechanism

Hagfish are jawless and thus lack the advantages provided by jaws with respect to prey capture and feeding (Clark and Summers, 2007). Instead, hagfishes feed by a "grasping and tearing" mechanism utilizing two hinged cartilaginous dental plates. These are lined with rows of grasping (i.e., not serrated) keratinous teeth. In the presence of food, these plates are protracted from the oral cavity, and as they extend from the mouth the plates unhinge, exposing the teeth. As the plates reach the fullest extent of protraction, they are in contact with the food, and then, as the plates retract, they clamp shut, embedding the teeth into the prey item. The food is then pulled into the body as retraction continues. Several cycles of protraction and retraction are necessary to push food into the esophagus (Clark and Summers, 2007). Although the "grasping" of this system of food acquisition lacks the bite forces of a jawed animal, the forces generated by the "tearing" of retraction can be equivalent (Clark and Summers, 2007).

The protraction and retraction of the dental plate are achieved by sliding up and down a central groove on a fixed basal plate, a hard cartilaginous structure that, during retraction, lies anterior and ventral to the dental plate (Figure 12.1). The motive force for dental plate motion is provided by a series of muscles. Protraction is achieved largely by the action of the deep protractor muscle, and retraction by the action of the clavatus, perpendicularis, and tubularis muscles (Clark et al., 2010). The clavatus muscle is the main driver of retraction, with the other muscles performing support functions. Consistent with an episodic heavy workload and a working environment inside hypoxic carrion falls, the clavatus muscle is highly tolerant of lactate accumulation (Baldwin et al., 1991). It is likely that an element of elastic recoil also aids in retraction, which would ease muscle work during feeding sessions (Clark et al., 2010).

Although gape (which determines the maximal size that can be swallowed whole) and tearing forces of hagfish are comparable to those of jawed fish (Clark and Summers, 2007), the major disadvantage of the hagfish feeding mechanism is the relatively slow speed of biting. Furthermore, this long cycle of protraction and retraction results in relatively low pressures generated within the oral cavity. In this regard, hagfish represent an intermediate phase of feeding biomechanics between the very low-pressure ciliary-based mechanisms of food acquisition in urochordates and cephalochordates and the high-pressure suction-based feeding mechanism of the true vertebrates (Clark and Summers, 2007).

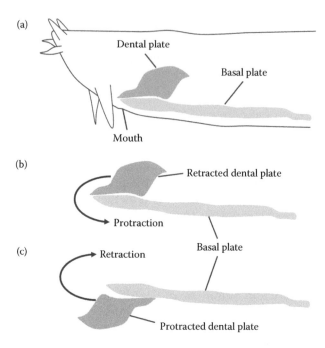

Figure 12.1 Overview of feeding apparatus of hagfishes, demonstrating the plates used to achieve tearing and grasping of food (a) and how these are protracted (b) and retracted (c). (Modified from Clark, A. J. and A. P. Summers. 2012. *J Fish Biol* 80:86–99.)

In teleost fish, there are a myriad of different mouth and digestive tract morphologies, reflecting adaptations to specific diets. Despite the variation in diet observed for hagfish (Table 12.1), the hagfish feeding apparatus appears to be conserved between species. The only notable differences between *E. stouti* and *M. glutinosa*, the two best studied species, relate to the number and size of teeth and differences in size of the feeding apparatus components (Clark and Summers, 2007). These differences in morphology do not, however, translate to any significant difference in feeding mechanics (Clark and Summers, 2007).

Behavioral modifications to aid feeding

A limitation of the hagfish feeding mechanism is an inability to perforate the integument of many of their known prey items. This is a consequence of the relatively low tooth stress (i.e., ability to maintain grasp on an item as the dental plate is retracted), which will thus constrain grasping and tearing (Clark and Summers, 2012). Consequently, when feeding on larger prey items, hagfish are thought to feed preferentially on softer tissues

(Shelton, 1978). This is most easily facilitated by entry via body openings (e.g., mouth, anus, and open wounds) to gain access to the inside of the animal (Martini, 1998).

Some behaviors also help facilitate food ingestion. Hagfish have been observed to shake their heads in a lateral motion and spin their bodies during dental plate retraction (Strahan, 1963; Martini, 1998), thus using body musculature to generate additional force (Clark and Summers, 2012). Furthermore, hagfish display a behavior whereby the body loops through itself forming a knot. The pressing of the knot against a feeding surface generates additional leverage for tearing flesh off carrion (Clark and Summers, 2012).

Digestion

Digestion is the breakdown of food into small, absorbable components for the provision of nutrients and energy and is a ubiquitous process across all multicellular organisms. The evolution of true digestive systems, with regionally distinct epithelia specialized in secretion and absorption, is considered a significant step in promoting enhanced processing of food (Yonge, 1937). The hagfishes display an intriguing strategy in which the digestive tract is relatively devoid of regional specialization, and digestive roles may be taken up by other epithelia. Despite this novel approach to digestion, this is a process that is poorly understood and has received little attention, particularly in recent years.

Digestion in the intestine

The digestive tract of hagfish is simple, consisting of a long intestinal tube with an associated liver and gallbladder. They lack a stomach, but do maintain a slightly acidic lumenal environment (pH ~5.5–5.8). This pH is generally lower than that of the intestine of other aquatic vertebrates and, in fact, decreases when digestion is occurring (Nilsson and Fänge, 1970). This decrease is not driven by gastric secretion of acid (as in elasmobranch and teleost fish; Wood et al., 2007; Bucking and Wood, 2008; Bucking et al., 2009) as, in addition to being stomachless, hagfish also lack detectable carbonic anhydrase activity (an enzyme crucial for the formation of HCl in higher vertebrates; Koelz, 1992). This therefore suggests that if acid-based digestion is occurring in hagfish, it is through a unique proton generation pathway (Nilsson and Fänge, 1970). There is evidence of a catheptic-type enzyme that is optimally active around pH 4 in the hagfish intestine, which may be analogous to the acid-activated pepsin in higher vertebrates (Nilsson and Fänge, 1970).

As would be predicted by their diets (Table 12.1), the hagfish intestine expresses or secretes enzymes for protein breakdown. Indeed, the presence of the proteolytic enzymes trypsin, chymotrypsin, carboxypeptidase

A, leucine aminopeptidase, and dipeptidase has been confirmed in hagfish gut tissue (Nilsson and Fänge, 1970). These enzymes appear to be biochemically similar to those found in higher vertebrates (Nilsson and Fänge, 1970). There is also evidence for chitinase in the hagfish intestine (Fänge et al., 1979), although the level of activity is low, supporting the observation that hagfishes may consume whole crustaceans, but they feed preferentially on their soft tissues (Shelton, 1978).

Adam (1963) suggested that lipids may act as a significant source of nutritional energy to hagfishes. Consistent with this hypothesis is a relatively high enzymatic activity of lipase in the hagfish intestine (Adam, 1963; Vigna and Gorbman, 1979). Lipase is an important intestinal enzyme involved in hydrolyzing dietary fats in the intestine, facilitating absorption. As hagfishes lack a true pancreas (Vigna and Gorbman, 1979; Thorndyke and Falkmer, 1998), the usual location for lipase synthesis, the enzyme is most likely produced by the diffuse pancreatic-like cells in the intestine (Barrington, 1972). Nevertheless, these pancreatic-like cells respond in a standard manner to hormonal stimulation. For example, porcine cholecystokinin (CCK), a potent hormone released from the intestine in response to a meal that stimulates pancreatic secretions (Lewis and Williams, 1990), increased the release of lipase into the hagfish gut (Vigna and Gorbman, 1979).

As a consequence of this reliance on lipids, the liver and intestinal cells of hagfish contain a marked quantity of intracellular fat-containing vacuoles (Spencer et al., 1966). Colipase, an enzyme required to stabilize and optimize lipase enzymatic functions in the presence of bile, has been detected in hagfish (Sternby et al., 1983). In contrast, colipase is not found in the invertebrate hepatopancreas (e.g., Cherif et al., 2007), suggesting that it evolved with the formation of exocrine pancreatic activity.

The gallbladder and bile

Bile, composed of bile alcohols or acids, salts, and water, allows for efficient emulsion of dietary fats for subsequent absorption. Indeed, bile solubilization of fats is required before lipase can begin to breakdown dietary lipid. Bile salts are unique to vertebrate species and are lacking in invertebrates with the exception of some species of sea squirts (e.g., *Ciona intestinalis*), which synthesize bile salt-like compounds, albeit for nondigestive purposes (sperm chemoattraction; Yoshida et al., 2002).

Bile alcohols are the primitive bile constituents and are found in hagfish and elasmobranchs, being largely supplanted by true bile salts in higher vertebrates such as reptiles and mammals (Haslewood, 1967; Hagey et al., 2010; Hofmann et al., 2010). In hagfish, 5α-myxinol is the only detected bile alcohol and, distinctively, it exists as a disulfate (Haslewood, 1966, 1967; Hagey et al., 2010). Elasmobranchs likewise have a single bile alcohol (5β-scymol), but radiations are apparent in teleost fish in which numerous bile acids and alcohols are present (Hagey et al., 2010).

Myxinol has several interesting structural properties, such as the longer carbon skeleton and pattern of hydroxylation, that confirm its primitive nature compared with more "advanced" bile alcohols and salts (Haslewood, 1967). Additionally, myxinol is structurally similar to the more primitive bile salt-like chemoattractant in *C. intestinalis*, suggesting a potential pathway for bile evolution (Yoshida et al., 2002). Structural reconstruction and analysis reveal that the ability of myxinol to act as a detergent may not be optimal compared with more advanced bile alcohols and salts (Haslewood, 1967). The resemblance to a primitive chemoattractant, and the predicted reduced ability to emulsify dietary fats, suggests that myxinol serves purposes other than digestion (Hofmann et al., 2010), a hypothesis consistent with the use of bile salts as pheromones in lampreys (e.g., Li et al., 2002) and other fishes (e.g., Zhang et al., 2001).

With the exception of potassium, the salt concentration of hagfish bile does not resemble the ionic composition observed in the plasma (Robertson, 1976; Figure 12.2). Indeed, calcium and magnesium are present at an order of magnitude greater than plasma and seawater levels (Robertson, 1976; Forster and Fenwick, 1994), whereas concentrations of sodium are twice, and chloride one-third, those observed in plasma (and surrounding seawater), despite the osmolarity being similar (Robertson, 1976, Figure 12.2). The resulting anion deficit is most likely compensated with myxinol disulfate. It has been suggested that bile is the primary route of divalent ion excretion, with biliary levels exceeding those of the urine (Rall and Burger, 1967). Comparing the ionic composition of bile with that found in elasmobranchs (which is similar to that of marine teleosts; Smith, 1930) reveals comparatively elevated sodium, calcium, and magnesium concentrations (Figure 12.2) in the hagfish bile, in accordance with biliary divalent ion excretion.

As in the true vertebrates, hagfish bile is stored in the gallbladder. Perhaps surprisingly, given evidence of a conserved CCK stimulation of pancreatic activity, the application of porcine CCK to the gallbladder of hagfish fails to induce contraction and bile secretion, in contrast to all vertebrates studied to date (Vigna and Gorbman, 1979). Contraction of vertebrate gallbladders via CCK is believed to be stimulated through two CCK receptors, CCK-A (CCK-AR) and CCK-B (CCK-BR), in the associated smooth muscle (e.g., Tang et al., 1996; Suzuki et al., 2001). The absence of an effect of CCK on gallbladder contraction would indicate that these receptors are either nonresponsive to porcine CCK, possibly due to a lack of homology (unlikely given the effect on lipase secretion), or lacking altogether. Hagfish gallbladders do, however, constrict in response to acetylcholine (Vigna and Gorbman, 1979), indicating that there is smooth muscle there to activate.

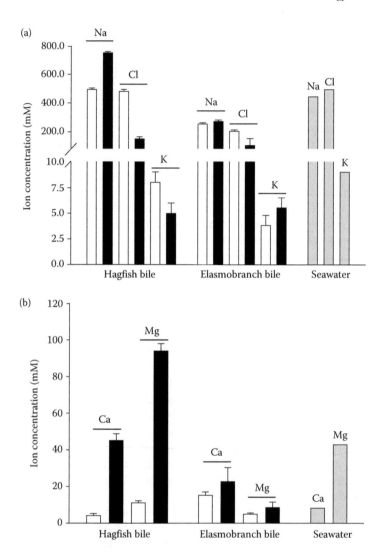

Figure 12.2 (a) Monovalent ion concentrations (mM) in plasma (white bars), gall-bladder bile (black bars), and seawater (gray bars) of hagfish and elasmobranchs. (b) Divalent ion concentrations (mM) in plasma (white bars), gallbladder bile (black bars), and seawater (gray bars) of hagfish and elasmobranchs. (Data compiled from Robertson, J. D. 1976. *J Zool* 178:261–277; Boyer, J. L. et al., 1976. *Am J Physiol* 230:970–973; Forster, M. E. and J. C. Fenwick. 1994. *Gen Comp Endocrinol* 94:92–103; Suzuki, N. et al., 1995. *Zool Sci* 12:239–242; Wood, C. M. et al., 2007. *J Exp Biol* 210:1335–1349; Sardella, B. A. et al., 2009. *J Comp Physiol B* 179:721–728; Anderson, W. G. et al., 2012. *Comp Biochem Physiol* 161A:27–35.)

There is clearly much that needs to be investigated regarding the hormonal control of digestive processes. To date, evidence suggests key similarities and differences with higher vertebrates. Whether these similarities represent conserved functions and processes in the evolution of vertebrate endocrine systems, or whether they are specializations to the unique aspects of hagfish biology, remains to be determined.

Skin digestion

The skin of the hagfish may be covered in one or two types of secretions. The continually secreted skin mucous is a product of epidermal or epithelial mucous cells, whereas the extruded slime is produced during feeding (see "Feeding and the role of slime"). Interestingly, examination of the skin mucous and extruded slime reveals the presence of protease activity (Subramanian et al., 2008). The presence of proteolytic enzymes is believed to play a role in defense against invading pathogens such as bacteria (Subramanian et al., 2008). In theory, however, these excreted enzymes could enhance the digestion of the carcass surrounding the scavenging hagfish immersed in a carrion fall, creating an enhanced opportunity for the absorption of dissolved nutrients across the epidermal surface (see "Nutrient absorption across the gill and skin epithelia"). The extruded slime has approximately a five-fold higher level of proteases relative to epidermal mucous (Subramanian et al., 2008), and considering that slime is extruded when feeding, this creates the potential to use slime as a digestive agent. Protease characterization using specific inhibitors showed that the extruded slime not only had higher levels of proteases, but also contained unique proteases that were not found in epidermal mucous. Evidence also suggested that approximately half the enzyme activity was trypsin-like (Subramanian et al., 2008).

Nutrient absorption

The assimilation of nutrients into the body is clearly a vital final component of the digestive process. In fact, given the potential limitations in feeding opportunities, the maximization of absorption efficiency is likely to be a critical factor in scavenging hagfish. Knowledge of nutrient absorption in hagfish is, however, greatly lacking.

Nutrient absorption across the intestine

Recent studies have demonstrated the ability of *E. stouti* gut to absorb inorganic phosphate, carbohydrates, and amino acids. Utilizing an *in vitro* double-perfused gut sac technique, glucose was shown to move from a mucosal perfusate into a serosal ringer perfused through major gut blood vessels (Bucking et al., 2011). A more detailed study using isolated gut sacs from the same species demonstrated concentration-dependent uptake kinetics of two amino acids. It was shown that both glycine and L-alanine transport were sodium-dependent and conformed to Michaelis–Menten uptake kinetics (Glover et al., 2011b). Inorganic phosphate uptake was also shown to occur across the gut (Schultz et al., 2014). This was also sodium-dependent, but was not saturable. Based on evidence of type II sodium phosphate cotransporter (Slc34a1) mRNA expression in intestinal tissue, it was proposed that a saturable component was present, but masked by the relatively greater flux of phosphate through diffusive pathways (Schultz et al., 2014).

Qualitatively, the observed patterns of amino acid uptake are consistent with findings in teleost fish. For example, on the basis of substrate specificity, it was proposed that amino acid absorption was achieved by transporters with similarity to those that facilitate glycine and L-alanine transport in teleosts (Glover et al., 2011b), although confirmation of this hypothesis requires molecular support. Quantitatively, however, amino acid transport characteristics differed. The affinity constant (K_m) values derived for hagfish were approximately an order of magnitude higher (i.e., an order of magnitude lower affinity) than those for amino acids in teleost fish and were thus more in line with those of mammalian gut (Glover et al., 2011b). The comparatively high teleost transporter affinities are usually rationalized in the context of a smaller gut transport surface area, and their poikilothermic existence (with higher affinities compensating somewhat for lower transport rates at low environmental temperatures; Karasov et al., 1985). That hagfish transport characteristics match those of mammalian gut might reflect the existence of a greatly enhanced absorptive surface area, with the ability of hagfish to utilize the gill and skin as additional routes of nutrient uptake (see "Nutrient absorption across the gill and skin epithelia").

Indicative of the lack of gross morphological differences in the intestinal structure, no regional differences in gut amino acid transport were observed (Glover et al., 2011b). This is distinct from the situation in most teleost fish, in which transport activities differ markedly depending on the gut region (Bakke et al., 2011). The explanation for this pattern requires further investigation, but it may reflect feeding strategy. Hagfishes gorge themselves on food, to the extent that recently ingested food can emerge virtually intact from the rear of the animal even as eating continues (Baldwin et al., 1991). Thus the amount of food ingested is not limited to the extent of the stomach, resulting in an enhanced capacity for ingestion. Consequently, however, regional specialization in digestive and transport functions down the length of the gut would be disadvantageous. Whether this simplistic, generalized gut function is a feature characteristic of early craniates, or a derived phenomenon characteristic of hagfish and their feeding behavior and physiology, remains unknown.

Nutrient absorption across the gill and skin epithelia

A key characteristic of hagfish scavenging behavior is the capacity of these animals to breach the internal cavities of carrion to access softer tissue. This will expose the hagfish to elevated levels of dissolved nutrients. Following observations by Stephens (1968) that *Eptatretus* was able to take up amino acids directly from seawater, the ability of the gills and skin of hagfish to act as nutrient absorptive surfaces was demonstrated (Bucking et al., 2011; Glover et al., 2011a). The immersion of hagfish in exposure chambers containing radiolabeled amino acids resulted in significant accumulation of the isotope in the animal. *In vitro* transport assays supported the findings of the *in vivo* work, with saturable uptake of L-alanine and glycine demonstrated across both gill and skin. Unlike the situation in the gut, transport of these amino acids seemed to occur via distinct (i.e., not shared) transport pathways and exhibited higher transport affinities than intestinal uptake (Figure 12.3). Skin from all regions of the body exhibited amino acid transport (Glover et al., 2013). More recently, Schultz et al. (2014) demonstrated that inorganic phosphate was also able to be absorbed by the gill and skin, suggesting that these epithelia may have a more general role in nutrient absorption.

It is important to recognize that the integument is a metabolically demanding tissue (Welsch and Potter, 1998). It thus remains to be determined whether the skin is a true absorptive surface, supplying building blocks and energy for the organism as a whole, or whether it is simply supplying its own metabolic and synthetic needs. The latter situation applies to most fish that display cutaneous oxygen uptake (for example, Glover et al., 2013).

Together, the gills, gut, and skin represent a very large surface area exposed to dissolved nutrients, and thus the use of these epithelia as nutrient transport loci is likely a mechanism that maximizes nutrient assimilation. To date, however, only a limited number of substrates (inorganic phosphate, glycine, and L-alanine) have been tested for their ability to traverse skin and gill, so it remains to be confirmed whether similar capacities for absorption exist for other nutrients. Furthermore, it is not known how widespread this phenomenon is among different hagfish species, and whether uptake patterns change with fed state or between species with different primary feeding strategies (i.e., scavenging vs. active predators). For example, as they are more inclined to burrow (Martini, 1998), *Myxine* skin may be exposed more frequently to a variety of organic and inorganic substrates via sediments. This could

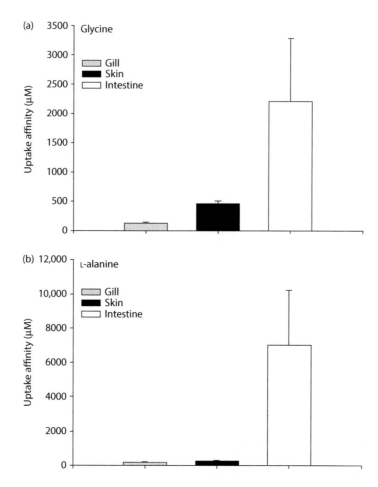

Figure 12.3 Comparison of uptake affinities for the transport of the amino acids glycine (a) and L-alanine (b) across gill, skin, and intestinal epithelia of the Pacific hagfish *E. stouti*. (Data from Glover, C. N. et al., 2011a. *Proc R Soc B* 278:3096–3101; Glover, C. N. et al., 2011b. *J Comp Physiol B* 181:765–771.)

have nutritional benefits, but could also increase susceptibility to toxi-cants (e.g., Chiu and Mok, 2011).

   Integumental nutrient absorption is common among marine inver-tebrates (e.g., Wright and Manahan, 1989), and the presence of cutaneous nutrient absorption may represent a primitive feature of hagfish. It has been proposed that the ability to utilize the skin surface is a function of the osmoconforming salt and water balance strategy of hagfish (Glover et al., 2011a). Using the skin as a transport surface is likely incompatible with its role as a barrier restricting water movements between the animal and its environment. Lacking any significant gradient for water exchange (Currie and Edwards, 2010) may therefore have allowed hagfishes to populate the skin with nutrient uptake transporters to maximize utilization of the infrequent opportunities for scavenging on carrion falls.

Nutrient absorption under duress

Feeding inside a decaying carcass is likely to offer significant challenges to metabolism and thus processes such as digestion and nutrient assimilation. Among these is the need to acquire, process, and absorb nutrients while hypoxic. Hagfish are highly hypoxia-tolerant (Forster, 1998), and one adaptation they display is an ability to maintain nutrient uptake despite oxygen limitations (Bucking et al., 2011). Both branchial and intestinal L-alanine transport *in vitro* was unimpacted by a 24 h pre-exposure to hypoxia, and in fact, glycine uptake via the gut and gill was significantly stimulated. The additional absorbed glycine accumulated at elevated levels in hagfish brain, where it is proposed to have functions in metabolic depression and/or cytoprotection (Bucking et al., 2011).

Conclusion

Despite the importance of feeding-related processes for understanding the basic biology of hagfishes, and the broader roles of these animals in areas such as benthic ecology, fisheries, and evolutionary physiology, this is an area of research in which significant questions remain. As with most aspects of hagfish biology, our knowledge of feeding and digestion is limited to a few relatively well-studied species. Although studies to date indicate that aspects of hagfish feeding, such as potential chemosensory modalities (Braun, 1998) and feeding apparatus morphologies (Clark and Summers, 2007), are conserved across the two most prominent genera (*Eptatretus* and *Myxine*), the occupation of distinct trophic niches between sympatric species (Zintzen et al., 2013) hints at subtle differences in key aspects of feeding and the ecological roles of hagfish.

   The biology and ecology of feeding are, however, well studied relative to the paucity of information regarding basic digestive and absorptive functions in hagfish. Of specific future interest is the potential interplay between the multiple epithelia that appear capable of performing roles in nutrient acquisition. Molecular characterization of the entities responsible will be crucial in this endeavor and will provide insight into the evolution of digestive systems. Studies that investigate the changes that occur in hagfish digestive tracts with changes in feeding modes (and the switching between feeding and fasting) will also be of value. Given their low metabolic rate and exothermic disposition, do hagfish up- and down-regulate intestinal functions and morphologies with episodic feeding events?

Two key areas are likely to drive further insight into hagfish feeding, digestion, and nutrient absorption. Innovations in observational and molecular methodologies for studying feeding ecology of benthic species *in situ* (Lindsay et al., 2012; Redd et al., 2014) are likely to find application in the study of hagfishes. At the physiological and biochemical levels, improved genomic information will provide a structural framework for investigating digestive and absorptive processes (e.g., Schultz et al., 2014), thus facilitating significant advances in our knowledge of these areas of hagfish biology.

References

Adam, H. 1963. Structure and histochemistry of the alimentary canal. In A. Brodal and R. Fänge (eds), *The Biology of Myxine*. University of Oslo Press, Oslo, pp. 256–288.

Anderson, W. G., C. M. Nawata, C. M. Wood, M. D. Piercey-Normore, and D. Weihrauch. 2012. Body fluid osmolytes and urea and ammonia flux in the colon of two chondrichthyan fishes, the ratfish, *Hydrolagus colliei*, and spiny dogfish, *Squalus acanthias*. *Comp Biochem Physiol* 161A:27–35.

Auster, P. J. and K. Barber. 2006. Atlantic hagfish exploit prey captured by other taxa. *J Fish Biol* 68:618–621.

Bakke, A. M., C. N. Glover, and Å. Krogdahl. 2011. Feeding, digestion and absorption of nutrients. In M. Grosell, A. P. Farrell, and C. J. Brauner (eds), *The Multifunctional Gut of Fish, Fish Physiology*, vol. 30. Academic Press, London, pp. 57–110.

Baldwin, J., W. Davison, and M. E. Forster. 1991. Anaerobic glycolysis in the dental plate retractor muscles of the New Zealand hagfish *Eptatretus cirrhatus* during feeding. *J Exp Biol* 260:295–301.

Barrington, E. J. W. 1972. The pancreas and intestine. In M. W. Hardisty and I. C. Potter (eds), *The Biology of Lampreys*, vol. 2. Academic Press, New York, pp. 135–169.

Boyer, J. L., J. Schwarz, and N. Smith. 1976. Biliary secretion in elasmobranchs. I. Bile collection and composition. *Am J Physiol* 230:970–973.

Braun, C. B. 1996. The sensory biology of the living jawless fishes. *Brain Behav Ecol* 48:262–276.

Braun, C. B. 1998. Schreiner organs: A new craniate chemosensory modality in hagfishes. *J Comp Neurol* 392:135–163.

Bucking, C., J. L. Fitzpatrick, S. R. Nadella, and C. M. Wood. 2009. Post-prandial metabolic alkalosis in the seawater-acclimated trout: The alkaline tide comes in. *J Exp Biol* 212:2159–2166.

Bucking, C., C. N. Glover, and C. M. Wood. 2011. Digestion under duress: Nutrient acquisition and metabolism during hypoxia in Pacific hagfish. *Physiol Biochem Zool* 84:607–617.

Bucking, C. and C. M. Wood. 2008. The alkaline tide and ammonia excretion after voluntary feeding in freshwater rainbow trout. *J Exp Biol* 211:2533–2541.

Catchpole, T. L., C. L. J. Frid, and T. S. Gray. 2006. Importance of discards from the English *Nephrops norvegicus* fishery in the North Sea to marine scavengers. *Mar Ecol Prog Ser* 313:215–226.

Cherif, S., A. Fendri, N. Miled, H. Trabelsi, H. Mejdoub, and Y. Gargouri. 2007. Crab digestive lipase acting at high temperature: Purification and biochemical characterization. *Biochimie* 89:1012–1018.

Chiu, K.-H. and H.-K. Mok. 2011. Study on the accumulation of heavy metals in shallow-water and deep-sea hagfishes. *Arch Environ Contam Toxicol* 60:643–653.

Clark, A. J., E. J. Maravilla, and A. P. Summers. 2010. A soft origin for a forceful bite: Motor patterns of the feeding musculature in Atlantic hagfish, *Myxine glutinosa. Zoology* 113:259–268.

Clark, A. J. and A. P. Summers. 2007. Morphology and kinematics of feeding in hagfish: Possible functional advantages of jaws. *J Exp Biol* 210:3897–3909.

Clark, A. J. and A. P. Summers. 2012. Ontogenetic scaling of the morphology and biomechanics of the feeding apparatus in the Pacific hagfish *Eptatretus stoutii. J Fish Biol* 80:86–99.

Cole, F. J. 1913. A monograph on the general morphology of the myxinoid fishes based on a study of *Myxine*. V. The anatomy of the gut and its appendages. *Trans R Soc Edin* 49:293–344.

Currie, S. and S. L. Edwards. 2010. The curious case of the chemical composition of hagfish tissues—50 years on. *Comp Biochem Physiol* 157A:111–115.

Davies, S., A. Griffiths, and T. E. Reimchen. 2006. Pacific hagfish, *Eptatretus stoutii*, spotted Ratfish, *Hydrolagus colliei*, and scavenger activity on tethered carrion in subtidal benthic communities off Western Vancouver Island. *Can Field-Nat* 120:363–366.

Dean, B. 1904. Notes on Japanese myxinoids. *J Coll Sci Imp Univ Tokyo* 19:1–24.

Derby, C. D. and P. W. Sorensen. 2008. Neural processing, perception, and behavioral responses to natural chemical stimuli by fish and crustaceans. *J Chem Ecol* 34:898–914.

Døving, K. B. and K. Holmberg. 1974. A note on the function of the olfactory organ of the hagfish *Myxine glutinosa. Acta Physiol Scand* 91:430–432.

Drazen, J. C., J. Yeh, J. Friedman, and N. Condon. 2011. Metabolism and enzyme activities of hagfish from shallow and deep water of the Pacific Ocean. *Comp Biochem Physiol* 159A:182–187.

Fänge, R., G. Lundblad, J. Lind, and K. Slettengren. 1979. Chitinolytic enzymes in the digestive system of marine fishes. *Mar Biol* 53:317–321.

Fernholm, B. 1974. Diurnal variations in the behaviour of the hagfish, *Eptatretus burgeri. Mar Biol* 27:351–356.

Forster, M. E. 1990. Conformation of the low metabolic rate of hagfish. *Comp Biochem Physiol* 96A:113–116.

Forster, M. E. 1998. Cardiovascular function in hagfishes. In J. M. Jørgensen, J. P. Lomholt, R. E. Weber, and H. Malte (eds), *Biology of the Hagfishes*. Chapman & Hall, London, pp. 237–258.

Forster, M. E. and J. C. Fenwick. 1994. Stimulation of calcium efflux from the hagfish, *Eptatretus cirrhatus*, gill pouch by an extract of corpuscles of Stannius from an eel (*Anguilla dieffenbachii*): Teleostei. *Gen Comp Endocrinol* 94:92–103.

Glover, C. N., C. Bucking, and C. M. Wood. 2011a. Adaptations to *in situ* feeding: Novel nutrient acquisition pathways in an ancient vertebrate. *Proc R Soc B* 278:3096–3101.

Glover, C. N., C. Bucking, and C. M. Wood. 2011b. Characterisation of L-alanine and glycine absorption across the gut of an ancient vertebrate. *J Comp Physiol B* 181:765–771.

Glover, C. N., C. Bucking, and C. M. Wood. 2013. The skin of fish as a transport epithelium: A review. *J Comp Physiol B* 183:877–891.

Gustafson, G. 1935. On the biology of *Myxine glutinosa* L. *Arkiv Zool* 28:1–8.

Hagey, L. R., P. R. Møller, A. F. Hofmann, and M. D. Krasowski. 2010. Diversity of bile salts in fish and amphibians: Evolution of a complex biochemical pathway. *Physiol Biochem Zool* 83:308–321.

Haslewood, G. A. D. 1966. Comparative studies of bile salts. Myxinol disulphate, the principal bile salt of hagfish (Myxinidae). *Biochem J* 100:233–237.

Haslewood, G. A. D. 1967. Bile salt evolution. *J Lipid Res* 8:535–550.

Heimberg, A. M., R. Cowper-Sallari, M. Semon, P. C. J. Donoghue, and K. J. Peterson. 2010. microRNAs reveal the interrelationships of hagfish, lampreys, and gnathostomes and the nature of the ancestral vertebrate. *Proc Natl Acad Sci USA* 107:19379–19383.

Hofmann, A. F., L. R. Hagey, and M. D. Krasowski. 2010. Bile salts of vertebrates: Structural variation and possible evolutionary significance. *J Lipid Res* 51:226–246.

Johnson, E. W. 1994. Aspects of the biology of Pacific (*Eptatretus stouti*) and Black (*Eptatretus deani*) hagfishes from Monterey Bay, California. M.S. thesis, California State University, Fresno, California.

Karasov, W. H., R. K. Buddington, and J. M. Diamond. 1985. Adaptation of intestinal sugar and amino acid transport in vertebrate evolution. In R. Gillies and M. Gilles-Baillien (eds), *Transport Processes, Iono- and Osmoregulation*. Springer, Berlin, pp. 227–239.

Kench, J. E. 1989. Observations on the respiration of the South African hagfish, *Eptatretus hexatrema* Mull. *Comp Biochem Physiol* 93A:877–892.

Kirino, M., J. Parnes, A. Hansen, S. Kiyohara, and T. E. Finger. 2013. Evolutionary origins of taste buds: Phylogenetic analysis of purinergic neurotransmission in epithelial chemosensors. *Open Biol* 3:130015.

Knapp, L., M. M. Mincarone, H. Harwell, B. Polidoro, J. Sanciangco, and K. Carpenter. 2011. Conservation status of the world's hagfish species and the loss of phylogenetic diversity and ecosystem function. *Aquat Conserv Mar Freshw Ecosyst* 21:401–411.

Koelz, H. R. 1992. Gastric acid in vertebrates. *Scand J Gastroenterol* 27:2–6.

Lesser, M. P., F. H. Martini, and J. B. Heiser. 1996. Ecology of the hagfish—*Myxine glutinosa* L., in the Gulf of Maine: I. Metabolic rates and energetics. *J Exp Mar Biol Ecol* 208:215–225.

Lewis, L. D. and J. A. Williams. 1990. Regulation of cholecystokinin secretion by food, hormones, and neural pathways in the rat. *Am J Physiol* 258:G512–G518.

Li, W., A. P. Scott, M. J. Siefkes, H. Yan, Q. Liu, S. S. Yun, and D. A. Gage. 2002. Bile acid secreted by male sea lamprey that acts as a sex pheromone. *Science* 296:138–141.

Lindsay, D. J., H. Yoshida, T. Uemura, H. Yamamoto, S. Ishibashi, J. Nishikawa, J. D. Reimer, R. J. Beaman, R. Fitzpatrick, K. Fujikura, and T. Maruyama. 2012. The untethered remotely operated vehicle PICASSO-1 and its deployment from chartered dive vessels for deep sea surveys off Okinawa, Japan and Osprey Reef, Coral Sea, Australia. *Mar Technol Soc J* 46:20–32.

MacAvoy, S. E., R. S. Carney, C. R. Fisher, and S. A. Macko. 2002. Use of chemosynthetic biomass by large, mobile, benthic predators in the Gulf of Mexico. *Mar Ecol Prog Ser* 225:65–78.

Martini, F. 1998. The ecology of the hagfishes. In J. M. Jørgensen, J. P. Lomholt, R. E. Weber, and H. Malte (eds), *Biology of the Hagfishes*. Chapman & Hall, London, pp. 57–77.

Martini, F., M. Lesser, and J. B. Heiser. 1997. Ecology of the hagfish *Myxine glutinosa* L., in the Gulf of Maine: II. Potential impact on benthic communities and commercial fisheries. *J Exp Mar Biol Ecol* 214:97–106.

McLeod, R. J. and S. R. Wing. 2007. Hagfish in the New Zealand fjords are supported by chemoautotrophy of forest carbon. *Ecology* 88:809–816.

Mittelbach, G. G. and L. Persson. 1998. The ontogeny of piscivory and its ecological consequences. *Can J Fish Aquat Sci* 55:1454–1465.

Nilsson, A. and R. Fänge. 1970. Digestive proteases in the cyclostome *Myxine glutinosa* (L.). *Comp Biochem Physiol* 32:237–250.

Rall, D. P. and J. W. Burger. 1967. Some aspects of renal secretion in *Myxine*. *Am J Physiol* 212:354–356.

Redd, K. S., S. D. Ling, S. D. Frusher, S. Jarman, and C. R. Johnson. 2014. Using molecular prey detection to quantify rock lobster predation on barrens-forming sea urchins. *Mol Ecol* 23:3849–3869.

Robertson, J. D. 1976. Chemical composition of the body fluids and muscle of the hagfish *Myxine glutinosa* and the rabbitfish *Chimaera monstrosa*. *J Zool* 178:261–277.

Sardella, B. A., D. W. Baker, and C. J. Brauner. 2009. The effects of variable water salinity and ionic composition on the plasma status of the Pacific hagfish (*Eptatretus stoutii*). *J Comp Physiol B* 179:721–728.

Schultz, A., S. C. Guffey, A. M. Clifford, and G. G. Goss. 2014. Phosphate absorption across multiple epithelia in the Pacific hagfish (*Eptatretus stoutii*). *Am J Physiol Regul Integr Comp Physiol* 307: R643–R652.

Shelton, R. G. J. 1978. On the feeding of the hagfish *Myxine glutinosa* in the North Sea. *J Mar Biol Assoc UK* 58:81–86.

Smith, H. W. 1930. The absorption and excretion of water and salts by marine teleosts. *Am J Physiol* 93:480–505.

Smith, C. R. and A. R. Baco. 2003. Ecology of whale falls at the deep-sea floor. *Oceanogr Mar Biol Ann Rev* 41:311–354.

Spencer, R. P., R. L. Scheig, and H. J. Binder. 1966. Observations on lipids of the alimentary canal of the hagfish *Eptatretus stoutii*. *Comp Biochem Physiol* 19:139–144.

Stephens, G. C. 1968. Dissolved organic matter as a potential source of nutrition for marine organisms. *Am Zool* 8:95–106.

Sternby, B., A. Larsson, and B. Borgström. 1983. Evolutionary studies on pancreatic collapse. *Biochim Biophys Acta* 750:340–345.

Strahan, R. 1962. Survival of the hag, *Paramyxine atami* Dean, in diluted seawater. *Copeia* 2:471–473.

Strahan, R. 1963. The behaviour of the myxinoids. *Acta Zool* 44:1–30.

Strahan, R. 1975. *Eptatretus longipinnis*, n. sp., a new hagfish (Family Eptatretidae) from South Australia, with a key to the 5–7 gilled Eptatretidae. *Aust Zool* 18:137–148.

Strahan, R. and Y. Honma. 1960. Notes on *Paramyxine atami* Dean (Family Myxinidae) and its fishery in Sado Strait, Sea of Japan. *Hong Kong Univ Fish J* 3:27–35.

Subramanian, S., N. W. Ross, and S. L. MacKinnon. 2008. Comparison of the biochemical composition of normal epidermal mucus and extruded slime of hagfish (*Myxine glutinosa* L.). *Fish Shellfish Immunol* 25:625–632.

Suzuki, N., T. Takagi, Y. Sasayama, and A. Kambegawa. 1995. Effects of ultimo-branchialectomy on the mineral balances of the plasma and bile in the sting-ray (Elasmobranchii). *Zool Sci* 12:239–242.

Suzuki, S., S. Takiguchi, N. Sato, S. Kanai, T. Kawanami, Y. Yoshida, and T. Noda. 2001. Importance of CCK-A receptor for gallbladder contraction and pan-creatic secretion: A study in CCK-A receptor knockout mice. *Jap J Physiol* 51:585–590.

Tamburri, M. N. and J. P. Barry. 1999. Adaptations for scavenging by three diverse bathyla species, *Eptatretus stouti*, *Neptunea amianta* and *Orchomene obtusus*. *Deep-Sea Res I* 46:2079–2093.

Tang, C. W., I. Biemond, and C. B. H. W. Lamers. 1996. Cholecystokinin recep-tors in human pancreas and gallbladder muscle: A comparative study. *Gastroenterology* 111:1621–1626.

Thorndyke, M. C. and S. Falkmer. 1998. The endocrine system of hagfishes. In J. M. Jørgensen, J. P. Lomholt, R. E. Weber, and H. Malte (eds), *Biology of the Hagfishes*. Chapman & Hall, London, pp. 399–412.

Vigna, S. R. and A. Gorbman. 1979. Stimulation of intestinal lipase secretion by porcine cholecystokinin in the hagfish, *Eptatretus stouti*. *Gen Comp Endocrinol* 38:356–359.

Wakefield, W. W. 1990. Patterns in the distribution of demersal fishes on the upper-continental slop off central California, with studies on the role of ontogenetic migration on particle flux. Ph.D. thesis, Scripps Institute of Oceanography, San Diego, California.

Welsch, U. and I. C. Potter. 1998. Dermal capillaries. In J. M. Jørgensen, J. P. Lomholt, R. E. Weber, and H. Malte (eds), *Biology of the Hagfishes*. Chapman & Hall, London, pp. 273–283.

Wood, C. M., M. Kajimura, C. Bucking, and P. J. Walsh. 2007. Osmoregulation, ionoregulation and acid–base regulation by the gastrointestinal tract after feeding in the elasmobranch (*Squalus acanthias*). *J Exp Biol* 210:1335–1349.

Worthington, J. 1905. Contribution to our knowledge of the myxinoids. *Am Nat* 39:625–663.

Wright, S. H. and D. T. Manahan. 1989. Integumental nutrient uptake by aquatic organisms. *Ann Rev Physiol* 51:585–600.

Yeh, J. and J. C. Drazen. 2011. Baited-camera observations of deep-sea megafaunal scavenger ecology on the California slope. *Mar Ecol Prog Ser* 424:145–156.

Yonge, C. M. 1937. Evolution and adaptation in the digestive tract of the metazoa. *Biol Rev* 12:87–114.

Yoshida, M., M. Murata, K. Inaba, and M. Morisawa. 2002. A chemoattractant for ascidian spermatazoa is a sulfated steroid. *Proc Natl Acad Sci USA* 99:14831–14836.

Zhang, C., S. B. Brown, and T. J. Hara. 2001. Biochemical and physiological evidence that bile acids produced and released by lake char (*Salvelinus namaycush*) function as chemical signals. *J Comp Physiol B* 171:161–171.

Zintzen, V., C. D. Roberts, M. J. Anderson, A. L. Stewart, C. D. Struthers, and E. S. Harvey. 2011. Hagfish predatory behaviour and slime defence mechanism. *Sci Rep* 1:131.

Zintzen, V., K. M. Rogers, C. D. Roberts, A. L. Stewart, and M. J. Anderson. 2013. Hagfish feeding habits along a depth gradient inferred from stable isotopes. *Mar Ecol Prog Ser* 485:223–234.

chapter thirteen

Hagfish slime
Origins, functions, and mechanisms

Douglas S. Fudge, Julia E. Herr, and Timothy M. Winegard

Contents

Introduction

The hagfishes are an ancient group of elongate, benthic craniates that inhabit mostly deep-sea habitats throughout the world's oceans with an antitropical distribution. Hagfishes are scavengers, but they are also known to take live prey such as soft-bodied benthic invertebrates (Martini, 1998). Recently, it was shown that hagfishes are capable of preying on burrow-dwelling teleosts (Zintzen et al., 2011), although it is not known how common this behavior is. Hagfishes possess several adaptations that make them supremely good scavengers. They have a low metabolic rate, which allows them to survive periods of low food availability and starvation (Lesser et al., 1997; Tamburri and Barry, 1999), and they also have an ability to find and take advantage of falling carrion faster than most other benthic scavengers, due to their exceptionally well-developed sense of smell and sensitivity to cues from dead organisms (Tamburri and Barry, 1999) and their moderately good swimming abilities (Lim, 2013). Once a hagfish arrives at a carrion fall, several adaptations allow it to take full advantage of the available food, including a knot-tying behavior that provides the leverage needed to tear off chunks of flesh with their modest jaws and keratinous teeth (Clark and Summers, 2007). Although hagfishes typically enjoy being some of the first animals to arrive at deep-sea carrion falls, other scavengers and predators inevitably show up. To fend off both competitors and predators, hagfishes possess a unique adaptation— the ability to release large amounts of fibrous slime from numerous epidermal slime glands (Ferry, 1941; Newby, 1946; Downing et al., 1981a, b; Fernholm, 1981; Fudge et al., 2005). Most aquatic animals, including hagfishes, produce some form of epidermal mucous as a barrier defense against microbial pathogens and parasites (Subramanian et al., 2009), but hagfish slime differs from epidermal mucous in its manner of production, its mode of release, and its function. In this chapter, we will describe what is known about the composition, cellular origins, and functions of the slime, as well as its mechanisms of deployment. We will conclude with some thoughts on the evolution of the slime, as well as some new questions that are inspiring further research in this area.

Slime function

Several researchers have speculated about the function of hagfish slime, but its most obvious function is defense against predators. Fernholm (1981) described observations of captive fishes suffocating after attacking hagfish in aquaria and getting slime on their gills. Lim et al. (2006) investigated the effects of hagfish slime on fish gills and found that the slime is remarkably good at disrupting respiratory flow in isolated fish heads (Lim et al., 2006). More recently, Zintzen et al. (2011) captured footage of hagfishes off the coast of New Zealand interacting with numerous fish predators near a baited camera (Zintzen et al., 2011). The researchers recorded fourteen species of fish, including both elasmobranchs and teleosts, attempting to feed on hagfish, and none were successful. While it cannot be ruled out that the predators were repelled by a noxious chemical released by the hagfish, no such chemicals have yet been identified, and the repeated attempts of the fishes to dislodge the slime from their gills suggest that it was the physical presence of slime on the gills that caused them to abort their attacks. These kinds of observations, as well as stomach content data from a variety of fishes in which hagfishes are conspicuously absent, suggest that the slime is an effective protection against fish predators (Martini, 1998). In contrast, many air breathers such as cetaceans, pinnipeds, and even some seabirds are known predators of hagfish from both stomach content data and direct observations (Martini, 1998). The susceptibility of hagfishes to air breathers and their apparent immunity to gill breathers may explain the general pattern that most hagfishes are found in habitats that are too deep to be reached by most air-breathing predators.

The slime may also be used to fend off competing scavengers while hagfishes feed on carrion falls (Isaacs and Schwartzlose, 1975; Tamburri and Barry, 1999; Zintzen et al., 2011), although this is an area that deserves more study. Some researchers have proposed that slime is used for reproduction, specifically in the localization of eggs (Koch et al., 1991a), although the origin of slime found on extruded eggs is more likely the cloacal glands, not the lateral slime glands (Tsuneki et al., 1985; Martini and Beulig, 2013). This issue will be discussed later in this chapter in more detail. Lastly, hagfish slime may also provide some defense against microbial pathogens, as higher levels of innate immune substances such as alkaline phosphatase, lysozyme, and cathepsin B are found in hagfish slime compared with hagfish epidermal mucous (Subramanian et al., 2008).

The slime glands

Hagfish slime originates within the slime glands, which exist in segmental pairs down both sides of the body, starting behind the head at the anterior end and ending near the end of the tail (Figure 13.1). The number of slime glands varies considerably among species, but is fairly constant within a species, which makes it a good taxonomic indicator (Fernholm, 1998). Two of the best-studied species, *Eptatretus stoutii* and *Myxine glutinosa*, have 158 and 194 glands, respectively. Out of all known species, *E. taiwanae* has the fewest slime glands (120), whereas *Nemamyxine elongata* has the most (400) (Fernholm, 1998). Little is known about whether the interspecies differences in slime gland number correlate with differences in traits other than the number of somites, such as how the slime is used. Our work on captive hagfishes suggests that *E. stoutii* slime glands contain larger volumes of slime than *M. glutinosa*, although the functional significance of this is not yet clear.

Slime gland morphology

The slime glands contain two main types of secretory cells, gland mucous cells (GMCs) and gland thread cells (GTCs), which produce the mucous and fibrous components of the slime, respectively (Figure 13.2). The gland is encased by a thin collagenous capsule, which itself is surrounded by a thin layer of striated muscle fibers, the *musculus decussates* (Lametschwandtner et al., 1986). These muscle fibers contribute to the forceful ejection of slime exudate from the gland, although contributions

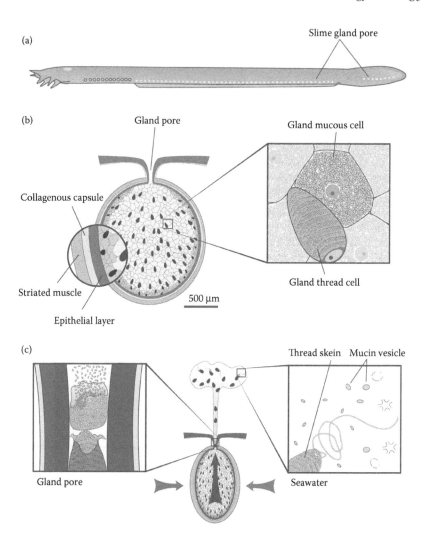

Figure 13.1 Anatomy of hagfish slime glands. (a) Pacific hagfish possess approximately 150 slime glands, which have a segmental distribution down both sides of the body. (b) In its full state, the hagfish slime gland has a spheroidal shape and contains two main kinds of secretory cells, GTCs, and gland mucous cells (GMCs). (c) Slime deployment from the slime glands involves contraction of the muscle layer surrounding the gland and subsequent holocrine expulsion of GTCs and GMCs through the narrow gland duct into seawater. During passage through the gland duct, GTCs lose their plasma membrane and release a coiled thread skein, and GMCs rupture and release numerous mucous vesicles. Mucous vesicles rupture and thread skeins unravel, and together they mix with seawater to form mature slime. (Adapted from Herr, J. E., A. Clifford, G. G. Goss and D. S. Fudge 2014. *J Exp Biol* 217:2288–2296.)

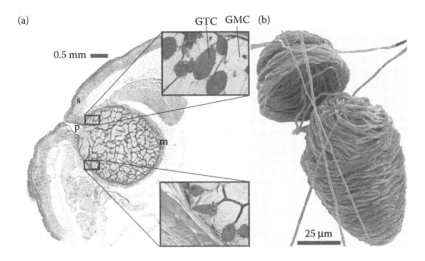

Figure 13.2 (**See color insert.**) Slime gland histology and skein structure. (a) H&E staining of a slime gland section from *E. stoutii* reveals two main cell types in the gland, GTCs and GMCs. The slime gland is located close to the skin (s) and is surrounded by a thin layer of muscle (m), which triggers slime exudate release when it contracts and forces GTCs and GMCs through the narrow gland pore (p). (b) GTCs lose their plasma membrane during holocrine secretion and release a single, elaborately coiled slime thread bundle, called a skein. This skein, viewed with SEM, has cracked open, revealing the staggered arrangement of loops (red) that spiral around the skein and form higher-order structures called conical loop arrangements (purple). (Adapted from Winegard, T., J. Herr, C. Mena, B. Lee, I. Dinov, D. Bird, M. Bernards, Jr. et al. 2014. *Nat Commun* 5:3534.)

from the more powerful surrounding myotomal musculature are also possible (Lim et al., 2006). Examinations of fixed tissue suggest that GMCs and GTCs arise from stem cells within a basal epithelial layer that abuts the capsule (Figure 13.2). As the cells grow and develop, they get pushed toward the center of the gland by the dividing and growing cells below (Newby, 1946). The result is a gradient of cell size and maturity during gland refilling, with the oldest, most mature cells in the center of the gland, and the youngest cells closest to the epithelium (Spitzer and Koch, 1998). GTCs develop with their pointed, apical ends toward the gland pore, which likely facilitates their passage through the gland duct during ejection. Based on the dimensions of the gland duct, mature GTCs and GMCs likely pass through it single file, with the plasma membranes getting sheared off during ejection (Spitzer and Koch, 1998). The result is the release of coiled threads (or "skeins") and countless mucous vesicles in a thick fluid referred to as slime gland exudate. Little is known about the structure and function of the gland pore, and whether it can dynamically regulate the jet of exudate that is released.

Aside from the variability that exists in the number of slime glands, little is known about how the slime glands themselves differ among the approximately 80 species of known hagfishes (Martini and Beulig, 2013). Leppi (1968) carried out histological studies of skin and slime glands from three species of hagfishes, but found no large differences in the staining properties of the glands (Leppi, 1968). Scanning electron microscopy of vascular corrosion casts in *E. stoutii* and *M. glutinosa* reveal a basket-like vascular pattern that extends throughout the gland capsule of both species. One main difference, however, is that capillary loops descend toward the center of the gland in *E. stoutii* but not *M. glutinosa* (Lametschwandtner et al., 1986). This pattern may correspond to differences in the recharge rate of the gland after sliming (Lametschwandtner et al., 1986), but at this point there is no evidence to support this claim.

Gland mucous cells

GMCs are large (~100–150 μm in diameter) cells that produce the mucous component of hagfish slime and occur in approximately equal numbers with GTCs (Figures 13.1 and 13.2). GMCs produce numerous disc-shaped mucous vesicles that are roughly 7 μm in diameter and are released into the environment via the holocrine secretion mode, meaning that the entire ruptured cell is ejected from the gland (Downing et al., 1981a; Luchtel et al., 1991). The glycoproteins within the mucous vesicles contain only 12–18% carbohydrate by weight (Salo et al., 1983), a value that is unusually low for mucous secretions, which typically have carbohydrate contents of about 85% (Gum, 1992). GMCs resemble the LMCs (large mucous cells) of the epidermis both in size and morphology, but it is not known whether LMCs release mucous via holocrine or merocrine (i.e., exocytotic) secretion, although some have suggested it is the latter based on the similarities between LMCs and teleost goblet cells (Whitear, 1986). While thread development and maturation in GTCs has been investigated in several studies (Fernholm, 1981; Winegard et al., 2014), little is known about GMC development. Like GTCs, newly differentiated GMCs undergo massive increases in size as they mature and move toward the center of the gland. Immature GMCs contain a prominent Golgi complex (Luchtel et al., 1991; Winegard, 2012), which produces the mucous vesicles that eventually fill up the vast majority of the cell volume.

Gland thread cells

GTCs produce the fibrous component of hagfish slime and are the most specialized epidermal cell type among lampreys and hagfishes (Quay, 1972). Like GMCs, GTCs likely originate from stem cells in the slime gland basal epithelium (Figure 13.2a). Following differentiation, GTCs undergo dramatic changes in size and morphology, starting as 10 μm diameter cells with large, round nuclei, and ending up as ellipsoidal cells that are about 150 μm long and up to 100 μm wide (Newby, 1946; Terakado et al., 1975; Downing et al., 1984; Winegard et al., 2014). In mature GTCs, a single, elaborately coiled thread skein occupies most of the cytoplasmic volume (Figure 13.2b). The mature slime thread is bidirectionally tapered, with a maximum diameter of about 3 μm near the middle, and less than 1 μm at the ends (Fudge et al., 2005). In the anterior of the cell, the conformation of the thread appears to be random, much like a soft rope would appear if dropped into a conical bucket without any attempt to coil it (Downing et al., 1981b; Fernholm, 1981; Fudge et al., 2005). In contrast to the randomly coiled cap, the rest of the slime thread is organized into repeating staggered loops that spiral around the long axis of the cell within higher-order structures called "conical loop arrangements." The pattern of thread coiling was originally described by Fernholm (1981) and Downing et al. (1981b), who were the first to observe both intact and partially unraveled skeins using SEM (Downing et al., 1981b; Fernholm, 1981).

More recently, the pattern of coiling was described in more detail using focused ion beam milling SEM and 3D modeling (Winegard et al., 2014). These data, along with new TEM images of GTCs at various stages of maturation, has led to a new understanding of how both individual loops and conical loop arrangements form in developing GTCs (Winegard et al., 2014) (Figure 13.3). According to this model, the slime thread originates and elongates from an area of intense protein production on the apical side of the nucleus that contains high concentrations of both mitochondria and ribosomes (Downing et al., 1984). As the thread elongates, it is constrained both by previous coils of thread and the surface of the nucleus, which undergoes dramatic changes in shape. The nucleus starts as a spheroid in young GTCs, elongates into a rounded cone in older cells, and elongates further into a narrow spindle with a wide base. In mature cells, the nuclear spindle retracts, leaving only a small hemispherical cap at the basal end of the cell. These changes in the shape of the nuclear template correspond with changes in the morphology of individual slime thread loops and the nested arrangement of the conical loop arrangements and ultimately determine the morphology of the mature skeins (Winegard et al., 2014) (Figure 13.2).

Slime thread synthesis within GTCs

The slime thread in GTCs consists mainly of two proteins, alpha (643 residues, 66.6 kDa) and gamma (603 residues, 62.7 kDa), which belong to the intermediate filament (IF) family of proteins (Spitzer et al., 1988; Koch et al., 1994, 1995). IFs are a diverse group of fibrous proteins that

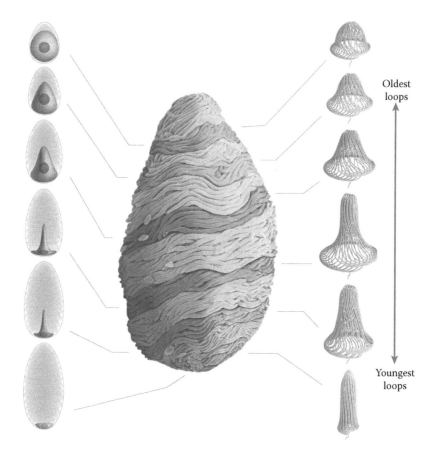

Figure 13.3 (**See color insert.**) Temporal and spatial model of thread assembly and coiling in GTCs. GTC growth and maturation is characterized by dramatic changes in nuclear size and morphology (left), which correspond with the shape of conical loop arrangements laid down during successive stages of thread production (right). Staggered loops are laid down in the space defined by previous loops and the apical surface of the nucleus, with their morphology changing as the nucleus becomes more spindle-shaped and retreats toward the basal end of the cell. The result is a mature thread skein that can be ejected through the gland pore and unravel to its full extended length of ~150 mm without tangling. (Adapted from Winegard, T., J. Herr, C. Mena, B. Lee, I. Dinov, D. Bird, M. Bernards, Jr. et al. 2014. *Nat Commun* 5:3534.)

form networks in the cytoplasm and nucleus (Aebi et al., 1988; Parry and Steinert, 1999). Alpha and gamma both have an unusually high threonine content (alpha, 13%; gamma, 10%) (Koch et al., 1994), which may be involved in posttranslational modifications that are believed to occur during slime thread maturation (Spitzer et al., 1984, 1988). A handful of independent studies on three different hagfish species have examined the changes in thread ultrastructure that occur in developing GTCs (Terakado et al., 1975; Downing et al., 1984; Winegard et al., 2014). Together, these studies reveal a fairly complex sequence of changes and a remarkable degree of similarity among the three species (Figure 13.4). In immature GTCs, thread production starts with a small bundle of IFs in a paranuclear region rich in mitochondria and ribosomes. The thread increases in both length and width via the addition of IFs and eventually via the incorporation of microtubules. The role of microtubules is unknown, but their function is most likely to deliver thread proteins and/or assembly intermediates to the growing thread (Winegard et al., 2014). The youngest threads are only 2–3 IFs in diameter (about 30 nm) and more mature ones are about 30 IFs in diameter (about 400 nm). The thread at this stage is wrapped with a filament that is about 12 nm in diameter, but it is unclear whether this filament exists in a low-pitch helix or as separate rings (Downing et al., 1984; Winegard et al., 2014). The most dramatic change to thread ultrastructure that occurs involves the simultaneous loss of the wrapping filaments, condensation of the constituent IFs, and the appearance of a

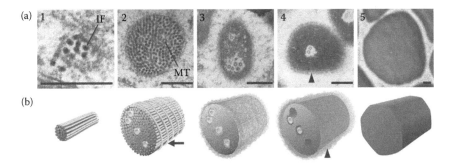

Figure 13.4 Thread ultrastructure changes during GTC maturation. (a) TEM sections of slime threads from immature up to fully mature GTCs. The thread in (1) consists of a bundle of about ten 12-nm diameter IFs. Threads increase in girth by the addition of more IFs, and eventually by the addition of MTs (2). In the next stage (3), 12-nm IFs become more tightly packed, which creates electron-lucent halos (asterisk) around the MTs. Further IF compaction is accompanied by the appearance of a fluffy rind (arrowheads) on the thread surface (4), which likely corresponds to the direct addition of IF subunits or proteins to the thread rather than the bundling of mature IFs. In the final stage, IF proteins compact further, MTs are absent, and the fluffy rind disappears. In fully mature GTCs, the thread takes up the vast majority of the cell volume and adjacent threads are packed so tightly that they conform to each other (5). Scale bars are all 200 nm. (b) Models of thread development corresponding to each of the stages depicted in (a). (Adapted from Winegard, T., J. Herr, C. Mena, B. Lee, I. Dinov, D. Bird, M. Bernards, Jr. et al. 2014. *Nat Commun* 5:3534.)

fluffy rind on the surface of the thread. Microtubules persist within the postcondensation thread and occupy electron-lucent spaces that presumably are created as the proteins within the IFs condense. At late stages of GTC maturation, these spaces are filled in, the fluffy rind is gone, and the thread has no discernable ultrastructure (Fan, 1965; Terakado et al., 1975; Downing et al., 1984; Winegard et al., 2014) due to the dense packing of IF proteins.

Gland interstitial cells

Winegard (2012) recently discovered and described a third cell type within the slime gland. Gland interstitial cells (GICs) occupy narrow spaces between GMCs and GTCs and were first discovered via fluorescent nuclear staining of paraffin-embedded hagfish slime gland sections (Figure 13.5a). GICs were likely overlooked for so long due to their exceedingly small volume compared with their GMC and GTC neighbors. TEM reveals that GICs possess a prominent nucleus and long bifurcating processes that form connections with other GICs, GTCs, and GMCs (Figure 13.5b). From TEM, it appears that each GTC and GMC borders one to three GICs, suggesting that GICs may be the most abundant cells in the slime glands. GICs exhibit large numbers of mitochondria, Golgi apparatus, and vesicles, with some GIC vesicles appearing in TEM to be in the process of fusing with neighboring GTCs and GMCs. The function of GICs is not known, but possibilities include acting as nurse cells for developing GTCs and GMCs and retaining immature GTCs and GMCs during sliming when mature cells are expelled from the gland.

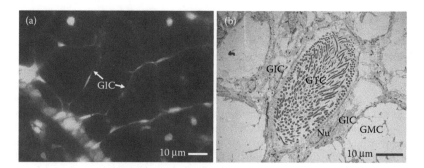

Figure 13.5 (**See color insert.**) Fluorescence and TEM images of GICs within the slime gland of *M. glutinosa*. (a) Nuclei were stained blue with DAPI and IFs were stained green with a pan-cytokeratin antibody. (b) TEM of a slime gland reveals a slender GIC with an elongated nucleus (Nu) wedged between a GTC and a GMC. In the upper left is likely another GIC sandwiched between the GTC and another GMC.

Slime composition, release, and setup

One of the remarkable things about hagfish slime is the vast volume that can be produced by a single hagfish. For example, a 150 g specimen of *E. stoutii* can produce a single mass of slime that is typically more than six times its own body volume and can repeat this behavior many times using stored slime exudate. Not surprisingly, the mucous and thread concentrations in hagfish slime are astoundingly low, with mucous concentrations about three orders of magnitude lower than typical mucous secretions (Fudge et al., 2005). In spite of the impressively economical nature of hagfish slime, hagfishes keep large amounts of it at the ready, with slime exudate making up approximately 3% of the wet mass of *E. stoutii* (Fudge et al., 2005). Based on observations of hagfishes sliming in aquaria using high-speed video, Lim et al. (2006) demonstrated that exudate is released from the slime glands in a forceful jet, which aids in the deployment process, but the jet alone is not sufficient for proper slime setup. Full deployment of the slime requires additional convective mixing, which could be supplied by the thrashing hagfish, the movements of an attacking predator, and/or flow caused by feeding movements by the predator. The same study showed that exudate release is local and only occurs from glands in the immediate vicinity of where the hagfish is physically attacked (Lim et al., 2006; Zintzen et al., 2011). These experiments, as well as videos of hagfishes being attacked by a variety of fish predators in the wild, demonstrate that hagfishes do not hide within their slime, but instead immediately try to escape after exudate is released (Lim et al., 2006; Zintzen et al., 2011) (Figure 13.6). How exactly a few tens of microliters of exudate transform into almost a liter of slime in a fraction of a second is still not well understood. Fernholm (1981)

Figure 13.6 Slime release by hagfishes successfully repels attacks by fish predators. Here a seal shark (*Dalatias licha*) bites a New Zealand hagfish (*E. cirrhatus*) and ends up with only a mouth full of slime and slime trailing from its gills (arrows). Aside from a small puncture wound (arrowhead), the hagfish appears to be mostly unharmed by the encounter. (Adapted from Zintzen, V., C. D. Roberts, M. J., Anderson, A. L., Stewart, C. D. Struthers and E. S. Harvey. 2011. *Sci Rep* 1:131.)

and Koch et al. (1991b) conducted *in vitro* experiments in which slime exudate was diluted with various solvents and solutes and examined under the microscope. These studies provided early glimpses into the processes of thread skein unraveling as well as mucin vesicle rupture and formed the basis of subsequent research in this area.

Mucous vesicle deployment

Building on the pioneering work of Downing et al. (1981a), Luchtel et al. (1991) explored mechanisms of mucin vesicle rupture by stirring slime exudate into a variety of solutions and assessing the degree of slime formation. From these experiments, they concluded that the interior of the vesicles has an osmolarity of about 900 mOsm, or close to that of seawater, which is consistent with the fact that hagfishes are osmo-conformers. Luchtel et al. (1991) also inferred that the vesicles rupture in all hyperosmotic salt solutions except ones made from di- and trivalent anions such as sulfate and citrate, although we now know this is not true for all the vesicles (Herr et al., 2014). Herr et al. (2010) measured the composition of the fluid component of slime exudate and found a distribution of ions similar to that of plasma, but with generally lower concentrations, presumably due to the high concentrations of organic osmolytes that are also present. The most common organic osmolytes in slime exudate fluid are the methylamines betaine, TMAO, and dimethylglycine, which have a cumulative concentration of 390 mM. Methylamines do not seem to be involved in vesicle stabilization, however, as even high concentrations of them can only prevent rupture of about half of the vesicles in *in vitro* assays (Herr et al., 2010). *In vitro* studies also reveal significant heterogeneity in the swelling behavior of mucin vesicles exposed to seawater, with some vesicles swelling slowly and others swelling very quickly, but only after a substantial delay (Herr et al., 2010, 2014).

Although the functional significance of having two kinds of vesicles in slime gland exudate is currently unknown, recent work has generated some clues about its possible molecular underpinnings. Herr et al. (2014) demonstrated that a substantial fraction of vesicles from *E. stoutii* requires Ca^{2+} for rupture, with the remaining vesicles rupturing in even Ca^{2+}-free solutions. It is likely that the Ca^{2+}-sensitive vesicles are also the vesicles with the fast, delayed swelling kinetics. Ca^{2+}-dependent rupture can be pharmacologically inhibited, suggesting a role for Ca^{2+}-activated membrane transporters, although these have yet to be identified. The same study also found evidence that aquaporins (AQP) are involved in vesicle swelling. Specifically, the AQP inhibitor mercuric chloride results in an order of magnitude decrease in the swelling rate of mucin vesicles, and PCR revealed the expression of two AQP genes in slime glands (Herr et al., 2014).

Thread skein deployment

One of the remaining mysteries of hagfish slime function is how thread skeins unravel from their coiled state to their full length in a fraction of a second. Early *in vitro* work described skein unraveling as "explosive" (Newby, 1946) although even explosive unraveling could not possibly account for the 1000-fold increase in length that must occur for a 150 μm skein to unravel to its full 150 mm length. Subsequent work on *M. glutinosa* revealed that hydrodynamic mixing as well as the presence of mucins are crucial for proper skein deployment (Winegard and Fudge, 2010). Later work on *E. stoutii* probed the phenomenon of spontaneous unraveling in seawater and demonstrated that it is sensitive to both ionic strength and temperature (Bernards et al., 2014). This study also showed that unraveling is mediated by a seawater-soluble protein adhesive, although the identity of this protein is not yet known.

Slime mechanics

On their own, hagfish slime mucous at its native concentration has almost no effect on the viscosity of seawater (Fudge et al., 2005), but in the presence of slime threads, the mucous acts synergistically to entrain large volumes of water (Koch et al., 1991b; Fudge et al., 2005). Ewoldt et al. (2011) quantified the mechanical properties of whole slime and concluded that hagfish slime is one of the softest biomaterials known, with an elastic modulus of 0.02 Pa, which is more than three orders of magnitude softer than gelatin. The slime also exhibits strain softening at large strains, with simultaneous local strain stiffening, which may correspond with the breaking of weak mucous-thread cross-links and the stretching of slime threads, respectively (Ewoldt et al., 2011).

Slime threads consist of IF proteins, yet their tensile mechanical properties differ radically from mammalian keratins such as wool, hair, and nail, which consist of IFs embedded in an amorphous keratin protein matrix (Fudge and Gosline, 2004). Slime threads are remarkably extensible, stretching to strains of 220% (more than three times their original length) before breaking (Fudge et al., 2003). In *E. stoutii*, whole slime threads break at an average length of about 34 cm. The elastic modulus of the threads is similar to that of rubber (about 6 MPa), or about 400 times softer than hydrated wool (Fudge et al., 2003). At strains of up to about 35%, the threads exhibit rubber-like mechanics, but at higher strains, they strain plastically and they exhibit dramatic strain stiffening at strains greater than 100%. Synchrotron x-ray diffraction experiments reveal that the transition between elastic and plastic behavior corresponds with protein chains in an alpha-helical conformation being pulled apart and reannealing into beta-sheets (Fudge et al., 2003). It is the formation of these stable beta-sheet structures that impart slime threads with their impressively high breaking stress. The study of slime thread mechanics has provided insights not only into the function of hagfish slime, but also has led to new insights into the behavior and function of IFs in living cells (Fudge et al., 2003, 2009), in mammalian keratins (Fudge and Gosline, 2004; Szewciw et al., 2010; Greenberg and Fudge, 2013), and in the quest to manufacture high-performance protein materials (Fudge et al., 2010; Negishi et al., 2012; Pinto et al., 2014).

Evolution of sliming in hagfishes

All extant hagfishes possess slime glands, but it is not clear when this trait appeared in the evolutionary history of the lineage. The one hagfish fossil (*Myxinikela siroka*) that has been described dates back over 300 Ma and shows no evidence of slime gland pores or slime glands (Bardack, 1991). Absence of evidence is not necessarily evidence of absence, but given the high quality of the specimen and some of the fine features that were preserved (branchial blood vessels and branchial pouches, for example), it is quite possible that *Myxinikela* lacked epidermal slime glands. If this is true, then it is tempting to speculate that the slime glands evolved in response to increasing predation pressure from gnathostomes later in their evolutionary history. Indeed, modern chondrichthyans and osteichthyans are both successfully repelled by hagfish slime (Zintzen et al., 2011), presumably because of the susceptibility of their gills to clogging with slime while trying to feed on hagfishes.

How epidermal slime glands evolved from ancestors that lacked them is not known, but there are currently two plausible hypotheses. One hypothesis is that the slime glands evolved as modifications of the cloacal glands, which produce both mucous and threads in association with egg and sperm release (Tsuneki et al., 1985), presumably for the localization of sperm and the anchorage of eggs (Koch et al., 1991a). The cloacal glands superficially resemble the epidermal slime glands and are often mistaken for them. Although a systematic comparison of epidermal and cloacal gland contents has not been done, the cloacal glands are known to stain positively with periodic acid–Schiff (PAS) stain, whereas the epidermal slime glands are PAS-negative (Tsuneki et al., 1985), suggesting that cloacal gland mucous is more heavily glycosylated and therefore more similar to typical mucins. Thread cells in the cloacal glands have not been examined in detail. One can imagine that cloacal glands capable of producing large volumes of fibrous mucous for reproduction might have conferred some protection against predators if the slime were released into a predator's mouth during an attack. Subsequent selection may then have led to the evolution of more glands and their specialization for defense. A cloacal gland origin for the slime glands would require the developmental program for cloacal glands to be eventually expressed in every segment and for the gland duct to be rerouted to drain to the external environment.

Another possible mechanism is that the slime glands evolved via invagination and specialization of the skin. According to this scenario, the slime glands evolved first, and subsequent modifications gave rise to the cloacal glands. If the epidermal invagination hypothesis is correct, then defensive sliming likely evolved before the appearance of the slime glands, unless unrelated selective pressures originally drove the invagination process, which is difficult to imagine. If epidermal slime production was used to deter predators, then invagination may have been selected for initially as a way to increase the surface area available for mucous production and storage, and eventually led to the ability to forcefully eject slime exudate. The invagination hypothesis is appealing in that the two main secretory cell types in the slime glands, GMCs, and GTCs, both have analogs in the epidermis: LMCs and epidermal thread cells (ETCs). LMCs are present in low numbers in the epidermis (they are far outnumbered by the small mucous cells), but they resemble GMCs in size and general morphology. ETCs are large, enigmatic cells of unknown function in the skin that, like GTCs, produce a keratin-like polymer within their cytoplasm (Schreiner, 1916; Blackstad, 1963). Interestingly, they also resemble "skein cells" found in lamprey epidermis, which produce a coiled cytoskeletal polymer, also of unknown function (Downing and Novales, 1971; Lane and Whitear, 1980; Spitzer and Koch, 1998). The presence of ETCs in hagfish skin favors the invagination hypothesis, but does not rule out a cloacal gland origin, especially since it is not known whether the tissue that gives rise to the cloacal gland contains its own kind of thread cell. Future work on hagfish embryology (Ota et al., 2014) and the discovery of new hagfish fossils will surely shed light on which of these hypotheses is more viable.

Conclusions and remaining questions

In 1941, the polymer physicist J.D. Ferry published a paper on hagfish slime in which he cautioned that "the heterogeneity of the original slime and its irreversible contraction render it unsuitable for study of mechanical properties in relation to its composition and structure" (Ferry, 1941). This warning from one of the great scientists of his day may have discouraged work on hagfish slime for several decades, but publications on hagfish slime started appearing again in significant numbers in the early 1980s. Since then, we have learned a tremendous amount about the genes and proteins involved in slime production, the histology and ultrastructure of cells within the slime glands, as well as the biophysics of the slime. These studies have answered many of the questions that Ferry was skeptical could be answered and the mechanisms they have revealed have raised new questions that continue to inspire research in this area.

We have made big strides in the last few years in understanding how the slime deploys in seawater, with several studies probing the behavior of isolated mucous vesicles and thread skeins under a variety of experimental manipulations (Herr et al., 2010, 2014; Winegard and Fudge, 2010; Bernards et al., 2014). These studies point to the existence of AQPs and much larger pore complexes in the mucin vesicle membrane as well as a seawater-soluble protein adhesive that mediates unraveling of ejected thread skeins, but these molecules have not been fully characterized (in the case of AQPs) or even identified (in the case of the pores and the glue). While *in vitro* studies have yielded important clues about deployment on the micro-scale, exactly how the slime exudate increases in volume by four orders of magnitude in a fraction of a second is still not understood.

Three recent papers on the structure and function of hagfish slime thread skeins have provided several new insights, but have also raised new questions. Our current model for how the slime thread is organized involves the patterning of thread loops on a nuclear template that morphs as the thread elongates and increases in diameter. While this partly explains how thread loops take on the shape they do, it raises deeper questions about how exactly the thread elongates, how constituent IFs are recruited to and bundled within the thread, and how the cell achieves such exquisite coiling and packing. Our recent work showing that skein deployment is triggered by solubilization of a seawater-soluble protein glue and the subsequent release of strain energy stored within the slime thread raises the interesting question of what the source of that strain energy is within the cell. While techniques for the culture of hagfish cells are currently lacking, the prospect of observing and probing GTC development using live cell imaging tools promises to one day reveal more wonders within these most remarkable of cells.

References

Aebi, U., M. Häner, J. Troncoso, R. Eichner and A. Engel. 1988. Unifying principles in intermediate filament (IF) structure and assembly. *Protoplasma* 145:73–81.

Bardack, D. 1991. First fossil hagfish (myxinoidea): A record from the Pennsylvanian of Illinois. *Science* 254:701–703.

Bernards, M. A. Jr., I. Oke, A. Heyland and D. S. Fudge. 2014. Spontaneous unraveling of hagfish slime thread skeins is mediated by a seawater-soluble protein adhesive. *J Exp Biol* 217:1263–1268.

Blackstad, T. W. 1963. The skin and the slime glands. In A. Brodal and R. Fänge (eds.), *The Biology of* Myxine. Universitetsforlaget, Oslo, 195 pp.

Clark, A. J. and A. P. Summers. 2007. Morphology and kinematics of feeding in hagfish: Possible functional advantages of jaws. *J Exp Biol* 210:3897–3909.

Downing, S. W. and R. R. Novales. 1971. The fine structure of lamprey epidermis. II. Club cells. *J Ultrastruct Res* 35:295–303.

Downing, S. W., W. L. Salo, R. H. Spitzer and E. A. Koch. 1981a. The hagfish slime gland: A model system for studying the biology of mucus. *Science* 214:1143–1145.

Downing, S. W., R. H. Spitzer, E. A. Koch and W. L. Salo. 1984. The hagfish slime gland thread cell. I. A unique cellular system for the study of intermediate filaments and intermediate filament-microtubule interactions. *J Cell Biol* 98:653–669.

Downing, S. W., R. H. Spitzer, W. L. Salo, J. S. Downing, L. J. Saidel and E. A. Koch. 1981b. Threads in the hagfish slime gland thread cells: Organization, biochemical features, and length. *Science* 212:326–328.

Ewoldt, R. H., T. M. Winegard and D. S. Fudge. 2011. Non-linear viscoelasticity of hagfish slime. *Int J Non-Linear Mech* 46:627.

Fan, W. 1965. Fine structure of thread cell differentiation in slime glands of Pacific hagfish *Polistotrema stoutii*. *Anat Rec* 151:348.

Fernholm, B. 1981. Thread cells from the slime glands of hagfish (Myxinidae). *Acta Zool* 62:137–145.

Fernholm, B. 1998. Hagfish systematics. In J. M. Jørgensen, J. P. Lomholt, R. E. Weber and H. Malte (eds.), *The Biology of Hagfishes*. Springer, The Netherlands, 33 pp.

Ferry, J. D. 1941. A fibrous protein from the slime of the hagfish. *J Biol Chem* 138:263–268.

Fudge, D. S., K. H. Gardner, V. T. Forsyth, C. Riekel and J. M. Gosline. 2003. The mechanical properties of hydrated intermediate filaments: Insights from hagfish slime threads. *Biophys J* 85:2015–2027.

Fudge, D. S. and J. M. Gosline. 2004. Molecular design of the alpha-keratin composite: Insights from a matrix-free model, hagfish slime threads. *Proc Biol Sci* 271:291–299.

Fudge, D. S., S. Hillis, N. Levy and J. M. Gosline. 2010. Hagfish slime threads as a biomimetic model for high performance protein fibres. *Bioinspir Biomim* 5:035002-3182/5/3/035002. Epub 2010 Aug 20.

Fudge, D. S., N. Levy, S. Chiu and J. M. Gosline. 2005. Composition, morphology and mechanics of hagfish slime. *J Exp Biol* 208:4613–4625.

Fudge, D. S., T. Winegard, R. H. Ewoldt, D. Beriault, L. Szewciw and G. H. McKinley. 2009. From ultra-soft slime to hard alpha-keratins: The many lives of intermediate filaments. *Integr Comp Biol* 49:32–39.

Greenberg, D. A. and D. S. Fudge, 2013. Regulation of hard alpha-keratin mechanics via control of intermediate filament hydration: Matrix squeeze revisited. *Proc Biol Sci* 280:20122158.

Gum, J. R., Jr. 1992. Mucin genes and the proteins they encode: Structure, diversity, and regulation. *Am J Respir Cell Mol Biol* 7:557–564.

Herr, J. E., A. Clifford, G. G. Goss and D. S. Fudge. 2014. Defensive slime formation in Pacific hagfish requires Ca^{2+} and aquaporin mediated swelling of released mucin vesicles. *J Exp Biol* 217:2288–2296.

Herr, J. E., T. M. Winegard, M. J. O'Donnell, P. H. Yancey and D. S. Fudge. 2010. Stabilization and swelling of hagfish slime mucin vesicles. *J Exp Biol* 213:1092–1099.

Isaacs, J. D. and R. A. Schwartzlose. 1975. Active animals of the deep-sea floor. *Sci Am* 233:84–91.

Koch, E. A., R. H. Spitzer and R. B. Pithawalla. 1991a. Structural forms and possible roles of aligned cytoskeletal biopolymers in hagfish (slime eel) mucus. *J Struct Biol* 106:205–210.

Koch, E. A., R. H. Spitzer, R. B. Pithawalla, F. A. Castillos, III and D. A. Parry. 1995. Hagfish biopolymer: A type I/type II homologue of epidermal keratin intermediate filaments. *Int J Biol Macromol* 17:283–292.

Koch, E. A., R. H. Spitzer, R. B. Pithawalla and S. W. Downing. 1991b. Keratin-like components of gland thread cells modulate the properties of mucus from hagfish (*Eptatretus stoutii*). *Cell Tissue Res* 264:79–86.

Koch, E. A., R. H. Spitzer, R. B. Pithawalla and D. A. Parry. 1994. An unusual intermediate filament subunit from the cytoskeletal biopolymer released extracellularly into seawater by the primitive hagfish (*Eptatretus stoutii*). *J Cell Sci* 107(Pt 11):3133–3144.

Lametschwandtner, A., U. Lametschwandtner and R. Patzner. 1986. The different vascular patterns of slime glands in the hagfishes, *Myxine glutinosa* Linnaeus and *Eptatretus stoutii* Lockington: A scanning electron microscope study of vascular corrosion cast*. *Acta Zool* 67:243–248.

Lane, E. and M. Whitear. 1980. Skein cells in lamprey epidermis. *Can J Zool* 58:450–455.

Leppi, T. J. 1968. Morphochemical analysis of mucous cells in the skin and slime glands of hagfishes. *Histochemie* 15:68–78.

Lesser, M. P., F. H. Martini and J. B. Heiser. 1997. Ecology of the hagfish, *Myxine glutinosa* L. in the Gulf of Maine. I. Metabolic rates and energetics. *J Exp Mar Biol Ecol* 208:215–225.

Lim, J. 2013. Kinematics and hydrodynamics of undulatory locomotion in hagfishes (Myxinidae) and hagfish-like robotic models. Ph.D. thesis, Harvard University, Cambridge, USA.

Lim, J., D. S. Fudge, N. Levy and J. M. Gosline. 2006. Hagfish slime ecomechanics: Testing the gill-clogging hypothesis. *J Exp Biol* 209:702–710.

Luchtel, D. L., A. W. Martin and I. Deyrup-Olsen. 1991. Ultrastructure and permeability characteristics of the membranes of mucous granules of the hagfish. *Tissue Cell* 23:939–948.

Martini, F. H. 1998. The ecology of hagfishes. In J. M. Jørgensen, J. P. Lomholt, R. E. Weber and H. Malte (eds.), *The Biology of Hagfishes*. Springer, The Netherlands, 57 pp.

Martini, F. H. and A. Beulig. 2013. Morphometrics and gonadal development of the hagfish *Eptatretus cirrhatus* in New Zealand. *PLoS ONE* 8:e78740.

Negishi, A., C. L. Armstrong, L. Kreplak, M. C. Rheinstadter, L. T. Lim, T. E. Gillis and D. S. Fudge 2012. The production of fibers and films from solubilized hagfish slime thread proteins. *Biomacromolecules* 13:3475–3482.

Newby, W. W. 1946. The slime glands and thread cells of the hagfish, *Polistrotrema stoutii*. *J Morphol* 78:397–409.

Ota, K. G., Y. Oisi, S. Fujimoto and S. Kuratani. 2014. The origin of developmental mechanisms underlying vertebral elements: Implications from hagfish evo-devo. *Zoology (Jena)* 117:77–80.

Parry, D. A. and P. M. Steinert. 1999. Intermediate filaments: Molecular architecture, assembly, dynamics and polymorphism. *Q Rev Biophys* 32:99–187.

Pinto, N., F. C. Yang, A. Negishi, M. C. Rheinstadter, T. E. Gillis and D. S. Fudge. 2014. Self-assembly enhances the strength of fibers made from vimentin intermediate filament proteins. *Biomacromolecules* 15:574–581.

Quay, W. B. 1972. Integument and the environment glandular composition function, and evolution. *Am Zool* 12:95–108.

Salo, W. L., S. W. Downing, W. A. Lidinsky, W. H. Gallagher, R. H. Spitzer and E. A. Koch. 1983. Fractionation of hagfish slime gland secretions: Partial characterization of the mucous vesicle fraction. *Prep Biochem* 13:103–135.

Schreiner, K. 1916. Zur Kenntnis der Zellgranula. *Arch Mikrosk Anat* 89:79–188.

Spitzer, R. H., S. W. Downing, E. A. Koch, W. L. Salo and L. J. Saidel. 1984. Hagfish slime gland thread cells. II. Isolation and characterization of intermediate filament components associated with the thread. *J Cell Biol* 98:670–677.

Spitzer, R. H. and E. A. Koch. 1998. Hagfish skin and slime glands. In J. M. Jørgensen, J. P. Lomholt, R. E. Weber and H. Malte (eds.), *The Biology of Hagfishes*. Springer, The Netherlands, 109 pp.

Spitzer, R. H., E. A. Koch and S. W. Downing. 1988. Maturation of hagfish gland thread cells: Composition and characterization of intermediate filament polypeptides. *Cell Motil Cytoskeleton* 11:31–45.

Subramanian, S., N. W. Ross and S. L. MacKinnon. 2008. Comparison of the biochemical composition of normal epidermal mucus and extruded slime of hagfish (*Myxine glutinosa* L.). *Fish Shellfish Immunol* 25:625–632.

Subramanian, S., N. W. Ross and S. L. MacKinnon. 2009. Myxinidin, a novel antimicrobial peptide from the epidermal mucus of hagfish, *Myxine glutinosa* L. *Mar Biotechnol (NY)* 11:748–757.

Szewciw, L. J., D. G. de Kerckhove, G. W. Grime and D. S. Fudge. 2010. Calcification provides mechanical reinforcement to whale baleen alpha-keratin. *Proc Biol Sci* 277:2597–2605.

Tamburri, M. N. and J. P. Barry. 1999. Adaptations for scavenging by three diverse bathyla species, *Eptatretus stoutii*, *Neptunea amianta* and *Orchomene obtusus*. *Deep Sea Res I Oceanogr Res Pap* 46:2079–2093.

Terakado, K., M. Ogawa, Y. Hashimoto and H. Matsuzaki. 1975. Ultrastructure of the thread cells in the slime gland of Japanese hagfishes, *Paramyxine atami* and *Eptatretus burgeri*. *Cell Tissue Res* 159:311–323.

Tsuneki, K., A. Suzuki and M. Ouji. 1985. Sex difference in the cloacal gland in the hagfish, *Eptatretus burgeri*, and its possible significance in reproduction. *Acta Zool* 66:151–158.

Whitear, M. 1986. The skin of fishes including cyclostomes. In J. Bereiter-Hahn, A. G. Matoltsy and K. S. Richards (eds.), *Biology of the Integument. 2. Vertebrates*. Springer, Berlin, 8 pp.

Winegard, T. M. 2012. Slime gland cytology and mechanisms of slime thread production in the Atlantic hagfish (*Myxine glutinosa*). M.Sc. thesis, University of Guelph, Guelph, Canada.

Winegard, T., J. Herr, C. Mena, B. Lee, I. Dinov, D. Bird, M. Bernards, Jr. et al. 2014. Coiling and maturation of a high-performance fibre in hagfish slime gland thread cells. *Nat Commun* 5:3534.

Winegard, T. M. and D. S. Fudge. 2010. Deployment of hagfish slime thread skeins requires the transmission of mixing forces via mucin strands. *J Exp Biol* 213:1235–1240.

Zintzen, V., C. D. Roberts, M. J., Anderson, A. L., Stewart, C. D. Struthers and E. S. Harvey. 2011. Hagfish predatory behaviour and slime defence mechanism. *Sci Rep* 1:131.

Index

T - #0215 - 111024 - C0 - 234/156/18 - PB - 9780367575519 - Gloss Lamination